Second Supplements to the 2nd Edition of

RODD'S CHEMISTRY OF CARBON COMPOUNDS

ELSEVIER SCIENCE PUBLISHERS B.V.,
Sara Burgerhartstraat 25
P.O. Box 211, 1000 AE Amsterdam, The Netherlands

Library of Congress Card Number 91-16129
ISBN 0-444 897585

© Elsevier Science Publishers B.V., 1992 All rights reserved.

No part of this publication may be reproduced, stored in a retrieval system, or transmitted, in any form or by any means, electronic, mechanical, photocopying, recording or otherwise, without the prior written permission of the publisher, Elsevier Science Publishers B.V., Copyright & Permission Department, P.O. Box 521, 1000 AM Amsterdam, The Netherlands.

Special regulations for readers in the USA - This publication has been registered with the Copyright Clearance Center Inc. (CCC), Salem, Massachusetts. Information can be obtained from the CCC about conditions under which photocopies of parts of this publication may be made in the USA. All other copyright questions, including photocopying outside of the USA, should be referred to the publisher.

No responsibility is assumed by the Publisher for any injury and/or damage to persons or property as a matter of products liability, negligence or otherwise, or from any use or operation of any methods, products, instructions or ideas contained in the material herein. Because of rapid advances in the medical sciences, the Publisher recommends that independent verification of diagnoses and drug dosages should be made.

This book is printed on acid-free paper.

Printed in The Netherlands

Second Supplements to the 2nd Edition of

RODD'S CHEMISTRY OF CARBON COMPOUNDS

VOLUME I

ALIPHATIC COMPOUNDS

★

VOLUME II

ALICYCLIC COMPOUNDS

★

VOLUME III

AROMATIC COMPOUNDS

★

VOLUME IV

HETEROCYCLIC COMPOUNDS

★

VOLUME V

MISCELLANEOUS

GENERAL INDEX

★

Second Supplements to the 2nd Edition of

RODD'S CHEMISTRY OF CARBON COMPOUNDS

A modern comprehensive treatise

Edited by
MALCOLM SAINSBURY
School of Chemistry, The University of Bath,
Claverton Down, Bath BA2 7AY, England

Second Supplement to

VOLUME I ALIPHATIC COMPOUNDS

Part C: Monocarbonyl Derivatives of Aliphatic Hydrocarbons, Their Analogues and Derivatives

ELSEVIER
Amsterdam – London – New York – Tokyo 1992

Cont 1-14-93

Contributors to this Volume

D.W. ANDERSON

Chemical Defence Establishment, Porton Down, Wiltshire SP4 0JQ, England

J. BERGMAN

Department of Organic Chemistry, Royal College of Technology,
S-100 44 Stockholm, Sweden

I.G.C. COUTTS

Department of Physical Sciences, Trent Polytechnic, Clifton Lane,
Nottingham NG11 8NS, England

A. FÜRSTNER

Institut für Organische Chemie der Technischen Universität,
Stremayrgasse 16, A-8010 Graz, Austria

R.B. MILLER

Department of Chemistry, University of California, Davis, CA 95616, U.S.A.

D.J. SIMMONDS

Department of Chemistry, Sheffield Polytechnic, Pond Street,
Sheffield S1 1WB, England

H. WEIDMANN

Institut für Organische Chemie der Technischen Universität,
Stremayrgasse 16, A-8010 Graz, Austria

QD251
R62
Suppl.2
vol.1
pt.C

Preface to Volume I C

The preparation, proof reading, and indexing of a book of this kind is a labour intensive process, which inevitably takes longer than expected. Volume I C of the second series of supplements to 'Rodd' is no exception and I would like to thank all the contributors for their industry, help, and patience in completing the tasks I placed before them three years ago. I have also received much encouragement and professional advice from Elsevier Science Publishers. I would like to record my thanks to their staff for assisting me as I take my first steps as editor of the second supplement.

Volume 1 of the Chemistry of the Carbon Compounds deals with all aspects of aliphatic chemistry. In this second sub-volume the carbonyl compounds in their various forms are updated from 1972, a year before the first supplement was published. It contains all the salient new discoveries in this diverse subject area. The fact that most of the individual topics are not easily delineated has made the task of searching the literature doubly difficult, and the authors of the six chapters are to be congratulated in the masterly way they have dealt with this problem.

Malcolm Sainsbury
July 1992

Contents
Volume 1 C

Aliphatic Compounds; Monocarbonyl Derivatives of Aliphatic Hydrocarbons, Their Analogues and Derivatives.

Chapter 9a. Monobasic Aliphatic Saturated Acids
by J. BERGMAN

Chapter 9b. Unsaturated Monobasic Acids
by R.B. MILLER

Chapter 10. Carbon Monoxide and its Derivatives, Carbonic Acid and its Derivatives
by I.G.C. COUTTS

Chapter 11. Carbamates and their Allies
by D.W. ANDERSON

List of Common Abbreviations and Symbols Used

A	acid
Å	Ångström units
Ac	acetyl
a	axial
as, asymm.	asymmetrical
at.	atmosphere
B	base
Bu	butyl
b.p.	boiling point
c, C	concentration
CD	circular dichroism
conc.	concentrated
D	Debye unit, 1×10^{-18} e.s.u.
D	dissociation energy
D	dextro-rotatory; dextro configuration
d	density
dec., decomp	with decomposition
deriv.	derivative
E	energy; extinction; electromeric effect
*E*1, *E*2	uni- and bi-molecular elimination mechanisms
E1cB	unimolecular elimination in conjugate base
ESR	electron spin resonance
Et	ethyl
e	nuclear charge; equatorial
f.p.	freezing point
G	free energy
GLC	gas liquid chromatography
g	spectroscopic splitting factor, 2.0023
H	applied magnetic field; heat content
h	Planck's constant
Hz	hertz
I	spin quantum number; intensity; inductive effect
IR	infrared
J	coupling constant in NMR spectra
J	Joule
K	dissociation constant
k	Boltzmann constant; velocity constant
kcal	kilocalories
M	molecular weight; molar; mesomeric effect
Me	methyl
m	mass; mole; molecule; *meta*-
m.p.	melting point
Ms	mesyl (methanesulphonyl)

[M]	molecular rotation
N	Avogadro number; normal
NMR	nuclear magnetic resonance
NOE	Nuclear Overhauser Effect
n	normal; refractive index; principal quantum number
o	*ortho-*
ORD	optical rotatory dispersion
P	polarisation; probability; orbital state
Pr	propyl
Ph	phenyl
p	*para-*; orbital
PMR	proton magnetic resonance
R	clockwise configuration
S	counterclockwise configuration; entropy; net spin of incompleted electronic shells; orbital state
S_N1, S_N2	uni- and bi-molecular nucleophilic substitution mechanism
S_Ni	internal nucleophilic substitution mechanism
s	symmetrical; orbital
sec	secondary
soln.	solution
symm.	symmetrical
T	absolute temperature
Tosyl	*p*-toluenesulphonyl
Trityl	triphenylmethyl
t	time
temp.	temperature (in degrees centrigrade)
tert	tertiary
UV	ultraviolet
α	optical rotation (in water unless otherwise stated)
$[\alpha]$	specific optical rotation
ϵ	dielectric constant; extinction coefficient
μ	dipole moment; magnetic moment
μ_B	Bohr magneton
μg	microgram
μm	micrometer
λ	wavelength
ν	frequency; wave number
χ, χ_d, χ_μ	magnetic; diamagnetic and paramagnetic susceptibilities
(+)	dextrorotatory
(–)	laevorotatory
–	negative charge
+	positive charge

Second Supplements to the 2nd Edition of Rodd's Chemistry of Carbon Compounds, Vol. 1C, edited by M. Sainsbury
© 1992 Elsevier Science Publishers B.V., Amsterdam

Chapter 8a

ALIPHATIC CARBONYL COMPOUNDS: ALIPHATIC ALDEHYDES

A. FÜRSTNER and H. WEIDMANN

1. Introduction

The recent literature reveals numerous improvements of earlier procedures and a prodigious number of new and quite effective methods for the synthesis of aliphatic aldehydes. A variety of monographs deals either comprehensively or with more or less specific areas of preparation of carbonyl compounds ("Aldehyde", Houben-Weyl, Methoden der Organischen Chemie, Vol. E3, J. Falbe, ed., G. Thieme, Stuttgart, New York, 1983; "Chromium Oxidation in Organic Chemistry", C. Cainelli, G. Cardillo, eds., Springer Verlag, Berlin, 1984; E.I. Negishi, "Organometallics in Organic Synthesis", Vol. 1, John Wiley & Sons, New York, 1980; A.I. Meyers, "Heterocycles in Organic Synthesis", John Wiley & Sons, London-New York-Sidney-Toronto, 1974; M. Peyreyre et al., "Tin in Organic Synthesis", Butterworth Ltd., London, 1987; P. Hodge, D.C. Sherrington, "Polymer Supported Reactions in Organic Synthesis", John Wiley & Sons, New York 1980; R.A. Shelton, J.K. Kochi, "Metal Catalyzed Oxidations of Organic Compounds", Academic Press, New York, 1981). In addition there are various review articles partly describing the respective transformations into aldehydes of alcohols, alkyl halides, amino, nitro, and carbonyl compounds, carboxylic acids and derivatives, alkenes, alkynes, epoxides and others by multifarious preparative procedures which will be cited in the corresponding chapters.

Irrespective of the structure of the aldehydes, and of the starting materials, the main sections are distinguished by the respective groups of procedures which either reduce or retain the number of carbon atoms or extend the carbon chains by one, two, three or more carbon atoms in the process of aldehyde formation. In the subsections the transformations into aldehydes of each of the different groups of precursors by various procedures are described.

2. Syntheses with reduction of the number of carbon atoms

Reactions of this type are accomplished as follows:

(i) from carboxylic acids by formation of their α-methylsulfenyl derivatives, followed by sodium periodate oxidation in methanol, which forms the nor-aldehyde dimethyl acetals. (B.M. Trost, Acc. Chem. Res., 1978, 11, 453).

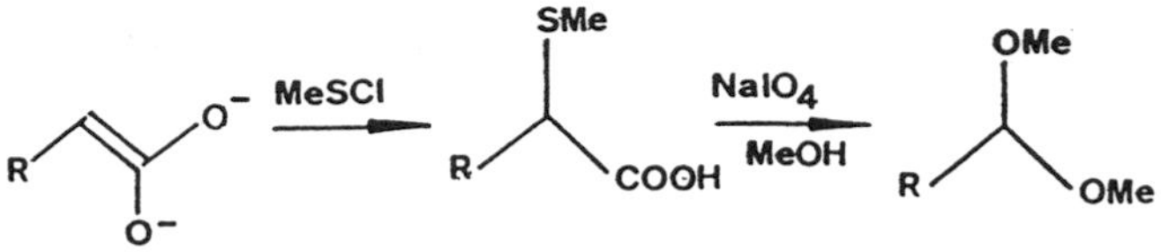

(ii) by ozonolysis of 1-alkenes, followed by reductive cleavage of the ozonides by thiourea. Cycloalkenes yield dials (D. Gupta, R. Soman, S. Dev, Tetrahedron, 1982, 38, 3013).

(iii) through enantioselective formation of 2-methyl-aldehydes by regioselective ring-opening of homochiral hydroxymethyl epoxides with trimethylaluminum, followed by periodate fission of the resulting 3-methyl-1,2-diols (W.R. Roush et al., Tetrahedron Lett., 1983, 1377). Best results are obtained with this reagent in the presence of wet silica gel (M. Daumas et al., Synthesis, 1989, 64).

3. Synthesis with retention of the number of carbon atoms

(a) *Oxidation reactions of primary alcohols*

$$RCH_2\text{-}OH \longrightarrow R\text{-}CHO$$

(for a review see: F.A. Luzzio, F.S. Guziec, "Recent Applications of Oxochromium (VI) - Amine Reagents", Org. Prep. Proc. Int., 1988, 20, 533)

Reactions of this type use a plethora of reagents and can be effected by:

(i) pyridinium chlorochromate (PCC, $C_5H_5N^+H(CrO_3Cl)^-$(E.J. Corey, J. W. Suggs, Tetrahedron Lett., 1975, 2647; review: G. Piancatelli, A. Scettri, M.D. Anria, Synthesis, 1982, 245) the acidity of which can be buffered by sodium acetate to save acid sensitive groups and to prevent (Z)-(E) isomerizations. Primary alcohols containing acid sensitive groups, or those requiring anhydrous conditions are converted into their borate esters by borane-dimethylsulfide prior to PCC oxidation (H.C. Brown, S.U. Kulkarni, C.G. Rao, Synthesis, 1979, 702). In this context the conversion of the substrates into trimethylsilyl ethers by triphenylmethenium tetrafluoroborate (M.E. Jung, J. Org. Chem., 1976, 41, 1479), or into tributylstannyl ethers by nitrosonium tetrafluoroborate (G.A. Olah, T.L. Ho, Synthesis, 1976, 609), is recommended.

(ii) the more stable and internally buffered 2,2'-bipyridinium chlorochromate, is equally mild and effective for the oxidation of saturated und unsaturated alco hols and allows simpler workup (F.S. Guziec, F.A. Luzzio, Synthesis, 1980, 691).

(iii) 4-(dimethylamino)pyridinium chlorochromate ($DMAPH^+$ $(CrO_3Cl)^-$). This reagent is preferred for the oxidation of primary 2-alkenols. Primary alcohols react very slowly (F.S. Guziec, F.A. Luzzio, J. Org. Chem., 1982, 47, 1787).

(iv) poly(vinyl(pyridinium chlorochromate)) (PVPCC), preferably in cyclohexane or heptane, oxidizes 2-alkenols faster than primary alkanols and less reagent is needed than in oxidations with PCC. Yields depend on the quality of the poly(vinylpyridine), but work-up of the reaction mixture is simple and the resin may be quantitatively recovered and regenerated (J.M.J. Frechet, J. Warnock, M.J. Farrall, J. Org. Chem., 1978, 43, 2618; ibid. 1981, 46, 1728). The Amberlyst A-26-chromate reagent (G. Cainelli et al., J. Am. Chem. Soc., 1976, 98, 6737) is similar to handle, but is less effective than PVPCC.

(v) tetrabutylammonium chlorochromate (TBACC) is a fairly stable, but light sensitive reagent for the selective oxidation of 2-alkenols (E. Santaniello, F. Milani, R. Casati, Synthesis, 1983, 749). The homogeneous solutions of methyl trialkylammonium dichromate, formed from Andogen 464 and potassium dichromate in benzene, also only oxidize 2-alkenols (R.O. Hutchins, N.R. Natale, W.J. Cook, Tetrahedron Lett., 1977, 4167) as does bis (benzyltriethylammonium) dichromate, a neutral oxidant (X. Huang, C. Chan, Synthesis, 1982, 1091). However, the seemingly similar tetrabutyl-ammonium chromate formed in situ in chloroform and only applicable in chlorohydrocarbon solvents is distincly less selective (S. Cacchi, F. LaTorre, D. Misti, Synthesis, 1979, 356).

(vi) the bulkier quinolinium chlorochromate. This reagent selectively oxidizes unbranched primary alkanols and alkenols (J. Singh et al., Chem. Ind. (London), 1986, 751).

(vii) pyridinium fluorochromate ($C_5H_5N^+H(CrO_3F)^-$) in acetonitrile.It is a less acidic and a more efficient oxidant than PCC (N.N. Bhattacharjee et al., Synthesis, 1982, 588).

(viii) pyridinium dichromate (PDC, $(PyH^+)_2(Cr_2O_7{}^{2-})$) (E.J. Corey, G. Schmidt, Tetrahedron Lett., 1979, 399), which needs to be handled with caution (J. Salmon, Chemistry in Britain, 1982, 703), exhibits solvent dependent selectivities. In the presence of excess of bis-trimethylsilyl peroxide as reoxidant, PDC can be more safely employed in catalytic amounts (K. Oshima et al., Tetrahedron Lett., 1983, 2185). Although 3- or 4-carboxy-pyridinium dichromates are safer than PDC, they are less effective oxidants (C. Lopez et al., Synth. Commun., 1985, 15, 1197).

(ix) pyridine-chromyl chloride-t-butanol, a remarkably effective oxidant for the large-scale preparation of simple alkanals and 2-alkenals, giving excellent yields (K.B. Sharpless, K. Alashi, J. Am. Chem. Soc., 1975, 97, 5927). The complex formed by slow addition of CrO_3 to hexamethylphosphoric triamide (HMPA) can be applied without additional solvent and is equally effective (G.Cardillo, M. Orena, S. Sandri, Synthesis, 1976, 394). This system is similar in action to CrO_3 in HMPA and with crown ethers (idem, Tetrahedron Lett., 1976, 3985), and to potassium dichromate in dimethylsulfoxide (DMSO) or in polyethylene glycols. It oxidizes 2-alkenols exclusively (E. Santaniello, P. Ferraboschi, Synthesis, 1980, 646).

(x) potassium dichromate/sulfuric acid in water/dichloromethane in the presence of tetrabutylammonium hydrogensulfate. The phase transfer conditions are quite compatible with acid sensitive alcohols and the resulting alkanals and alkenals, but are restricted to lipophilic substrates (D. Landini, F. Montanara, F. Rolla, Synthesis, 1979, 134). See also G. Gelbard, T. Brunelet, C. Jouitteau (Tetrahedron Lett, 1980, 4653).

(xi) pyridinechromium peroxide (PCP) ($PyCrO_5$, pyridine-bis(peroxy)-oxo-chromium (VI)), prepared either from pyridine, CrO_3 and H_2O_2 (G.W.J. Fleet, W. Little, Tetrahedron Lett., 1977, 3749) or in crystalline form as for 2,2'-bipyridylchromium peroxide (BPCP, 2,2'-bipyCrO_5) (H. Firouzabadi et al., Tetrahedron, 1986, 42, 719). The latter reagent is the more effective, allowing oxidations of various kinds of primary alcohols under neutral conditions at ambient temperatures. Yields are high. In contrast to most oxochromium (VI) reagents, the oxidation with both of these complexes succeeds with equimolar amounts of substrate and oxidant.

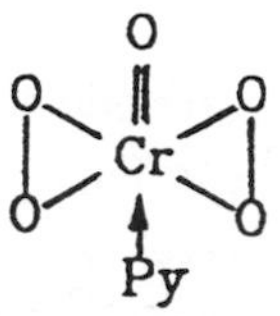

(xii) 3,5-dimethylpyrazole-CrO_3, which is also useful for larger scale oxidations of saturated and unsaturated primary alcohols and affords very good yields (E.J. Corey, G.W.J. Fleet, Tetrahedron Lett., 1973, 4499).

(xiii) diisopropyl sulfide/N-chlorosuccinimide (NCS)/triethylamine (K.S. Kim et al., J. Chem. Soc., Chem. Commun., 1984, 762). Unlike the original reagent combination described by E.J. Corey, C.U. Kim (Tetrahedron Lett., 1974, 287), this reagent distinguishes between saturated and unsaturated primary alcohols and secondary alcohols merely by selection of an appropriate reaction temperature. An analogous procedure employs the dimethylselenide-NCS complex. The intermediate chlorodimethylselenium alkoxides ($Me_2Se(Cl)OCH_2R$) undergo base-induced elimination with formation of aldehydes (T. Takaki et al., J. Org. Chem., 1983, 48, 54).

Oxochromium (VI) reactants adsorbed on inorganic supports (xiv - xx). They form part of the important group of polymer-bound reagents and are particularly advantageous (reviews: A. McKillop, D.W. Young, Synthesis, 1979, 401 and 481).

(xiv) silica gel supported chromic acid, prepared and employed under various conditions (E. Santaniello, F. Ponti, A. Manzocchi, Synthesis, 1978, 534; J.D. Lou, Y.Y. Wu, Chem. Ind. (London), 1987, 531).

(xv) chromic acid adsorbed on Kieselguhr, a stable and efficient oxidant (J.D. Lou, Synth. Commun., 1989, 19, 1841).

(xvi) pyridinium chromate on silica gel containing a small amount of acetic acid. This reagent shows solvent dependent oxidation rates (R.P. Singh, H.N. Subbarao, S. Dev, Tetrahedron, 1979, 35, 1789).

(xvii) pyridinium chlorochromate/amine with ultrasonication. This renders PCC less acidic, simplifies the workup and gives excellent results. Good compatibility with acid sensitive groups is also noted (L.L. Adams, F.A. Luzzio, J. Org. Chem., 1989, 54, 5387). PCC on aluminum oxide gives equally good results (Y.S. Cheng, W.L. Liu, S.H. Chen, Synthesis, 1980, 223).

(xviii) chromic acid on aluminum silicate in nonpolar solvents (D.D. Lou, Y.Y. Wu, Synth. Commun., 1987, 17, 1717).

(xix) chromyl chloride (CrO_2Cl_2) on silica gel under neutral, anhydrous conditions. While C-C-double bonds are cleaved, carbon-halogen bonds, esters, ethers and cyano groups are not affected (J.S. Filippo, C.I. Chern, J. Org. Chem., 1977, 42, 2182). The reagent is more selective than that described by K.B. Sharpless et al. (J. Am. Chem. Soc., 1977, 99, 3120).

(xx) CrO_3/graphite obtained at 200°C under reduced pressure, then successively washed with water, hydrochloric acid and acetone. Despite long reaction times yields of aldehydes are good (J.M. Lalancette, G. Rollin, P. Dumas, Can. J. Chem, 1972, 50, 3058). Graphite nitrate ($C_{24}{}^{+}NO_3^{-}$) is also recommended (H.B. Cagan, Chem. Tech., 1976, 6, 510).

(xxi) sodium hypochlorite and 2,2,6,6-tetramethylpiperidine-1-oxyl radical as catalyst, cocatalyzed by bromide ion in dichloromethane-buffered water under phase transfer conditions. Oxidation of 1-alkanols only is very rapid and quantitative at 0°C (P. Anelli et al., J. Org. Chem., 1987, 52, 2559).

(xxii) N-iodosuccinimide and tetrabutylammonium iodide gives excellent yields of alkanals and 2-alkenals (S. Hanessian, D.H. Wong, M. Therien, Synthesis, 1981, 394).

The combination of t-butyl hydroperoxide and bis-(2,4,6-trimethylphenyl)-diselenide is also a very effective oxidant (M.Shimizu, I. Kuwajima, Tetrahedron Lett., 1979, 2801).

(xxiii) the combination of nitro ethylacetate/diethyl azodicarboxylate/triphenyl-phosphine in refluxing tetrahydrofuran is also claimed to be a useful oxidant system (O. Mitsunobu, N. Yoshida, Tetrahedron Lett., 1981, 2295).

Activated dimethylsulfoxide (DMSO) mediated oxidations of alcohols play an essential role in the preparation of carbonyl compounds (review: A.J. Mancuso, D. Swern, Synthesis, 1981, 165) as exemplified by the following procedures:

(xxiv) DMSO/oxalyl chloride (OC)/triethylamine (Et_3N). This combination has been extensively investigated and has wide application (A.J. Mancuso, D.S. Brownfain, D. Swern, J. Org. Chem., 1979, 44, 4148). The actual reagent in "Swern oxidations" was found to be chloro dimethylsulfonium chloride ($Me_2S^+Cl\ Cl^-$) also resulting from dimethyl sulfide and chlorine (E.J. Corey, C.U. Kim, J. Am. Chem. Soc., 1972, 94, 7586) or N-chlorosuccinimide (idem, Tetrahedron Lett., 1974, 287). The "Swern oxidation" even allows the preparation of highly reactive aldehydes which are interceptable by Wittig reactions (R.E. Ireland, D.W. Norbeck, J. Org. Chem., 1985, 50, 2198).

(xxv) DMSO/trichloromethyl chloroformate (phosgene dimer)-Et_3N gives equally good results as DMSO/OC (S. Takano, Tetrahedron Lett., 1988, 6619).

(xxvi) DMSO-thionyl chloride/Et_3N (K. Omura, D. Swern, Tetrahedron, 1978, 34, 1651). This requires precautions to prevent potential Pummerer rearrangement.

(xxvii) by DMSO/phenyl dichlorophosphate (PDCP, $PhOP(O)Cl_2$)-Et_3N (H.J. Liu, J.M. Nyangulu, Tetrahedron Lett., 1988, 3167).

The reagents described above are invariably employed at temperatures between -60 to -30°C.

(xxviii) DMSO/carbodiimide incorporated in a polystyrene matrix as activator (N.M. Weinshenker, C.H. Shen, Tetrahedron Lett., 1972, 3285).

In the following group of procedures transition metal reagents are used in alcohol to aldehyde oxidation reactions (reviews: A.P. Kozikowski, H.F. Wetter, Synthesis, 1976, 561; R.A. Sheldon, Bull. Soc. Chim. Belg., 1985, 94, 651):

(xxix) silver carbonate/celite, a remarkably effective supported reagent, oxidizing primary alcohols, and particularly 2-alkenols, to aldehydes in excellent yields (A. McKillop, D.W. Young, Synthesis, 1979, 401).

(xxx) palladium acetate or tetrakis(triphenylphoshine)palladium/triphenyl-phosphine (TPP)/arylbromide/potassium carbonate in dimethoxyethane (DME), both forming with primary alcohols the complex $RCH_2OPd(Ar)(TPP)_2$ which expels the aldehyde (Y. Tamaru et al., J. Org. Chem., 1983, 48, 1286). Also by the combination of palladium acetate/sodium hydrogencarbonate/tetra-n-butyl-ammoniumbromide/iodobenzene in DMF under solid-liquid phase transfer conditions. The oxidations proceed rather slowly, but alkanals are formed in good yields (B.M. Chondary et al., Tetrahedron Lett., 1985, 6257).

(xxxi) tetrakis(pyridine)silver dichromate ($Py_4Ag_2Cr_2O_7$), which oxidizes 2-alkenols selectively but only in refluxing benzene (H. Firouzabadi et al. Synth. Commun., 1984, 14, 89), or barium ferrate which shows similar selectivity under the same conditions (idem, ibid., 1986, 16, 723).

(xxxii) catalytic amounts of nonacarbonyl diiron ($Fe_2(CO)_9$) in benzene, which invariably isomerizes various kinds of primary alkenols, alkenol ethers and esters with formation of aldehyde enols, enol ethers and esters. All of which are readily convertible into aldehydes (N. Iranpoor, A. Imanich, Synth. Commun., 1989, 19, 2955).

(xxxiii) ruthenium dioxide/oxygen, which forms aldehydes from 2-alkenols only. It does not effect primary alkanols (M. Matsumoto, N. Watanabe, J. Org. Chem., 1984, 49, 3436).

(xxxiv) ruthenium dioxide/triphenylphosphine/ phenyliodosodiacetate ($PhI(OAc)_2$) which has limited selectivity. Thus carboxylic acids are formed together with alkanals (P. Müller, J. Codry, Tetrahedron Lett., 1981, 2361).

(xxxv) N-methylmorpholine-N-oxide/ruthenium dichloride·tristriphenylphosphine (TPP). This combination oxidizes primary alkanols and 2-alkenols quite selectively (K.B. Sharpless, K. Akashi, K. Oshima, Tetrahedron Lett., 1976, 2503). Quite acceptable selectivities are also achieved with the ruthenium dichloride tris-TPP complex without (H. Tomioka et al., Tetrahedron Lett., 1981, 1605), or with addition of bis-trimethylsilyl peroxide (S. Kanemoto et al., Tetrahedron Lett., 1983, 2185), or with ruthenium trichloride hydrate with calcium hypochloride or sodium periodate as reoxidants (J.R. Genet, D. Pons, S. Juge, Synth. Commun., 1989, 19, 1721).

(xxxvi) the trinuclear ruthenium acetate catalyst ($Ru_3O(OAc)_6L_3$, L = H_2O or TPP) and oxygen under pressure at 100°C (C. Bilgrien, S. Davis, R.S. Drago, J. Am. Chem. Soc., 1987, 109, 3786).

(xxxvii) tetrapropylammonium perruthenate/N-methylmorpholine-N-oxide. This self re-generating combination of reagents is very efficient and can be used under mild conditions. It allows simple workup ,even in the presence of various acid sensitive groups (W.P. Griffith, S.V. Ley, J. Chem. Soc., Chem. Commun., 1987, 1625; idem, Aldrichimica Acta, 1990, 23(1),143).

(xxxviii) dicyclopentadienyl zirconium hydride (Cp_2ZrH_2) or isopropoxide ($Cp_2Zr(O\text{-}iPr)_2$) with benzophenone (T. Nakano et al., Synthesis, 1986, 774) or benzaldehyde or cyclohexanone as hydrogen acceptors (T. Nakano, Y. Ishi, M. Ogawa, J. Org. Chem., 1987, 52, 4855). These reagent combinations oxidize di-primary and primary-secondary diols selectively to hydroxyaldehydes. Propargylic alcohols are unaffected.

(b) *Oxidation of primary halides*

$$RCH_2{-}X \longrightarrow R{-}CHO$$

Again a variety of reagent conditions can be employed including:

(i) dimethyl sulfoxide (DMSO)/sodium hydrogen carbonate/sodium iodide at elevated temperature with alkyl chlorides and bromides (R. Engel et al., Synth. Commun., 1986, 16, 1343).

(ii) DMSO/silver tetrafluoroborate. The silver salt considerably accelerates this Kornblum-type oxidation, with primary allylic halides even reacting at ambient temperature (B. Ganem, R.K. Boeckmann, Tetrahedron Lett., 1974, 917).

(iii) dimethylformamide(DMF)/N-ethylmorpholine-N-oxide. This is recommended for the preparation of 2-alkenals (S. Suzuki et al., Bull. Chem. Soc. Jpn., 1986, 59, 3287), or 4-dimethyl-aminopyridine-N-oxide/base for the synthesis of alkanals (S. Mukaiyama, S. Inaga, M. Yamaguchi, Bull. Chem. Soc. Jpn., 1981, 54, 2221).

(c) *Oxidation of amines*

(i) with dry air, which reacts with the lithium salts of N-alkyl-N-trimethylsilyl metal amides yielding hydrolyzable aldoximes (H.G. Chen, P. Knochel, Tetrahedron Lett., 1988, 6701).

$$R{-}CH_2NHSi(Me)_3 \xrightarrow{R'Li} R{-}CH_2\bar{N}{-}Si(Me)_3Li^+ \xrightarrow{O_2} R{-}CH{=}NOH$$

(ii) the formation of N,N-dimethylallylamine-N-oxides, which give 2-alkenals by a redox reaction with acetic anhydride (K. Takabe et al., Chem. Lett., 1982, 1987).

$$R\text{-}CH{=}CH\text{-}CH_2\text{-}\overset{+}{N}Me_2(\to O^-) \xrightarrow{Ac_2O} R\text{-}CH{=}CH\text{-}CHO$$

(d) *From alkenes, alkynes and their derivatives*

Here an addition-oxidation sequence is commonly used and may be achieved by:

(i) regioselective addition of diisoamyl borane to 1-alkenes, followed by PCC oxidation and hydrolysis of the diisoamylborinates (H.C. Brown, S.U. Kulkarni, C.G. Rao, Synthesis, 1980, 151; idem, J. Organomet. Chem., 1979, 172, C20).

(ii) regioselective addition of dibromoborane dimethylsulfide (DMS) to 1-alkynes, followed by oxidative hydrolysis of the alkenyl-B,B-dibromoborane dimethylsulfide complex (H.C. Brown, J.B. Campell, J. Org. Chem., 1980, 45, 389).

$$R\text{-}C{\equiv}C\text{-}H \xrightarrow{HBBr_2.DMS} R\text{-}CH{=}CH\text{-}BBr_2.DMS \xrightarrow[H_2O]{Ox.} R\text{-}CH_2\text{-}CHO$$

(iii) reactions of 1-alkynes with boron tribromide, followed by treatment with buffered methanolic potassium acetate and dihydrogen peroxide. This leads to the formation of 2-bromoaldehydes (Y. Satoh et al., Synthesis, 1985, 406). In another approach 2-bromoaldehydes are formed by the addition of bromine to aldehyde trimethylsilyl enolethers, followed by spontaneous loss of bromotrimethylsilane (R.H. Reuss, A. Hassner, J. Org. Chem., 1974, 39, 1785).

(iv) rearrangement of propargylamines into allenylamines. These form 2-alkenals on hydrolysis (J.C. Craig, N.N. Ekwuribe, Tetrahedron Lett., 1980, 2587). This synthesis allows labelling of both alkenic carbon atoms.

(v) treatment of 1-trimethylsilyl-1-alkenes with osmium tetroxide, followed by dehydration and concomittent C-O migration of the trimethylsilyl group. Hydrolysis of the resulting aldehyde trimethylsilyl enolethers gives the corresponding aldehydes (P.F. Hudrlik, A.M. Hudrlik, A.M. Kulkarni, J. Am. Chem. Soc., 1985, 107, 4260). α,β-Epoxysilanes were found to be inferior precursors.

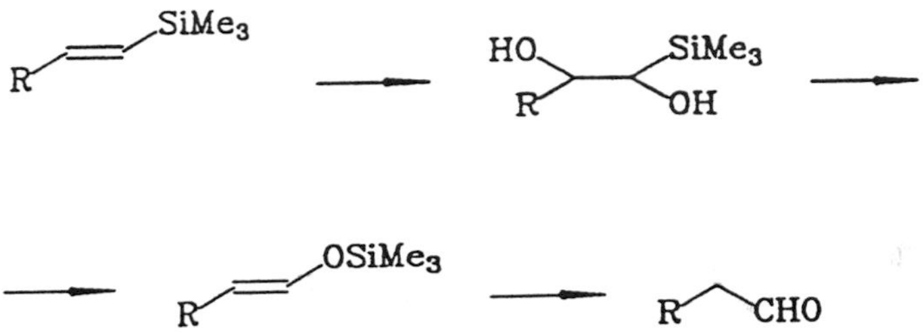

(vi) epoxidation of 1-alkenes, followed by isomerization of the oxiranes by reaction with nickel bromide, or its bis-triphenylphosphine complex in tetrahydrofuran, or toluene (A. Miyashita et al., Chem. Lett., 1986, 1323).

(vii) "Carbometallation" (P. Knochel, J.F. Normant, Tetrahedron Lett., 1986, 1039) of 1-metallo alkenes, by allylzinc halides in the presence of chlorotrimethyl-silane or -stannane, followed by oxidation of the resulting 1,1-dimetallo-4-alkenes. This leads to formation of 4-alkenals (P.Knochel, C. Xiao, M.C.P. Yeh, Tetrahedron Lett., 1988, 6697).

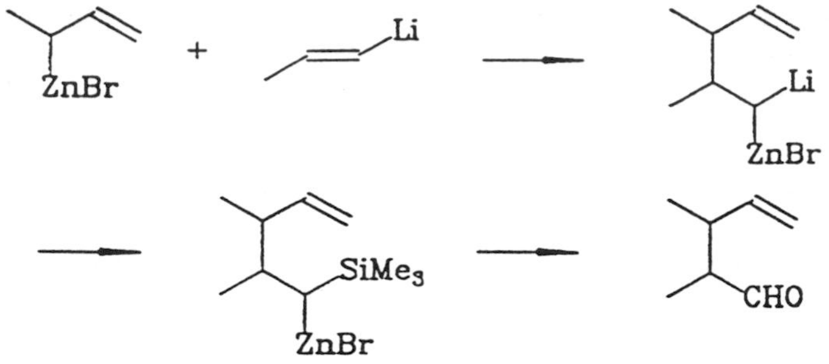

(viii) regio- and stereoselctive oxidation of allylic chlorides by 2-propane nitronate in the presence of tetrakis(triphenylphosphine) palladium. This method is suitable for variously substituted substrates (S. Suzuki et al., Synth. Commun., 1985, 15, 1123).

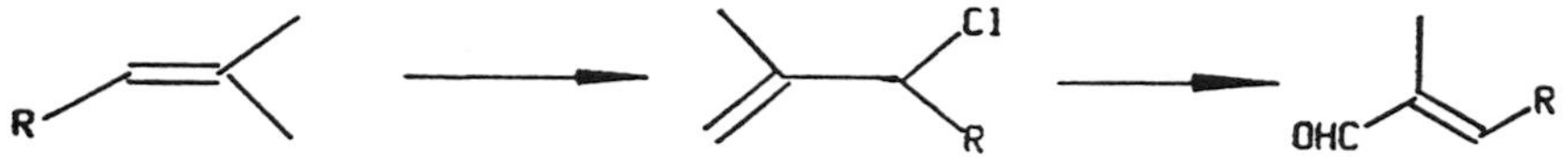

(e) *Reduction of carboxylic acids and derivatives*

(Review: J.S. Cha, "Recent developments in the synthesis of aldehydes by reduction of carboxylic acids and their derivatives with metal hydrides", Org. Prep. Proc. Int., 1989, 21, 451).

Carboxylic acids and their allies can be reduced by:

(i) isobutylmagnesium bromide in the presence of dichloro-bis-(cyclopentadienyl)titanium (for saturated carboxylic acids only) (F. Sato, T. Jiubo, M. Sato, Synthesis, 1981, 871). The reaction mechanism is unresolved.

(ii) reduction of O-acyl thexyl borinic acids, formed from carboxylic acids and excess hexylbromoborane. This procedure exhibits a high degree of compatibility with various functional groups. Diacids similarly form dials (J.S. Cha, J.E. Kim, K.W.

Lee, J. Org. Chem., 1987, 52, 5030). Thexyl chloroborane can also be used (H.C. Brown et al., J. Org. Chem., 1987, 52, 5400).

(iii) either by treatment of acyloxy-9-borabicyclo [3.3.1] nonanes (**1**) obtained from 9-BBN and carboxylic acids, with lithium 9-borabicyclo[3.3.1]nonane (Li-9BBNH) (J.S. Cha et al., Tetrahedron Lett., 1987, 4575), or by stepwise treatment of (**1**) with t-butyllithium and 9-BBN (idem, ibid., 1987, 6231). Final hydrolysis of the products from both procedures affords aldehydes.

RC(=O)OBBN (1) — 1. Li^{+}BBN, 2. H_2O → RCHO

RC(=O)OBBN (1) — 1. t-BuLi, 2. 9-BBN, 3. H_2O → RCHO

(iv) reduction of carboxylic acids by borane/dimethylsulfide with formation of trialkylboroxines, which are then oxidized by PCC (H.C. Brown, C.G. Rao, S.U. Kulkarni, Synthesis, 1979, 704).

3 RCO_2H → trialkylboroxine ($(RCH_2BO)_3$) —PCC→ 3 RCHO

(v) reduction with tri-t-butoxy-lithiumaluminum hydride of the products resulting from the reaction of carboxylic acids and chloro-N,N-dimethylformamidium chloride in the presence of copper(I)iodide at -78°C. Excellent compatibility with various functional groups is observed (T. Fujisawa et al., Tetrahedron Lett., 1983, 1543).

(vi) reduction of unsaturated esters by lithium aluminium hydride in pentane in the presence of an excess of diethylamine. This is an exceptionally facile procedure (J.S. Cha, S.S. Kim, J.Org.Chem., 1987, 52, 5486). A similar method using bis(N-methylpiperazinyl)lithium aluminium hydride is also recommended (T.D. Hubert, D.P. Eyman, D.F. Wiemer, J.Org.Chem., 1984, 49, 2279).

(vii) reduction of esters and lactones by sodium bis(2-methoxyethoxy)aluminum hydride (SMEAH) in the presence of either N-methylpiperazine or -morpholine. Very good yields are claimed (R. Kanagawa, T. Tokoroyama, Synthesis, 1976, 526).

(viii) reduction of N-methoxy-N-methylamides by lithiumaluminum hydride, or by diisobutylaluminum hydride (DIBAL) (S.Nahm, S.M. Weinreb, Tetrahedron Lett., 1981, 3815).

(ix) hydrogenolysis of acid chlorides in acetone, with diisopropylethylamine as acid scavenger (J.A. Peters, H. van Bekkum, Rec. Trav. Chim. Pays-Bas, 1971, 90, 1323).

(x) hydrogenolysis of acid chlorides in dilute solutions in the presence of equimolar amounts of 2,6-dimethylpiperidine (A.W. Burgstahler, L.O. Weigel, C.G. Schaefer, Synthesis, 1976, 767).

(xi) sodium borohydride reduction of acid chlorides in dimethylformamide/pyridine at -70°C (J.H. Babler, Synth. Commun., 1982, 12, 839).

(xii) rapid reduction of acid chlorides by Amberlyst A-26 supported borohydride (K.Y. Cordeev et al., Zh. Org. Khim., 1985, 21, 2615).

(xiii) reaction of acid chlorides, or esters, with S-methyl-1,4-diphenylthiosemicarbazide with the formation of thiazolines, followed by reduction of the corresponding thiazolinium chlorides and hydrolysis (G. Doleschall, Tetrahedron, 1976, 32, 2549).

(xiv) simply by mixing acid chlorides and group 6B anionic hydrides ($HM(CO)_4L^-$; M = Cr, W; L = (CO), PR_3) (S.C. Kao et al., Organometallics, 1984, 3, 1601).

(xv) reduction of saturated acid chlorides (only) by tetramethylammonium hydridotetracarbonylferrate ($Me_4N^+HFe(CO)_4^-$) in solution (T.E. Cole, R. Pettit, Tetrahedron Lett., 1977, 781), or in a polymer-bound form (J.P. Collman, Acc. Chem. Res., 1975, 8, 342).

(xvi) reduction of acid chlorides by μ-bis(cyantrihydroborato)-tetrakis-(triphenylphosphine)dicopper (I) ($(Ph_3P)_2CuBH_3(CN)_2$) under neutral conditions (R.O. Hutchins, M. Markowitz, Tetrahedron Lett., 1980, 813), or by bis(triphenylphosphine)copper(I)borohydride ($(Ph_3P)_2CuBH_4$) (G.W.J. Fleet, P.J.C. Harding, Tetrahedron Lett., 1979, 975).

(xvii) reduction of acid chlorides by tri-n-butylstannane catalyzed by tetrakis (triphenylphosphine) palladium, or palladium dichloride/triphenylphosphine complex. These reactions are very mild, selective and allow the preparation of unsaturated, halo- and nitro aldehydes (P. Four, F. Guibe, J. Org. Chem., 1981, 46, 4439).

(xviii) reduction of carboxylic acids (R.J.R.Corriu, G.F. Lanneau, M. Perrot, Tetrahedron Lett., 1987, 3941), or acid chlorides by hypervalent silicon hydrides (idem, ibid., 1988, 1271).

SiH_2 / NR_2 —RCOOH→ $SiH(OOCR)$ / NR_2 → RCHO

SiH_2 / NR_2 —RCOCl→ $SiH(Cl)$ / NR_2 + RCHO

(xix) an indirect procedure consisting of the formylation by ethylformate of carboxylic acid dianions, followed by decarboxylation (G.K. Koch, J.M.M. Kop, Tetrahedron Lett., 1974, 603).

$RCH{=}C(O^-Li^+)_2$ → → RCH_2CHO

(xx) the reduction of N-ethyl nitrilium tetrafluoro borates, obtained from nitriles by reaction with triethyloxonium tetrafluoroborate, by trimethylsilane, followed by hydrolysis of the resulting aldimines (J.L. Frey, J. Chem. Soc., Chem. Commun., 1974, 45).

(f) *Substituted aldehydes from aldehydes*

The following procedures are recommended:

(i) synthesis of α-phenylseleno aldehydes by treatment of alkanals with phenylselenyl chloride in ethyl acetate at ambient temperature (K.B. Sharpless et al., J. Am. Chem. Soc., 1973, 95, 6137).

(ii) preparation of α-phenylseleno aldehydes by reaction of aldehyde enamines from aldehydes with piperidine, followed by treatment with phenylselenyl chloride. Oxidation forming the selenoxide elimination and hydrolysis finally affords 2-alkenals (D.R. Williams, K. Nishitani, Tetrahedron Lett., 1980, 4417).

(iii) potassium aldehyde enolates are very good nucleophiles which can be transformed into α-iodo-, α-phenylsulfenyl-, α-alkyl- and α-allyl aldehydes (P. Groenewegen, H. Kallenberg, A. van der Gen, ibid., 1979, 2817). Direct bromination of aldehydes is accomplished by 5,5-dibromo-2,2-dimethyl-4,6-dioxo-1,3-dioxane (dibromo Meldrum's acid) (R. Bloch, Synthesis, 1978, 140). Potassium enolates of 2-alkenals are readily allylated, diallylated and prenylated with formation of α-branched aldehydes (P. Groenewegen, H. Kallenberg, A. van der Gen, Tetrahedron Lett., 1978, 491).

(iv) enantioselective synthesis of α-hydroxy aldehydes by reaction of aldehydes with a homochiral hydrazine derivative (H_2N-NH-RAMP or -SAMP), followed by metal enolate formation and oxidation (D. Enders, V. Bhushan, Tetrahedron Lett., 1988, 2437).

N–RAMP (SAMP) H R Li^+ → 1. Ox 2. E^+ → N–RAMP (SAMP) H R OE → O H R OE

ee>96%

(v) from 2-alkenals, which are subject to exclusive 1,4-reduction with the formation of alkanals by reaction with sodium dithionite ($Na_2S_2O_4$) under phase transfer conditions in water/benzene at 80°C (O. Louis-Andre, G. Gelbard, Tetrahedron Lett., 1985, 831).

4. Syntheses with extension by one carbon atom

This and the following sub-sections deal with the preparation of aldehydes by carbon-carbon coupling reactions, considered to be "the most useful carbonyl syntheses" (W.C. Still et al., J. Am. Chem. Soc., 1974, 96, 5561). (Reviews: S.F. Martin, Synthesis, 1979, 633; B.T. Gröbel, D. Seebach, Synthesis, 1977, 357; O.W. Lever, Tetrahedron, 1976, 32, 1943; P.W. Hickmott, Tetrahedron, 1982, 38, 1975; N.H. Werstink, Tetrahedron, 1983, 39, 205; T.A. Hase, J.K. Koskimies, Aldrichimica Acta, 1981, 14, 73).

(a) *From alkyl halides*

Alkyl halides can be converted into aldehydes either directly or indirectly by:

(i) an improved Bouveault-reaction which involves the treatment of primary, secundary or tertiary alkyl halides with lithium sand in mineral oil in the presence of dimethylformamide with ultrasonication and final acidic hydrolysis (C. Petrier, A.L. Gemal, J.L. Luche, Tetrahedron Lett., 1982, 3361).

(ii) reaction of Grignard reagents with anhydrous formic acid (F. Sato et al., Tetrahedron Lett., 1980, 2869). In an improved version lithium or sodium formate is used (M.Bogavac et al., ibid., 1984, 1843).

(iii) reactions of alkyl, alkenyl or alkinyl Grignard or organolithium reagents with N-formylpiperidine (G.A. Olah, M. Arvanaghi, Angew. Chem., 1981, 93, 925).

(iv) the formylation of alkyl, alkenyl or alkynyl Grignard reagents by N-(N-formyl-N-methyl)aminopiperidine (D.L. Comins, A.I. Meyers, Synthesis, 1978, 463; W. Amaratunga, J.M.J. Frechet, Tetrahedron Lett., 1983, 1143).

(v) the preparation of metallo aldimines from isonitriles and Grignard reagents, followed by hydrolysis (H.M. Walborsky et al., J. Org. Chem., 1974, 39, 600).

$$RMgX + {}^{-}C{\equiv}N^{+}{-}R' \longrightarrow \underset{XMg}{\overset{R}{>}}{=}N{-}R' \longrightarrow RCHO$$

(vi) reductive formylation through reactions of alkyl halides with polymer bound tetraalkylammonium tetracarbonylferrate (P-$PhCH_2N^+Me_3$ $HFe(CO)_4^-$) (G. Cainelli et al., J. Org. Chem., 1978, 43, 1598). This is superior to the corresponding reactions in solution (M.P. Coke, J. Am. Chem. Soc., 1970, 92, 6080).

(vii) catalytic formylation of allylic halides in the presence of bis (triphenylphosphine) palladium dibromide complex and carbon monoxide/hydrogen under pressure at 100°C. (A. Kasahara, T. Izumi, H. Yanai, Chem. Ind. (London), 1983, 898).

(viii) catalytic reductive formylation of allyl and alkenyl chlorides, bromides and iodides by tributylstannane/tetrakis(triphenylphosphine) palladium and carbon monoxide at 50°C under low pressure (V.P. Baillargeon, J.K. Stille, J. Am. Chem. Soc., 1983, 105, 7175). Formation of 2-and 3-alkenals.

(ix) alkylation at the nucleophilic carbon atom of the 2,4-N,N-diphenyl-5-methylsulfenyl-1,2,4-triazolinium ion, followed by sodium borohydride reduction. This leads to the formation of the 3-alkyltriazolinium halide, the hydrolysis of which liberates the aldehyde (G. Doleschall, Tetrahedron Lett., 1975, 1889).

MeS, Ph, N, ⊕, N–N, Ph, ⊖ $\xrightarrow[KI]{RX}$ $\xrightarrow{NaBH_4}$ MeS, Ph, N, N–N, Ph, –R $\xrightarrow{H_3O^+}$ RCHO

(x) alkylation of the dilithium salt of methylsulfenylacetic acid, followed by sodium periodate treatment in methanol affording aldehyde dimethylacetals (B.M. Trost, Acc. Chem. Res., 1978, 11, 453).

$$MeS{-}CH{=}C(O^-)_2 + RX \longrightarrow MeS(R)CH{-}COOH \xrightarrow[MeOH,\ -CO_2]{NaIO_4} R{-}CH(OMe)_2$$

(xi) reaction of alkyl halides with lithiophenylthiomethyltrimethylsilane, followed by oxidation, thermal rearrangement of the resulting sulfonyl derivative and final hydrolysis (D.J. Ager, R.C. Cookson, Tetrahedron Lett., 1980, 1677), or by a reverse approach (idem, ibid., 1981, 587).

$$(Me_3Si)(PhS)CH^- \ Li^+ \xrightarrow[2.\ Ox]{1.\ RX} (Me_3Si)(PhS{=}O)CH{-}R \longrightarrow (Me_3SiO)(PhS)CH{-}R \longrightarrow RCHO$$

An analogous reaction sequence employs phenylselenylmethyltrimethylsilane (K. Sachdev, H.S. Sachdev, ibid., 1976, 4223).

(xii) alkylations of the lithium salt of dithiomethane monosulfoxide, followed by hydrolysis with mercuric chloride in hydrochloric acid/tetrahydrofuran (J.E. Richman, J.L.Herrmann, R.H. Schlessinger, Tetrahedron Lett., 1973, 3267).

$$(EtS{=}O)(EtS)C^- \ Li^+ \xrightarrow{RX} (EtS{=}O)(EtS)C{-}R \longrightarrow RCHO$$

This reaction sequence exemplifies the principle of formyl anion equivalents by "Umpolung" reviewed by B.T. Gröbel, D. Seebach (Synthesis, 1977, 357).

(b) *From nitro compounds*

Nitro compounds are infrequently used as substrates, however, conversion to aldehydes can be achieved by:

(i) reaction of an excess of a primary nitro compound with trialkyl orthoformate/zinc chloride, followed by the elimination of alcohol and hydrolysis of the resulting 2-nitro enol ether with formation of 2-nitro-aldehydes (L. Rene, R. Royer, Synthesis, 1981, 878).

(c) *From carboxylic acid derivatives*

Reduction of carboxylic acids is achieved by:

(i) high temperature, pressurized dicobalt octacarbonyl catalyzed, trialkylsilane-induced carbonylations of alkyl acetates affording trialkylsilyl aldehyde enolethers which form aldehydes on hydrolysis. Lactones give ω-formyl alkanoic acids (N. Chatani, S. Murai, N. Sonoda, J. Am. Chem. Soc., 1983, 105, 1370).

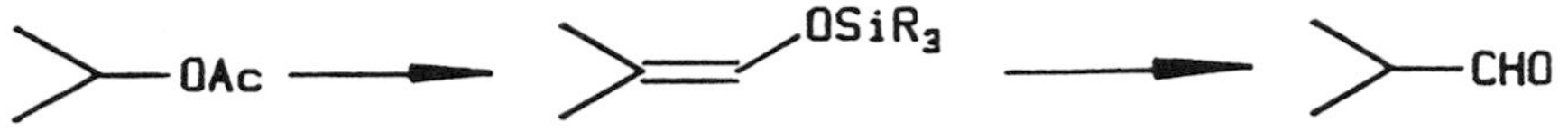

(ii) conversion of methyl alkanoates into methyl trimethylsilyl keteneacetals, which are treated with N-t-butylforminidoyl cyanide yielding 2-methoxycarbonyl-N-t-butylaldimines, the hydrolysis of which forms 2-formyl alkanoates (K. Okano et al., J. Chem. Soc., Chem. Commun., 1985, 119).

(d) *From alkenes, alkynes and derivatives*

Aldehydes may be obtained by:

(i) a sequence of reactions starting with the addition of dibromoborane to 1-alkenes, conversion into alkylboronates, reaction with methoxy(phenylthio)-

methyllithium (MPML) to form alkylated borates, their rearrangement into chain-extended boronates by mercuric chloride, oxidation and hydrolysis (H.C. Brown, T. Imai, J. Am. Chem. Soc., 1983, 105, 6285). The syntheses of α-homochiral aldehydes is also possible by this sequence (H.C. Brown et al., J. Am. Chem. Soc., 1985, 107, 4980; H.C. Brown, B. Singaram, Acc. Chem. Res., 1988, 21, 287).

R ⟶ R–CH₂CH₂–BBr_2.DMS ⟶ R–CH₂CH₂–B(OO) ⟶ R–CH₂CH₂–B(OO)–CH(SPh)(OMe) ⟶ R–CH₂CH₂–CH(OMe)–B(OO) ⟶ R–CH₂CH₂–CHO

Similar homologation reactions can be performed by employing dichloro- or bromochloromethyllithium instead of MPML (D.S. Matteson, D. Majumdar, J. Am. Chem. Soc., 1980, 102, 7588; idem, Organometallics, 1983, 2, 1529; H.C. Brown et al., ibid., 1985, 4, 1925; review: D.S. Matteson, Tetrahedron, 1989, 45, 1859).

(ii) low temperature reactions of alkenyltrimethylsilanes with 1,1-dichloromethylether/titanium tetrachloride, followed by hydrolysis forming (E)-alkenals (K. Yamamoto, O. Nunokawa, J. Tsuji, Synthesis, 1977, 721).

(iii) hydrozirconation of different kinds of alkenes, invariably forming 1-zirconyl alkanes, followed by carbonylation with carbon monoxide under pressure and finally hydrolysis (C.A. Bertelo, J.Schartz, J. Am. Chem. Soc., 1975, 97, 228).

(iv) hydrozirconation of alkenes or alkynes, followed by isocyanide insertion and hydrolysis (E.I. Negishi, D.R. Swanson, S.R. Miller, Tetrahedron Lett., 1988, 1631).

R_1–CH=CH–R_2 $\xrightarrow{Cp_2ZrClH}$ Cp_2ClZr–CH(R_1)–CH₂–R_2 ⟶

$\xrightarrow{R_3-NC}$ Cp_2ClZr–C(=NR_3)–CH(R_1)–CH₂–R_2 ⟶ OHC–CH(R_1)–CH₂–R_2

(e) *From carbonyl compounds*

Carbonyl compounds and their derivatives are commonly used and chain extension can be effected by:

(i) homologation of aldehydes as described in section 4a (ix) for alkyl halides. Functional groups in aldehydes such as keto, ester, nitro, halo or double bonds are unaffected (G. Doleschall, Tetrahedron Lett., 1980, 4183).

(ii) dimethylformamide mediated formylation (superior to the Vilsmeier reaction) of alkenyl anions obtained from toluenesulfonyl hydrazones by reaction with butyllithium. This procedure affords 2-alkenals (P.C. Traas, H. Boelens, H.J. Takken, Tetrahedron Lett., 1976, 2287).

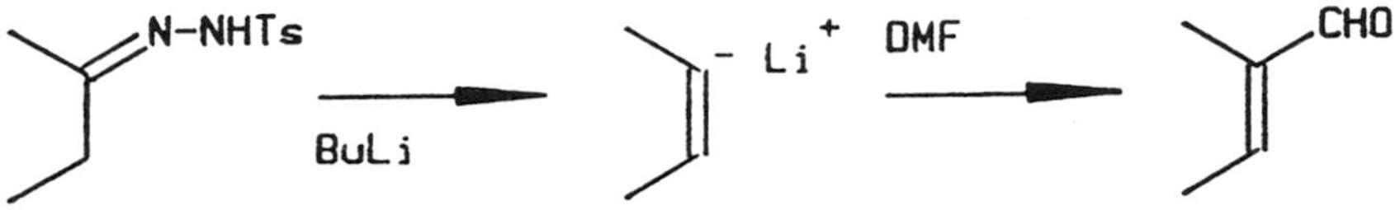

N-NHTs BuLi $^{-}$ Li^{+} DMF CHO

(iii) Wittig reactions of ketones with formation of intermediates readily subject to fragmentation (K. Schönauer, E. Zbiral, Tetrahedron Lett., 1983, 573).

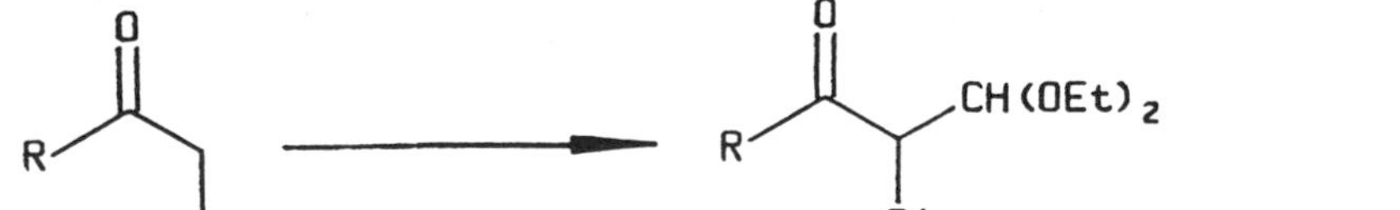

$=O + PH_3P{=}CH{-}O{-}CH_2CH_2{-}SiMe_3 \longrightarrow$ CHO

(iv) 1,3-carbonyl transposition of ketones on reaction with diethoxy carbenium tetrafluoroborate, followed by sodium borohydride reduction and treatment with acid (R. Dasgupta, U.R. Ghatak, Tetrahedron Lett., 1985, 1581).

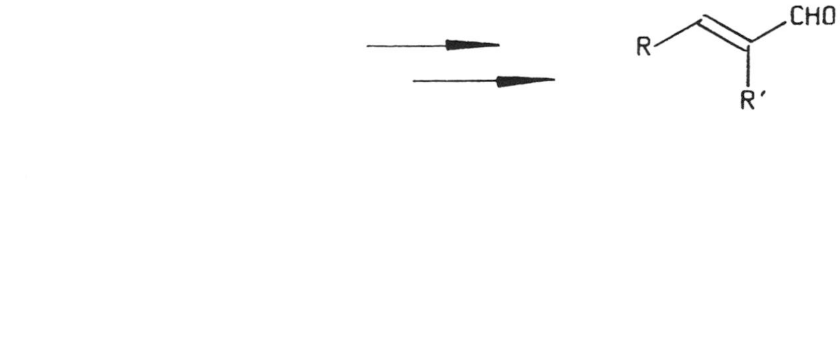

O R R' → O R CH(OEt)$_2$ R'

→ R CHO R'

(v) conversion of carbonyl compounds with tosylmethylisocyanide (TOSMIC) and thallium ethylate in ethanol into oxazoline derivatives, affording 2-hydroxyaldehydes on hydrolysis (O.H. Oldenziel, A.M. van Leusen, Tetrahedron Lett., 1974, 167).

R, R'C=O → [oxazoline: R, R', EtO, O, N] → R, R'C(OH)CHO

(vi) dealkylselenolithiation of seleno acetals, followed by reactions with dimethylformamide yielding 2-seleno alkanals (J.N. Denis, W.Dumont, A. Krief, Tetrahedron Lett., 1976, 453).

R, R'C(SeR)SeR —BuLi→ R, R'C(SeR)⁻Li⁺ —DMF→ R, R'C(SeR)CHO

(vii) reaction of carbonyl compounds with lithiochloromethyltrimethylsilane with the formation of versatile α,ß-epoxytrimethylsilanes. Their treatment with acids yields aldehydes, with methanol methylacetals, and with 1,3-propanedithiol 1,3-dithians (C. Burford et al., J. Am. Chem. Soc., 1977, 99, 4536).

(viii) "tandem" reactions of lithio methoxy(phenylthio)(trimethylsilyl)methane (T. Mandai et al. Tetrahedron Lett., 1985, 2675) and alkyl halides to give α,ß-unsaturated carbonyl compounds, followed by desilylation and liberation of the β-alkoxy-γ-oxo-aldehydes (J. Otera, Y. Niibo, H. Nozaki, J. Org. Chem., 1989, 54, 5003).

O + SPh–C(–OMe)–SiMe$_3$ + RX → O, R, SPh, OMe, Me$_3$Si → O, R, CHO

(ix) reactions of aldehydes with lithio N-morpholinomethyl diphenylphosphinic oxide (I) (A. van der Gen et al.,Tetrahedron Lett., 1979, 2433), or of ketones with lithio N-methylanilinomethyl diphenylphosphinic oxide (II) (N.Boekhof, F.J. Jonkers, A. van der Gen, Tetrahedron Lett., 1980, 2671), followed by elimination of diphenylphosphinic acid and hydrolysis of the resulting enamines.

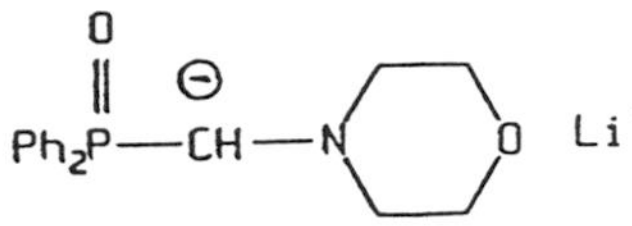

(x) reaction of carbonyl compounds with lithio bis(ethylenedioxyboryl)methide with the formation of alkenyl boronates, followed by the usual oxidative conversion into aldehydes (D.S. Matteson, R.J. Moody, J. Org. Chem., 1980, 45, 1091).

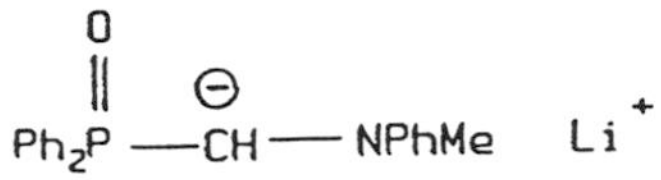

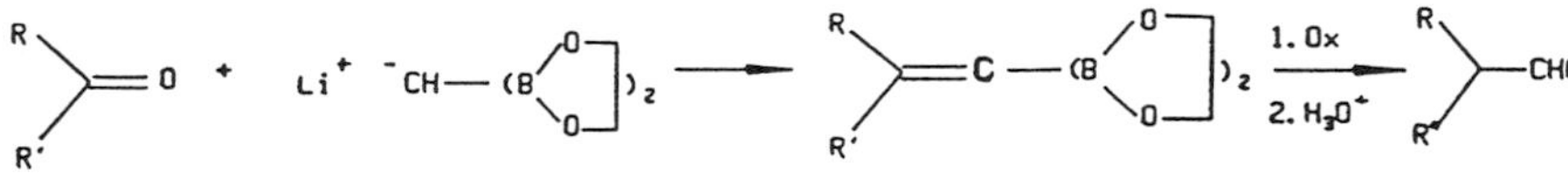

(xi) reaction of numerous different electrophiles with 2-trimethylsilylthiazole (2-TST, Dodoni's Thiazole) making available a wealth of differently and differentially substituted aldehydes through re-iterable and stereoselective syntheses (A. Dodoni et al., Gazetta Chim. Ital., 1988, 118, 211; idem, ibid., J. Org. Chem., 1988, 53, 1748).

(xii) reactions of even sterically hindered, particularly cyclic ketones with 1,1-diphenylphosphonio-1-methoxymethyllithium and iodomethane with the formation of methyl enol ethers. These suffer hydrolysis on treatment with trichloroacetic acid releasing the corresponding aldehydes (E.J. Corey, M.A. Tius, Tetrahedron Lett., 1980, 3535).

$$RR'C{=}O + Ph_2P{-}\overset{Li^+}{CH^-}{-}OMe \xrightarrow{MeI} \longrightarrow RR'CH{-}CHO$$

(xiii) samarium diiodide-induced cross coupling of ketones and dioxolane with formation of α-hydroxy aldehydes (M. Matsukawa, J. Inanga, M. Yamaguchi, Tetrahedron Lett., 1987, 5877).

$$\text{dioxolane} \xrightarrow[THF/HMPT]{SmI_2,\ PhI} \text{dioxolan-2-yl}^{\bullet} \xrightarrow{RC(O)R'} \text{dioxolanyl}{-}C(OH)RR' \longrightarrow OHC{-}C(OH)RR'$$

5. Syntheses with extension by two carbon atoms

(a) *From alkyl halides*

Three procedures are reported which are effected by:

(i) alkylation of lithiated 2-methyl thiazoline, an acetaldehyde equivalent, with alkyl halides, followed by aluminum amalgam reduction and hydrolysis with aqueous mercuric chloride. This method allows the formation of long chain aldehydes (A.I. Meyers, J.L. Durandetta, J. Org. Chem., 1975, 40, 2021).

(ii) reactions of ß-lithio ethoxyethene with 1-iodoalkanes in HMPT, followed by hydrolysis of the enolethers formed (R.H. Wollenberg, K.F. Albizati, R. Periers, J. Am. Chem. Soc., 1977, 99, 7365). For applications in natural product synthesis see R.H. Wollenberg, R. Periers (Tetrahedron Lett., 1979, 297).

(iii) two carbon homologation of alkyl halides with formation of 1-phenyl-2-hydroxyalkanes. These products suffer carbon-carbon bond fission on treatment with lead tetraacetate/iodine in refluxing benzene (K. Shankaran et al., Indian J. Chem., 1982, 21B, 408).

(b) *From alkenes*

(i) Reactions of terminal trialkylboranes with 1,2-dimethoxyethenyllithium lead to the corresponding borates, which rearrange with the formation of chain extended new boranes. Their hydrolysis yields aldehydes extended by two carbon atoms (J. Koshino et al., Synth. Commun., 1983, 13, 1149; E.I. Negishi, "Organometallics in Organic Synthesis", John Wiley & Sons, New York, Vol. 1, 1980, p 286).

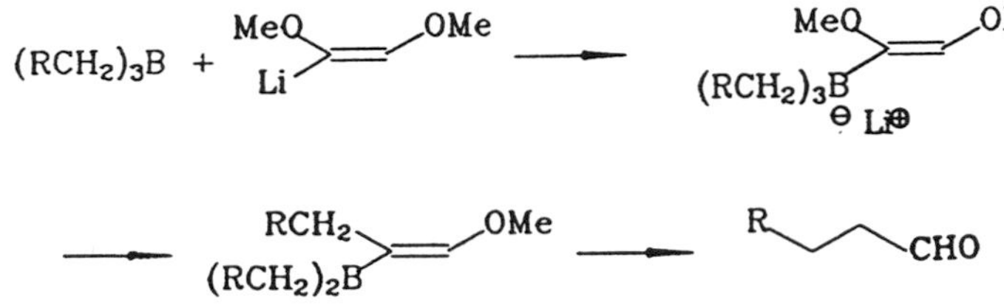

$(RCH_2)_3B$ + MeO(Li)C=CH–OMe → MeO[$(RCH_2)_3B^{\ominus}$]C=CH–OMe $Li^{\oplus}$

→ RCH_2[$(RCH_2)_2B$]C=CH–OMe → R–CH_2–CH_2–CHO

(c) *From carbonyl compounds*

A variety of procedures are advanced which require:

(i) reactions of ketones with ß-lithio ethoxyethene (c.f. subsection 5a (ii)), obtained from ß-tributylstannyl ethoxyethene, followed by mild hydrolysis with formation of α,ß-unsaturated aldehydes (J. Ficini et al., Tetrahedron Lett., 1977, 3589).

(ii) 1,3-functional group transposition through oxidation of t-allylalcohols, which results from reactions between carbonyl compounds and vinylmagnesium halides. Interestingly, PCC (c.f. section 3a, (i)) forms α,ß-unsaturated aldehydes, however,

Collins reagent (Org. Synth., 1972, 52, 5), gives α,ß-epoxy-aldehydes (P. Sundararaman, W. Herz, J. Org. Chem., 1977, 42, 813; J.H. Babler, M.J. Coghlan, Synth. Commun., 1976, 6, 469).

(iii) a similar sequence of reactions via the formation of secondary or tertiary propargylic alcohols, addition of thiophenol and Meyer-Schuster rearrangement of the ß-phenylsulfenyl allylalcohols yields α,ß-unsaturated aldehydes (M. Julia, C. Lefebre, Tetrahedron Lett., 1984, 189).

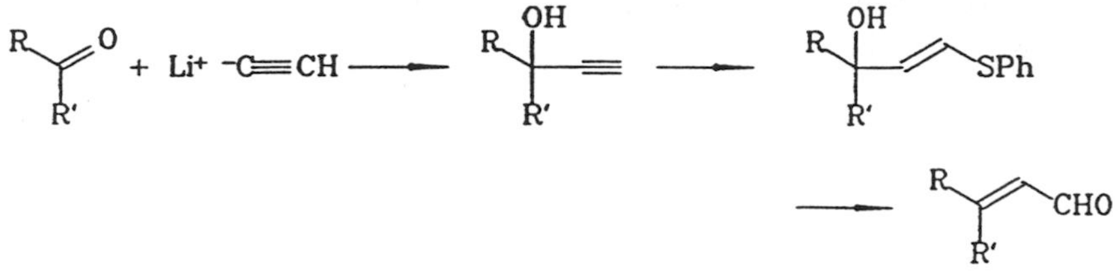

(iv) reactions of 2-(ethoxy)-1-(alkylthio)vinyl lithium with either carbonyl compounds, epoxides or alkyl halides, resulting in different intermediates, each yielding α-alkylthioaldehydes in a single step (I. Vlattas et al., J. Am. Chem. Soc., 1976, 98, 2008).

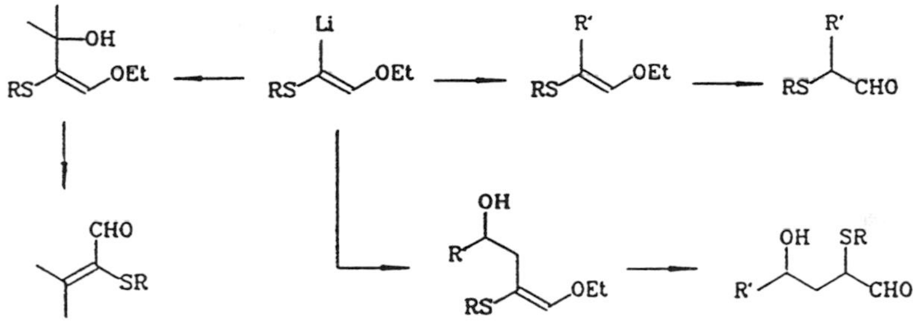

(v) the cycloaddition of aldehyde nitrones and trimethylvinylsilane, followed by a ring opening and elimination reaction with the formation of α,ß-unsaturated aldehydes (P. DeShong, M. Leginus, J. Org. Chem., 1984, 49, 3421).

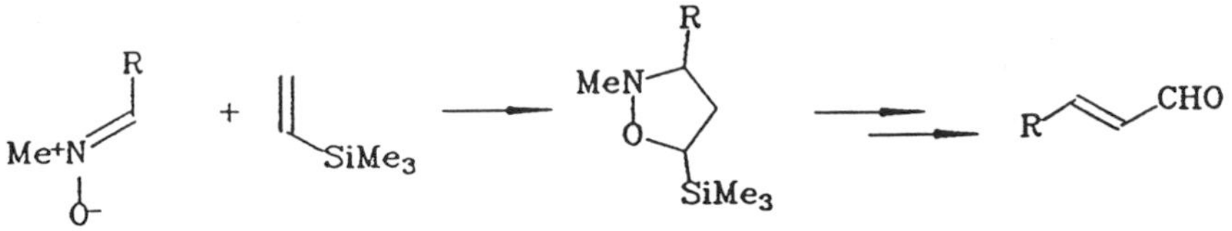

(vi) a procedure comprising the following steps: trimethylsilylation of N-cyclohexylaldimines with the formation of 2-trimethylsilyl enamines, lithiation and Peterson olefination and reaction with an aldehyde. Finally hydrolysis with trifluoroacetic acid yields α,ß-unsaturated aldehydes (S.G. Mills et al., Tetrahedron Lett., 1988, 3895).

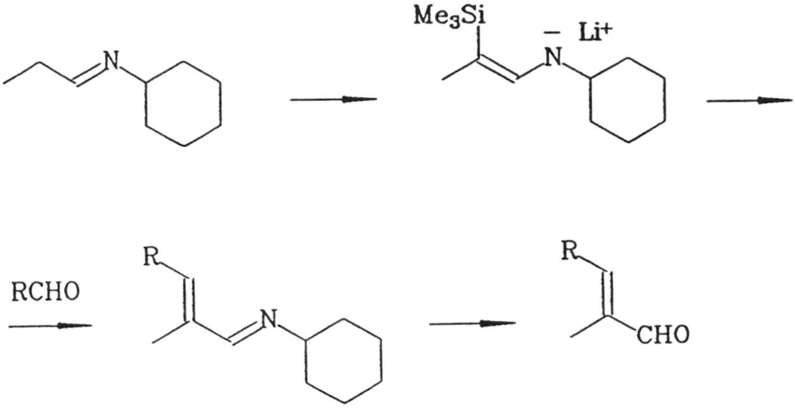

RCHO

(vii) the reaction of ketones with phenylsulfenyl cyclopropyllithium via a spiro epoxide that rearranges with the formation of a cyclobutanone derivative. Ring opening yields ß-carboxyaldehydes (B.M. Trost et al., J. Am. Chem. Soc., 1975, 97, 5873).

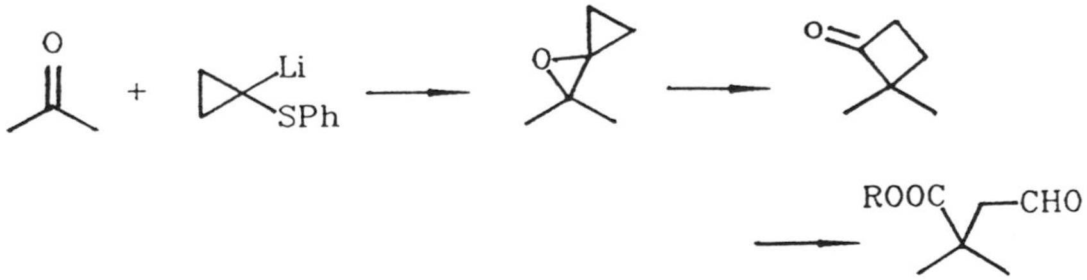

(viii) the direct formylolefination of aldehydes employing 3-formylmethyl-triphenylarsonium bromide. This is found to be superior to the corresponding Wittig reagent (Y. Huang, L.Shi, J. Yang, Tetrahedron Lett., 1985, 6447).

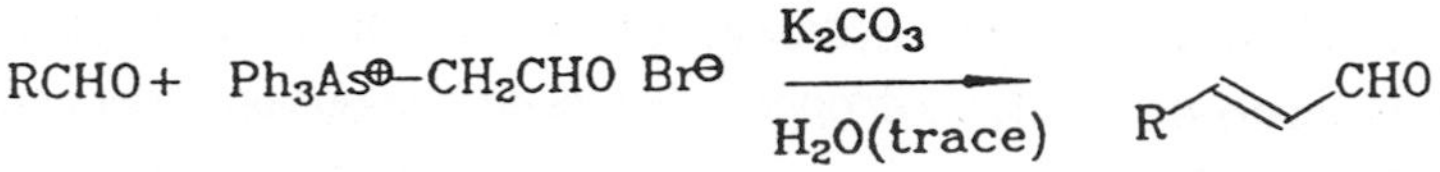

$$RCHO + Ph_3As^{\oplus}\text{–}CH_2CHO\ Br^{\ominus} \xrightarrow[H_2O(trace)]{K_2CO_3} R\text{–}CH{=}CH\text{–}CHO$$

(ix) catalytic alkenyl cuprate addition to α,ß-unsaturated homochiral cyclic acetals resulting in an asymmetric synthesis of γ,δ-alkenals (P. Mangeney <u>et al</u>., Tetrahedron Lett., 1987, 2363).

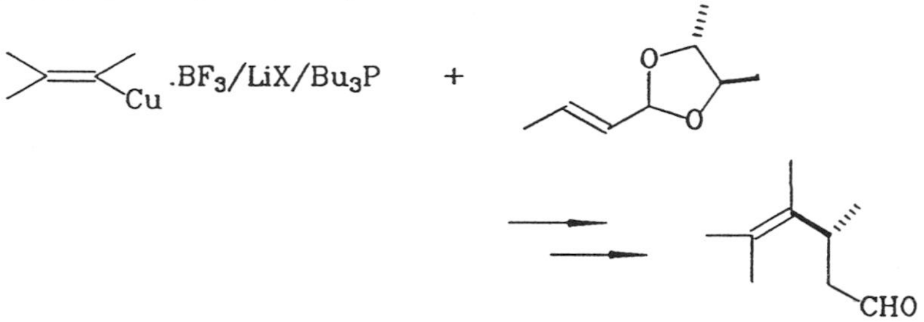

Cu $.BF_3/LiX/Bu_3P$ +

O

O

CHO

6. Syntheses with extensions by three carbon atoms

(Review: N.H. Werstiuk, "Homoenolate Anions and Homoenolate Anion Equivalents", Tetrahedron, 1981, <u>39</u>, 205)

(a) *From alkyl halides*

There is a wealth of routes from alkyl halides which include:

(i) homologation by a d^3-reagent, a 3-lithioacrolein equivalent, is accomplished by reactions of alkyl halides with the lithium salt of a 2-phenylsulfonylcyclo-propanolether. O-Deprotection, followed by a ring opening elimination reaction then gives α,ß-unsaturated aldehydes (M.Pomakotr, S. Pisutjarocupong, Tetrahedron Lett., 1985, 3613).

MEMO, Li, RX, SO_2Ph → HBF_4 → HO, R, SO_2Ph → base, H_2O → R–CH=CH–CHO

(ii) reactions of Grignard reagents with either 3-trimethylsilyloxy- or 3-ethoxyacroleins with formation of α-hydroxyenolethers, which on hydrolysis yield α,ß-unsaturated aldehydes. Thus, oxopropenylation of carbon nucleophiles occurs via intermediate vinylogous formates (review: E. Breitmeier et al., Synthesis, 1987, 1; C.W. Spangler et al., Synth. Commun., 1985, 15, 371).

RMgX + OHC–CH=CH–OR' → R–CH(OH)–CH=CH–OR' → R–CH=CH–CHO

(iii) through reactions of alkyl iodides with the allyloxy carbanionic reagents derived from the triethylsilyl-, or phenyl- ether of allylalcohol result in the formation of the respective vinyl ethers. These are valuable intermediates which give aldehydes on hydrolysis (W.C. Still, T.L. Macdonald, J. Am. Chem. Soc., 1974, 96, 5561).

Li—OR, R'I → R'–CH=CH–OR → R'–CH₂–CH₂–CHO

R = Ph; $SiEt_3$

(iv) alkylation of lithiated 1-trimethylsilyl allenyl ethers, followed by fluoride mediated desilylation and hydrolysis with formation of the 2-alkenals (J.C. Clinet, G. Linstrumelle, Tetrahedron Lett., 1980, 3987).

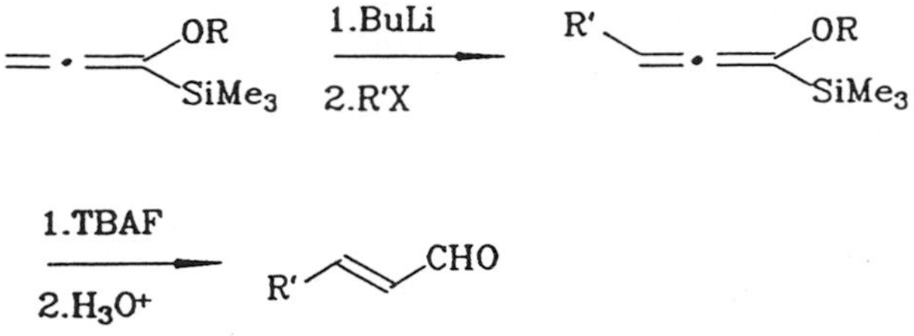

(v) alkylation of allyldimesitylborane in the presence of mesityllithium, followed by oxidation of the resulting alkenylborane (A. Pelter, B. Singaram, J.W. Wilson, Tetrahedron Lett., 1983, 631).

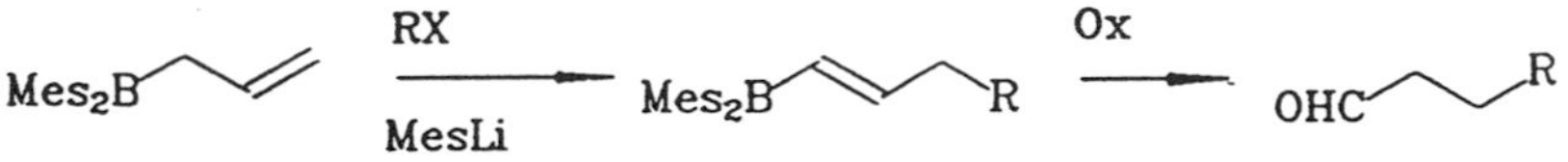

(vi) homochiral ß-alkylaldehydes are obtained by alkylation of enantiomerically pure metalloallyl amines (which are homoenolate equivalents), followed by hydrolysis of the resulting enamines (H. Ahlbrecht et al., Tetrahedron Lett., 1980, 3175).

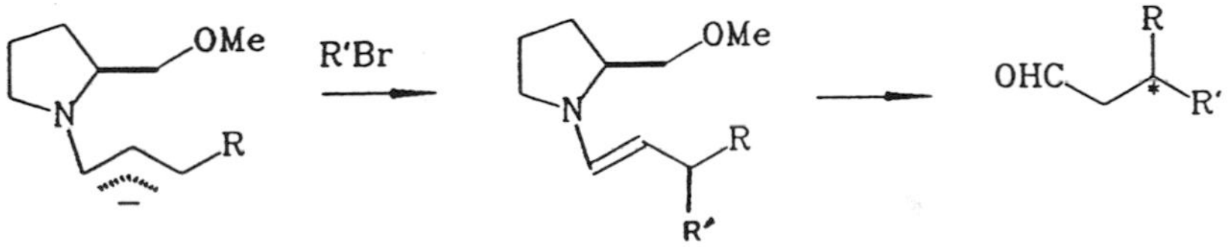

(vii) α-triisopropylsilylaldehydes which are rarely accessible by other methods, are synthesized by regioselective α-alkylation of the sterically demanding lithio allyltriisopropylsilane, followed by epoxidation of the intermediate enesilane and hydrolyic rearrangement (J.M. Muchowski, R. Naef, M.L. Maddox, Tetrahedron Lett., 1985, 5375).

(viii) by reactions of lithio 3-phenylsulfenyl-3-(trimethylsilyl)methoxypropene, a 3-trimethylsilylacrolein equivalent, with alkyl halides, followed by oxidation and elimination yields 3-trimethylsilyl-2-alkenals. (T. Mandai et al., Tetrahedron Lett., 1985, 2677).

(ix) via reaction of acrolein with 1,3-propanedithiol, triphenylphosphine and hydrobromic acid, followed by treatment with base yields 2-(2-triphenylphosphorylideneethyl)dithiane which adds alkyl halides in the presence of silver nitrate. Electroreduction liberates the aldehyde group and final treatment with triethylamine gives 2-alkenals (H.J. Christau, B. Chabaud, C. Niangoran, J. Org. Chem., 1983, 48, 1527).

(x) 1*H*-2-vinyl-5,6-dihydro-1,3-oxazine in tandem reactions with Grignard reagents and alkyl iodides, followed by hydrolysis of the intermediates gives α-alkylaldehydes (A.I. Meyers et al., J. Org. Chem., 1973, 38, 36).

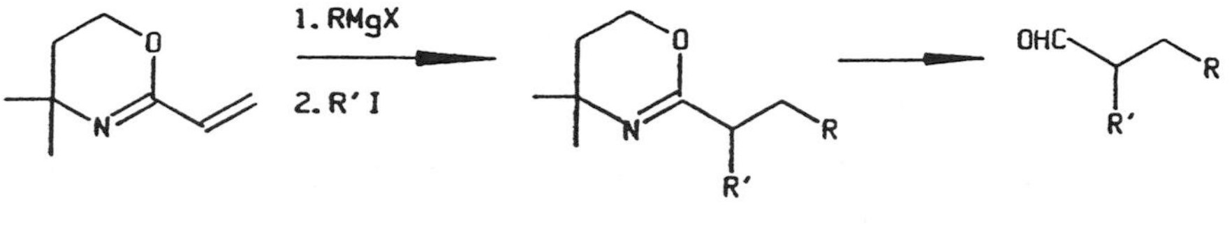

1. RMgX
2. R'I
O N R R' OHC R R'

(xi) 3-hydroxy-1,5-alkadienes, readily accessible by reactions of allyl Grignard reagents with 2-alkenals, are subject to oxy-Cope rearrangement with the formation of 5-enals on treatment with potassium hydride in DMSO at ambient temperature (D.A. Evans, D.J. Baillargeon, J.V. Nelson, J. Am. Chem. Soc., 1978, 100, 2242).

HO CHO

(b) *From carbonyl compounds*

(i) Reactions of aldehydes with (3,3-diisopropoxypropyl)triphenylarsonium ylide, a ß-formylvinylanion equivalent, followed by hydrolysis and elimination yield 4-hydroxy-2-alkenals (P. Chabert, J.B. Ousset, C. Mioskowski, Tetrahedron Lett., 1989, 179).

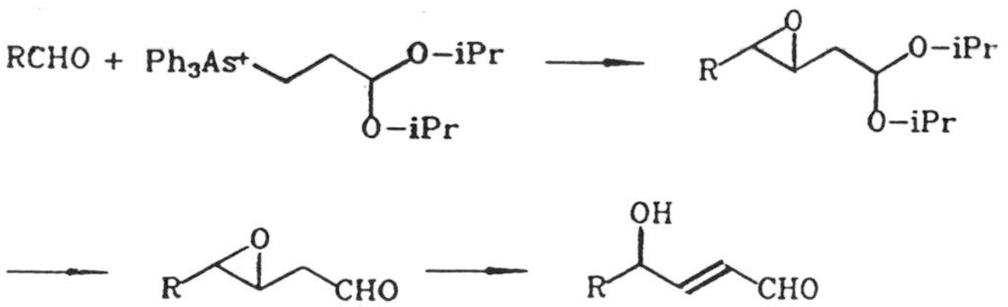

RCHO + Ph_3As^+ O–iPr O–iPr R O O–iPr O–iPr

R O CHO R OH CHO

(ii) Reactions of carbonyl compounds with 2-methoxy-cyclopropyllithium, followed by mesylation of the intermediate alcohols and methanolysis, results in the formation of 3-alkenal dimethylacetals yielding 3-alkenals on hydrolysis (E.J. Corey, P. Ulrich, Tetrahedron Lett., 1975, 3685).

O Li OMs
+ OMe OMe

MeOH H_3O^+ CHO

(c) *From carboxylic acid chlorides*

(i) α-Silyloxyallylsilanes are acylated in the presence of titanium tetrachloride and the resulting 3-oxo-silylenolethers hydrolyzed with the formation of 3-oxoalkanals (H. Sakurai, A. Hosomi, H. Hashimoto, J. Org. Chem., 1978, 43, 2551).

O $SiMe_3$ $TiCl_4$ O
R Cl + $OSiMe_3$ R $OSiMe_3$

O
R CHO

7. Syntheses with extensions by four or more carbon atoms

(a) *From carbonyl compounds*

(i) In an extension of the two-carbon homologation reaction described above (c.f. section 5c, (viii)) 3-formylallyl-triphenylarsonium bromide, (a "formyl-enyl olefination" reagent), is employed in the four carbon prolongation of aldehydes (Y. Wu, Y. Huang, Tetrahedron Lett., 1986, 4583).

RCHO + $PhAs^{+}$ ... CHO Br^{-} → R ... CHO

(b) *From alkyl halides*

(i) Alkylations of either vinylthioallyl- or vinylthio-(2-ethoxyallyl)-lithium with alkyl halides, followed by Thio-Claisen rearrangement result in five carbon extensions with the formation of 4-alkenals or 4-oxoalkanals (K. Oshima et al., J. Am. Chem. Soc., 1973, 95, 2693, 4446).

Li, S, Y → R, S, Y → Y=H: R ... CHO; Y=OEt: R ... O ... CHO

(c) *From alkenes*

(i) 2,5,7-Trienals can be obtained by direct Vilsmeier formylations of conjugated trienes (P.C. Traas et al., Tetrahedron Lett., 1977, 2129).

Second Supplements to the 2nd Edition of Rodd's Chemistry of Carbon Compounds, Vol. 1C, edited by M. Sainsbury
© 1992 Elsevier Science Publishers B.V., Amsterdam

Chapter 8b

ALIPHATIC KETONES

DEREK J SIMMONDS

1. INTRODUCTION

Since the publication of the first supplement (G.Pattenden in "Rodd's Chemistry of Carbon Compounds", 2nd Edition, Supplement to 1c/1d, M.F. Ansell (Ed.), pp. 57-107), a thorough account of ketone chemistry has appeared (A.J. Waring in "Comprehensive Organic Chemistry, Vol. 1", J.F. Stoddart (Ed.), Pergamon, Oxford, 1979, p. 1017). Sections dealing with ketones appear in "The Chemistry of the Carbonyl Group, Vol. 2", J. Zabicky (Ed.), Interscience, London, 1970, and in "The Chemistry of Functional Groups, Supplement A", Part 1, S. Patai (Ed.), John Wiley and Sons, New York, 1989.

In this review, the growing importance of cyclic ketones, enol derivatives, silicon-containing ketones, and ketene acetals has been recognised. As a result, other aspects may have suffered from down-grading or omission. Spectroscopic identification is well covered elsewhere (eg "Spectroscopic Methods in Organic Chemistry", 4th Edition, D.H. Williams and I. Fleming, McGraw-Hill, London, 1987), and comprehensive reagent lists for classical ketone transformations appear in "Compendium of Organic Synthetic Methods, Vol. 4", L.G. Wade, John Wiley and Sons, New York, 1980, and in "Advanced Organic Chemistry", 3rd Edition, J. March, John Wiley and Sons, New York, 1985, which also examines mechanistic aspects of the reactions. Furthermore this chapter is complementary to the first supplement (G. Pattenden, <u>op. cit.</u>) and transformations covered there are re-introduced only if there have been significant developments.

2. SATURATED KETONES

a) **Methods of formation and preparation**

(1) By oxidation of secondary alcohols

$R_2CHOH \longrightarrow R_2C{=}O$

A vast number of reagents is available for this transformation (see "Oxidation in Organic Chemistry", Part D, W.S. Trahanovsky (Ed.), Academic Press, New York, 1982; cf. J. March, op. cit., p. 1057, and L. G. Wade, op. cit. p. 257), and the list continues to grow. However, most of the "new" reagents are variants of a limited number of oxidising systems, (a) to (e), and selection can be made on the basis of particular requirements (scale, convenience, toxicity, sensitivity of other functional groups etc.).

(a) Dimethyl sulphoxide, the basis of the Pfitzner-Moffat protocol (see H-J. Liu and J.M. Nyangulu, Tetrahedron Lett., 1988, 29, 3167 and references cited therein), can be activated by oxalyl chloride to give the Swern reagent (K. Omura and D. Swern, Tetrahedron, 1978, 34, 1651). Swern oxidation (reviews: A.J. Mancuso and D. Swern, Synthesis, 1981, 165; T.T. Tidwell, Org. React., 1990, 39, 297) is probably the most reliable procedure and the best for large-scale preparations (cf. R.E. Ireland and D.W. Norbeck, J. Org. Chem., 1985, 50, 2198).

(b) Chromium (VI) oxide, precursor of the Jones reagent (see R.N. Warrener et al, Aust. J. Chem., 1978, 31, 1113), is parent to many useful chromate (VI) reagents (see "Chromium Oxidations in Organic Chemistry", G. Cainelli and G. Cardillo, Springer-Verlag, Berlin-Heidelberg, 1984). The most valuable of these reagents are pyridinium chlorochromate (review: G. Piancatelli, A. Scettri and M. D'Auria, Synthesis, 1982, 245), pyridinium dichromate

(E.J. Corey and G. Schmitt, Tetrahedron Lett., 1979, 399), polymer-supported chromate reagents (T. Brunelet, C. Jouitteau and G. Gelbard, J. Org. Chem., 1986, J1, 4016; but see A. McKillop and D.W. Young, Synthesis, 1979, 401 for other useful supported reagents), and variants designed to provide mild conditions (eg. J.M. Aizpurua et al, Tetrahedron, 1985, 41, 2903; H. Firouzabadi et al, ibid, 1986, 42, 719). These reagents may provide the method of choice for small-scale oxidations.

(c) Ruthenium (VIII) oxide is an effective reagent (review: J.L. Courtney, "Organic Synthesis by Oxidation with Metal Compounds", W.J. Meijs and C.R.H. de Jonge (Eds), Plenum, New York, 1986, p.445), but is expensive and potentially hazardous. It can, however, be generated in situ from ruthenium(III) chloride/sodium perbromate (Y. Yamamoto, H. Suzuki and Y. Moro-oka, Tetrahedron Lett., 1985, 26, 2107), or ruthenate (VI) salts can be useful, in catalytic quantities, with a co-oxidant (G. Green et al, J. Chem. Soc., Perkin Trans. 1, 1984, 681). Excellent reagents based on ruthenium (VII), and used with morpholine-N-oxide co-oxidant, are tetrabutyl- and tetrapropylammonium perruthenate (W.P. Griffith et al, J. Chem. Soc., Chem. Commun., 1987, 1625; review: W.P. Griffith and S.V. Ley, Aldrichimica Acta, 1990, 23, 13).

(d) Peroxides provide good reagents when used with co-oxidants. The usual peroxide species are hydrogen peroxide (eg. K. Yamawaki et al, Synth. Commun., 1988, 18, 869; O. Bortolini et al, J. Org. Chem., 1986, 51, 2661; B.M. Trost and Y. Masuyama, Tetrahedron Lett., 1984, 25, 173), or t-butyl hydroperoxide (eg. S. Kanemoto et al, ibid, 1984, 25, 3317; K. Yamawaki et al, Synthesis, 1986, 59, and Synth. Commun., 1986, 16, 537), or peracetic acid (E.J. Corey et al, Tetrahedron Lett., 1985, 26, 5855).

(e) Oxyanions (other than chromates and ruthenates) are useful oxidants. Permanganate is perhaps the best known (see D.G. Lee et al, J.

Org. Chem., 1979, 44, 3446, and 1978, 43, 1532, and Tetrahedron Lett., 1981, 22, 4889; c.f. S.L. Regen and C. Koteel, J. Amer. Chem. Soc., 1977, 99, 3837) but sodium bromate (H. Tomioka, K. Oshima and H. Nozaki, ibid, 1982, 23, 539), zinc (II) or copper (II) nitrate on silica gel (T. Nishigushi and F. Asano, J. Org. Chem., 1989, 54, 1531), and barium or silver ferrate are also useful (H. Firouzabadi et al, Synth. Commun., 1986, 16, 723 and 211; cf. K.S. Kim et al, Tetrahedron Lett, 1986, 27, 2875). Sodium hypochlorite in acetic acid has been advocated for large-scale ketone preparations (R.V. Stevens, K.T. Chapman and H.N. Weller), J. Org. Chem., 1980, 45, 2030).

(f) Palladium (II) acetate is a good oxidant for 2° alcohols. It can be used in catalytic quantities with CCl_4 or $CBrCl_3$ as reoxidant (H. Nagashima, K. Sato and J. Tsuji, Tetrahedron, 1985, 41, 5645; cf. B.M. Choudary et al, Tetrahedron Lett., 1985, 26, 6257).

(g) Secondary alcohols, R_2CHOH, are readily converted into allylcarbonate esters from which the ketone, R_2CO, is produced in a mild procedure using a palladium or ruthenium catalyst (J. Tsuji et al, ibid, 1984, 25, 2791, and 1986, 27, 1805).

(h) Certain enzyme systems are useful for the oxidation of secondary alcohols, eg horse liver alcohol dehydrogenase immobilised on glass beads (see J. Grunwald et al, J. Amer. Chem. Soc., 1986, 108, 6732 and references cited therein).

(2) From alkenes

(a) By hydroboration followed by oxidation, eg in a one-pot procedure using a chromate (VI) oxidant (H.C. Brown et al, Tetrahedron, 1986, 42, 5511 and 5515; cf. "Organic Syntheses via Boranes", H.C. Brown, John Wiley and Sons, New York, 1975):

$$RCH{=}CHR \longrightarrow (RCH_2CHR)_3B \longrightarrow RCH_2COR$$

(b) By hydroboration using an appropriate monoalkyl-t-hexylborane followed by either carbonylation or cyanidation or iodoalkynation, and then oxidation (H.C. Brown, R.K. Bakshi and B.Singaram, J. Amer. Chem. Soc., 1988, 110, 1529) eg:

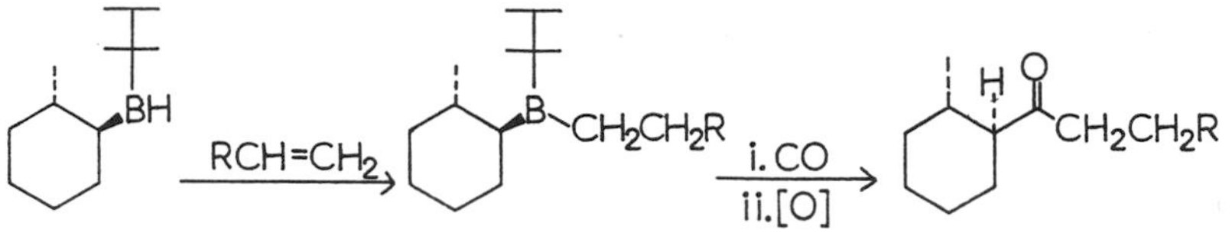

(c) By oxidation using palladium (II) salts (see "Organic Synthesis with Palladium Compounds", J. Tsuji, Springer-Verlag, New York, 1980; cf. 1st Suppl., p.58):

$$RCH{=}CHR \xrightarrow[H_2O]{PdCl_2} RCH_2COR$$

(3) From vicinal diols

(a) By oxidative cleavage into two carbonyl compounds using thallium (III) salts (A. McKillop, R.A. Raphael and E.C. Taylor, J. Org. Chem., 1972, 37, 4204), or NBS/triphenylbismuth (D.H.R. Barton et al, Tetrahedron, 1986, 42, 5627), or pyridinium chlorochromate (A. Cisneros, S. Fernandez and J.E. Hernandez, Synth. Commun., 1982, 12, 833):

$$R_2C(OH)C(OH)R_2 \longrightarrow 2R_2CO$$

(b) By a pinacol rearrangement (cf. First supplement, p.59) (review: M. Bartok and A. Molnar in "The Chemistry of Functional Groups, Supplement E", S. Patai (Ed.), John Wiley and Sons, New York, 1980, p.722); a related approach is possible using certain α-iodoethers as precursors for benzyl ketones (A. Citterio and M. Gandolfi, Tetrahedron Lett., 1985, 26, 1665):

$$RC(OMe)(Ph)CH_2I \xrightarrow{\text{peracid}} RCOCH_2Ph$$

(4) From epoxides

(a) Appropriate epoxides give ketones in a rearrangement catalysed by Lewis acids among other promoters. With Lewis acids the migratory aptitude of R^1 is aryl>acyl>H>ethyl>methyl (review: J. Gorzynski-Smith, Synthesis, 1984, 629):

$$R_2C(-O-)C(R)(R^1) \longrightarrow R_2C(R^1)COR$$

(b) α,β-Epoxysilanes can be hydrolysed to ketones in a synthesis based on carbonyl precursors (cf. preparation 6) (F. Cooke, G. Roy and P. Magnus, Organometallics, 1982, 1, 893):

$$RR^1CO \xrightarrow{Me_3SiC(Li)(Cl)Me} R-\underset{O}{C(R^1)-C(Me)}-SiMe_3 \xrightarrow{H_2O} RR^1CHCOMe$$

(5) From silyl ethers. Jones reagent with KF oxidises trimethylsilyl ethers or t-butyldimethylsilyl ethers (H-J. Liu and I-S. Han, Synth. Commun., 1985, 15, 759), whereas the Swern reagent $[DMSO/(COCl)_2]$ is selective for trimethylsilyl ethers (C.M. Alfonso, M.T. Barros and C.D. Maycock, J. Chem. Soc., Perkin Trans. I, 1987, 1221):

$$R_2CHOSiR^1_3 \longrightarrow R_2CO$$

Treatment of an alkylcyclopropyl silyl ether with an aryl triflate, in the presence of a Pd-allyl complex as catalyst, affords a β-arylketone (S. Aoki et al, J. Amer. Chem. Soc., 1988, 110, 3296):

$$R-C_3H_4(OSiR^1_3) + ArOTf \xrightarrow{[Pd]} RCOCH_2CH_2Ar$$

(6) From vinylsilanes and vinyl halides

$$R^1CH{=}C(X)R^2 \longrightarrow R^1CH_2COR^2$$

Vinyl silanes ($X = SiR_3$) are oxidised to ketones either in the Stork-Colvin reaction using peracids (G. Stork and E. Colvin, ibid, 1971, 93,

2080), or, if X = SiR $(OEt)_2$, using H_2O_2/KHF_2 (cf. preparation from alkynes below) (K. Tamao, M Kumada and K. Maeda, Tetrahedron Lett., 1984, 25, 321). Vinyl halides (X = hal) give the ketones after treatment with mercury (II) acetate and TFA (S.F. Martin and T. Chou, ibid, 1978, 1943).

(7) From N-phenylthioimidates by sequential treatment with two organometallic reagents, eg organolithiums (M.E. Jazouli, S. Masson and A. Thuillier, J. Chem. Soc., Chem. Commun., 1985, 1598):

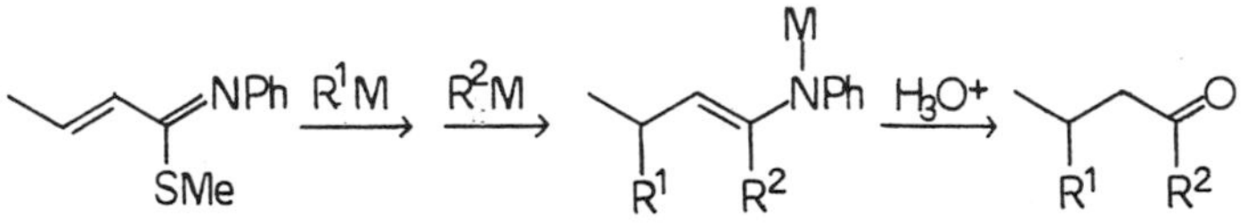

(8) From alkyl halides

(a) By treatment with acyl anion equivalents ("RCO"), eg. the nucleophiles obtained by base treatment of 2-alkyl-1,3-dithianes (D. Seebach and E.J. Corey, J. Org. Chem., 1975, 40, 231) or of aldehyde cyanohydrin acetals, RCH(CN)OCH(Me)OEt (G. Stork and L. Maldonado, J. Amer. Chem. Soc., 1971, 93, 5286; review: J.D. Albright, Tetrahedron, 1983, 39, 3207):

Hydrolysis of the alkylation product affords the ketone, $RCOR^1$, in either case. The use of reagents of this sort, that serve as synthons for a structural entity but with a polarity that is the reverse of what would be expected of that entity, is called '<u>Umpolung</u>' and is an important strategy in synthesis (review: D. Seebach, <u>Angew. Chem. Int. Ed. Engl.</u>, 1979, <u>18</u>, 239). The examples given above are usually prepared from aldehydes, RCHO, using thioacetalisation and cyanohydrin formation respectively (for other examples see A.I. Meyers <u>et al</u>, <u>J. Org. Chem.</u>, 1985, <u>50</u>, 1019; K. Ogura <u>et al</u>, <u>Chem. Lett.</u>, 1986, 1597). Other <u>Umpolung</u> reagents utilise alkyl halides (RX and R^1X) as sources of both alkyl groups either directly or, in the anion-generating step, <u>via</u> an organometallic derivative, eg. (D. Seebach <u>et al</u>, <u>Justus Liebigs Ann. Chem.</u>, 1977, 830):

CH_2=(1,3-dithiane) $\xrightarrow{RLi}$ $RCH_2^{\ominus}$(1,3-dithiane) $\xrightarrow{R^1X}$ 2-RCH_2-2-R^1-1,3-dithiane

The extensive range of reagents that serve as acylanion equivalents has been reviewed (T.A. Hase and J.K. Koshimies, <u>Aldrichimica Acta</u>, 1981, <u>14</u>, 73 and 1982, <u>15</u>, 35; cf. "Umpoled Synthons", T.A. Hase (Ed.), John Wiley and Sons, New York, 1987).

(b) With dihydro-1,3-oxazines in the Meyers synthesis of ketones (A.I. Meyers and E.M. Smith, <u>J. Org. Chem.</u>, 1972, <u>37</u>, 4289):

2-Me-oxazine $\xrightarrow[RX]{BuLi}$ 2-CH_2R-oxazine $\xrightarrow{MeI}$ N-Me oxazinium (CH_2R) $\xrightarrow{R^1MgX}$ 2-R^1-2-CH_2R-N-Me-oxazine $\xrightarrow{H_3O^+}$ RCH_2COR

(c) With sodium tetracarbonylferrate (Collman's reagent) and triphenylphosphine (J.P. Collman *et al*, *J. Amer. Chem. Soc.*, 1972, 94, 1788, and 1973, 95, 2689; cf. L.S. Hegedus and R.J. Perry, *J. Org. Chem.*, 1985, 50, 4955):

$$RX + Na_2Fe(CO)_4 \xrightarrow{Ph_3P} [RCOFe(CO)_3PPh_3]^- \xrightarrow{R^1X} RCOR^1$$

(d) By direct oxidation of secondary halides using tetrabutylammonium dichromate (D. Landini and F. Rolla, *Chem. Ind.*, 1979, 213).

(9) From acyl chlorides with organometallic reagents.

$$R^1COCl + \text{"RM"} \longrightarrow R^1COR$$

(a) Using a preformed alkylaluminium halide and Pd(O) catalyst (K. Wakamatsu *et al*, *Bull. Chem. Soc. Jpn.*, 1985, 58, 5425), or a tetraalkylsilane with $AlCl_3$ (G.A. Olah *et al*, *Synthesis*, 1977, 677), or (b) using a dialkyl cuprate reagent (G.H. Posner, C.E. Whitten and P.E. McFarland, *J. Amer. Chem. Soc.*, 1972, 94, 5106; cf. C.R. Johnson and D.S. Dhanoa, *J. Chem. Soc., Chem. Commun.*, 1982, 358; R.M. Wehmeyer and R.D. Dieke, *Tetrahedron Lett.*, 1988, 29, 4513), or (c) using a Grignard reagent with an Fe(III) catalyst (V. Fiandenese *et al*, *ibid*, 1984, 25, 4805, and 1987, 28, 2053), or (d) using an alkylmanganese halide (G. Friour *et al*, *Tetrahedron*, 1984, 40, 683, and *Synthesis*, 1985, 50), or (e) using an organonickel reagent, from nickel metal with RX (S. Inaba and R.D. Rieke, *J. Org. Chem.*, 1985, 50, 1373), or (f) using an alkylrhodium (I) complex, from $[RhCl(CO)(Ph_3P)_2]$ with RLi or RMgX (L.S. Hegedus *et al*, *J. Amer. Chem. Soc.*, 1975, 97, 5448), or (g) using a tetraalkyl tin, R_4Sn, with a palladium complex, eg. $[PdCl(Bz)(Ph_3P)_2]$ (D. Milstein and J.K. Stille, *ibid*, 1978, 100,

3036; cf. J.W. Labadie and J.K. Stille, ibid, 1983, 105, 669 and 6129), or (h) using an alkylvanadium dichloride, from VCl_3 with RMgX (T. Hirao et al, Tetrahedron Lett., 1986, 27, 929), or (i) using an alkylzinc chloride, from $ZnCl_2$ with either RLi (E. Negishi et al, ibid, 1983, 24, 5181) or RMgX (Y. Tamara et al, ibid, 1985, 26, 5529; R.A. Grey, J. Org. Chem., 1984, 49, 2288), or (j) using the zirconium reagent $RZrcp_2Cl$, from $Zrcp_2Cl_2$ with RLi (D.B. Carr and J. Schwartz, J. Amer. Chem. Soc., 1979, 101, 3521), or (k) using a tetraalkyl lead, R_4Pb (J. Yamada and Y. Yamamoto, J. Chem. Soc., Chem. Commun., 1987, 1302).

(10) From carboxylic acids,

$$R^1COOH \longrightarrow R^1COR$$

(a) Using a Grignard reagent, RMgX, in the presence of either an α-chloroenamine or dichlorotriphenylphosphorane as condensing agent (T. Fujisawa et al, Chem. Lett., 1983, 1791 and 1267), or (b) via formation of the 2-pyridyl ester and reaction with a suitable lithium dialkylcuprate (S. Kim and J.I. Lee, J. Org. Chem., 1983, 48, 2608):

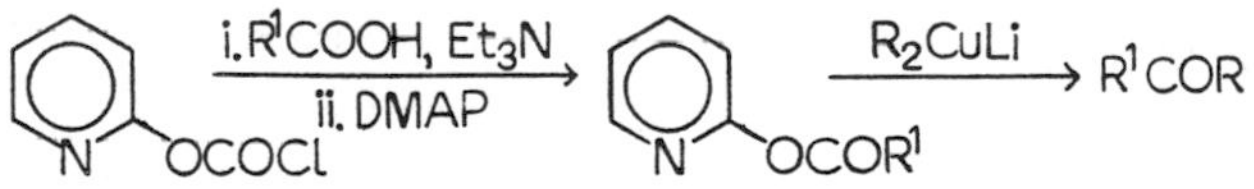

2-pyridyl–OCOCl $\xrightarrow[\text{ii. DMAP}]{\text{i. } R^1COOH,\ Et_3N}$ 2-pyridyl–$OCOR^1$ $\xrightarrow{R_2CuLi}$ R^1COR

(c) Suitable, α-branched acids give ketones via thiomethylation and decarboxylation (B.M. Trost and Y. Tamaru, J. Amer. Chem. Soc., 1977, 99, 3107):

$$RR^1CHCOOH \xrightarrow[\text{ii. MeSSMe}]{\text{i. LDA}} \left[\begin{array}{c} RR^1CCOO^- \\ | \\ SMe \end{array} \right] \xrightarrow[NaHCO_3]{\text{NCS, EtOH}} RCOR^1$$

(d) By formation of the corresponding bis-anion (LDA), alkylation with an ester, and decarboxylation (Y. Kuo, J.A. Yahner and C. Ainsworth, <u>ibid</u>, 1971, <u>93</u>, 6321):

$$R\bar{C}HCOO^- \xrightarrow{R^1COOMe} \begin{array}{c} RCHCOO^- \\ | \\ COR^1 \end{array} \longrightarrow RCH_2COR^1$$

(e) By alkylation of the lithium carboxylate followed by hydrolysis (G. Rubottom and C. Kim, <u>J. Org. Chem.</u>, 1983, <u>48</u>, 1550):

$$RCOOLi \xrightarrow{R^1Li} \begin{array}{c} RC(OLi)_2 \\ | \\ R^1 \end{array} \xrightarrow{H_2O} RCOR^1$$

(f) By treatment of the lithium carboxylate with dichlorotriphenylphosphorane followed by a Grignard reagent (G. Palumbo, C. Ferreri and R. Caputo, <u>Tetrahedron Lett.</u>, 1983, <u>24</u>, 1307):

$$RCOOLi \xrightarrow{Ph_3PCl_2} RCOO\overset{+}{P}Ph_3 \xrightarrow{R^1MgX} RCOR^1 + Ph_3PO$$

(11) <u>From acid anhydrides</u>, using a Grignard reagent with HMPT (F. Huet, G. Emptoz and A. Jubier, <u>Tetrahedron</u>, 1973, <u>29</u>, 479) or, better, using an organomanganese reagent as described for acyl chlorides above.

(12) From esters

(a) Using either a Grignard reagent and LDA, or an alkyl lithium, in either case in the presence of a silicon trapping reagent to give an enol derivative (C. Fehr, J. Galindo and R. Perret, Helv. Chim. Acta, 1987, 70, 1745; M.P. Cooke, J. Org. Chem., 1986, 51, 951):

$$RCH_2COOMe \xrightarrow[\text{or } R^1Li, Me_3SiCl]{R^1MgX, LDA, Me_3SiCl} RCH{=}C(R^1)OSiMe_3 \longrightarrow RCH_2COR^1$$

(b) via conversion to a hindered α-silyl ester and treatment with a Grignard reagent eg. (G.L. Larson et al, ibid, 1985, 50, 5260):

$$RCH_2COOMe \longrightarrow RCH(SiMe_2Ph)COOMe \xrightarrow[\text{ii. } H^+]{\text{i. } R^1MgX} RCH_2COR^1$$

(c) Methyl ketones can be prepared via methylenation of esters using the Tebbe reagent (F.N. Tebbe, G.W. Parshall and G.S. Reddy, J. Amer. Chem. Soc., 1978, 100, 3611):

$$RC(OR^1){=}O + Cp_2Ti(\mu\text{-}CH_2)(\mu\text{-}Cl)AlMe_2 \longrightarrow RC(OR^1){=}CH_2 \longrightarrow RCOCH_3$$

(d) γ,δ-unsaturated ketones can be prepared by exposure of the ester to a dienylzirconium complex (H. Yasuda et al, Chem. Lett, 1983, 217):

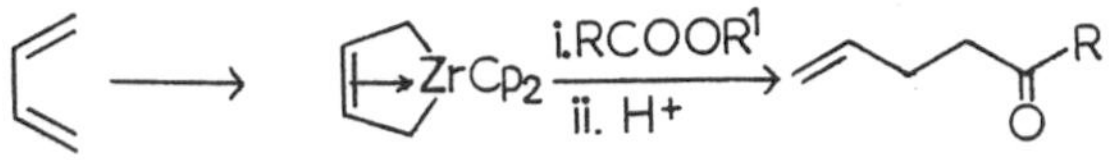

$$\text{butadiene} \longrightarrow \text{ZrCp}_2\ \text{complex} \xrightarrow[\text{ii. H}^+]{\text{i. RCOOR}^1} \text{CH}_2\text{=CHCH}_2\text{CH}_2\text{COR}$$

(13) From amides

(a) Alkyl lithiums convert tertiary amides (S. Wattanasin and F.G. Kathawala, Tetrahedron Lett., 1984, 25, 811), or N-acylaziridines (idem, ibid, 1984, 25, 4805), or tertiary thioamides (Y. Tominaga, S Kohra and A. Hoxomi, ibid, 1987, 28, 1529) into the corresponding ketones. Alkyllanthanum triflates, $R^1La(OTf)_2$, can also be used with tertiary carboxamides (S. Collins and Y. Hong, ibid, 1987, 28, 4391):

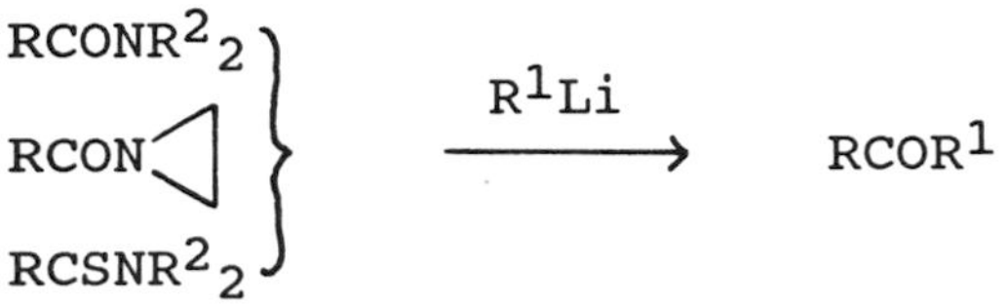

$$\left.\begin{array}{l}\text{RCONR}^2{}_2\\ \text{RCON}\triangleleft\\ \text{RCSNR}^2{}_2\end{array}\right\}\xrightarrow{\text{R}^1\text{Li}}\text{RCOR}^1$$

(14) From nitriles

(a) Using a Grignard reagent with benzene as solvent but in the presence of stoichiometric ether (P. Canonne, G.G. Foscolos and G. Lemay, ibid, 1980, 155).

(b) By degradation of α-branched nitriles via peroxide formation (R.W. Freerksen et al, J. Org. Chem., 1983, 48, 4087):

$$\text{RR}^1\text{CHCN}\xrightarrow[\text{ii. O}_2]{\text{i. LDA}}\text{RR}^1\text{C(OO}^-\text{)CN}\xrightarrow[\text{ii. }^-\text{OH}]{\text{i. H}^+/\text{Sn}^{2+}}\text{RR}^1\text{CO}$$

(c) γ,δ-unsaturated ketones can be prepared by exposure of the nitrile to a dienylzirconium complex as described for the preparation from esters above (H. Yashuda et al, Chem. Lett., 1981, 671):

$$RCN + \text{(diene)}ZrCp_2 \longrightarrow CH_2{=}CHCH_2CH_2C(O)R$$

(15) From secondary nitroalkanes

(a) By a Nef reaction, involving acid hydrolysis of the conjugate base of the nitro group (K. Steliou and M-A. Poupart, J. Org. Chem., 1985, 50, 4971; cf. R.S. Varma, M. Varma and G.W. Kabalka, Tetrahedron Lett., 1985, 26, 3777; review: H. Pinnick, Org. React., 1990, 38, 655), or (b) by direct treatment with aqueous $TiCl_3$ (J.E. McMurry, Acc. Chem. Res., 1974, 7, 281), or with activated silica gel (E. Keinan and Y. Mazur, J. Amer. Chem. Soc., 1977, 99, 3861), or with H_2O_2/K_2CO_3 (G.A. Olah et al, Synthesis, 1980, 662), or (c) by oxidative cleavage of the conjugate base of the nitroalkane using ceric ammonium nitrate, or permanganate (G.A. Olah and B.G.B. Gupta, ibid, 1980, 44; N. Kornblum et al, J. Org. Chem., 1982, 47, 4534):

$$RR^1CHNO_2 \longrightarrow \left[RR^1C{=}\overset{+}{N}(O^-)_2 \right] \longrightarrow RR^1CO$$

(d) by direct reaction with the cobalt-Schiff's base complex, [Co(Salpr)] (A. Nishinaga et al, Nippon Kagaku Kaishi, 1988, 487).

(16) From alkynes

(a) By hydrosilylation followed by oxidation of the vinyl(bisalkoxy)silane intermediate (as described for the preparation from vinyl silanes) (K. Tamao et al, Organometallics, 1983, 2, 1694):

$$R{-}{\equiv}{-}R^1 \xrightarrow[\text{ii. } H_2O_2/KHF_2]{\text{i. } MeSiH(OEt)_2)/H_2PtCl_6} RCH_2COR^1$$

(b) By mercuration, eg. mercuric oxide/Nafion-H, followed by hydrolysis (G. Olah and D. Meidar, Synthesis, 1978, 671; cf. N.X. Hu et al, Tetrahedron Lett., 1986, 27, 6099).

(c) By hydroboration using a dialkylborane, eg. 9-borobicyclo[3.3.1]nonane, followed by oxidation (H.C. Brown, C.G. Scouten and R. Liotta, J. Amer. Chem. Soc., 1979, 101, 96):

$$R{-}{\equiv}{-}R^1 \xrightarrow{\text{9-BBN}} RCH{=}C(R^1){-}B(C_8H_{14}) \xrightarrow[H_2O_2]{NaOH} RCH_2COR^1$$

(d) Via formation of an alkynyl trialkylborate:

$$R_3B + Li{-}{\equiv}{-}R^1 \longrightarrow [R_3\bar{B}{-}{\equiv}{-}R^1]Li^+$$

One alkyl group (R) will be incorporated into the ketone product so that an alkene, from which R_3B is prepared (as described for the preparation of ketones from alkenes), is a relevant precursor here also. In the presence of an electrophilic reagent the alkynyl trialkylborate is converted into a vinyl borane, eg. E=Me (A. Pelter et al,

J. Chem. Soc., Perkin Trans. 1, 1976, 2419 and 2428), from which the ketone is obtained by oxidation:

$$R_2\bar{B}(R)\text{-}{\equiv}\text{-}R^1 \xrightarrow[E=Me]{\text{eg. } Me_2SO_4} R^2B(R)C{=}C(E)R^1 \xrightarrow[H_2O_2]{NaOH} RCOCH(E)R^1$$

By using trimethylsilylmethyl triflate (in place of Me_2SO_4) the ketone E = CH_2SiMe_3 is produced (K.K. Wang, K.E. Yang and S. Dhumrongvaraporn, Tetrahedron Lett., 1987, 28, 1003 and 1007), while the use of tributyltin hydride gives a vinyl borane, E = $SnBu_3$, from which the unbranched ketone, E = H, is obtained after oxidation (K.K. Wang and K-H. Chu, J. Org. Chem., 1984, 49, 5175).

(17) From aldehydes

(a) *Via* conversion into epoxysilanes or into acyl anion equivalents, and further reactions as described in preparations 4(b) and 8(a) above.

(b) By reaction with the appropriate alkylvanadium dichloride (prepared from vanadium (III) chloride using RMgX or RLi) (T. Hirao, D. Misu and T. Agawa, J. Amer. Chem. Soc., 1985, 107, 7179):

$$R^1CHO + RVCl_2 \longrightarrow R^1COR$$

(c) By conversion into enol derivatives or nitrogen derivatives, and the alkylation of these derivatives using reactions analogous to those described in Sections 2(e) and 2(f) below.

(d) By conversion into an α-acyloxyphosphonium

salt followed by Wittig reaction with a second aldehyde and hydrolysis of the enol ester so formed (E. Anders and T. Gassner, Chem. Ber., 1984, 117, 1034):

$$R^1CHO \xrightarrow{PPh_3/RCOCl} R^1CH(\overset{+}{P}Ph_3)OCOR \xrightarrow[ii.\ R^2CHO]{i.\ Base} R^1C(OCOR)=CHR_2 \longrightarrow R^1COCH_2R^2$$

(e) By Wittig reaction with a lithio-α-diphenylphosphinoylamine, and hydrolysis of the resultant enamine (N.L.J.M. Broekhof et al, Recl.: J.R. Neth. Chem. Soc., 1984, 103, 317):

$$R^1CHO + Ph_2POC(R^2)(NR_2)Li \longrightarrow R^1CH=C(R^2)NR_2 \longrightarrow R^1CH_2COR^2$$

(f) Methyl ketones are formed by the reaction of aldehydes with trimethylsilyldiazomethane followed by hydrolysis (T. Aoyama and T. Shiori, Synthesis, 1988, 228).

$$RCHO \xrightarrow[ii.\ H_3O^+]{i.\ Me_3SiCHN_2} RCOMe$$

(18) From ethers, R_2CHOMe, by oxidation using either uranium hexafluoride (G.A. Olah and J. Welch, J. Amer. Chem. Soc., 1978, 100, 5396), or nitronium tetrafluoroborate (T-L. Ho and G.A. Olah), J. Org. Chem., 1977, 42, 3097).

(19) By fragmentation reactions involving either the pyrolysis of tertiary alkoxides, $R_2C(R^1)O^-$ (E.M. Arnett et al, ibid, 1978, 43, 815), or acid treatment of tertiary hydroperoxides, $R_2C(R^1)OOH$ (V.A. Yablokov, Russ. Chem. Rev., 1980, 49, 833), to give the ketone R_2CO when R^1 is a branched alkyl group.

(20) From sulphides by photolysis of their phenacyl derivatives (E. Vedejs and D.A. Perry, J. Org. Chem., 1984, 49, 573):

$$R_2CH{-}SH \longrightarrow R_2CH{-}S{-}CH_2COPh \xrightarrow{h\nu} R_2CO$$

(21) From Grignard reagents by carbonylation and alkylation (M. Yamashita and R. Suemitsu, Tetrahedron Lett., 1978, 761):

$$RMgBr \xrightarrow[\text{ii. } R^1I]{\text{i. } Fe(CO)_5} RCOR^1$$

(22) From acyloins by dehydroxylation using lithium diphenylphosphide (A. Leone-Bay, J. Org. Chem., 1986, 51, 2378):

$$RC(OH){=}C(OH)R \longrightarrow RCH_2COR$$

(23) From isonitriles via the formation of lithio-imines, alkylation and hydrolysis (M.J. Marks and H.M. Walborsky, J. Org. Chem., 1981, 46, 5405 and 1982, 47, 52):

$$R\text{-}\overset{+}{N}{\equiv}\overset{-}{C} \xrightarrow{R^1Li} R\text{-}N{=}\underset{\underset{Li}{|}}{C}\text{-}R^1 \xrightarrow[\text{ii. } H_2O]{\text{i. } R^2X} R^1COR^2$$

(24) <u>From keteneimines</u> <u>via</u> alkylation and hydrolysis (A.I. Meyers, E.M. Smith and M.S. Ao, <u>ibid</u>, 1973, <u>38</u>, 2129):

$$R^3\text{-}N{=}C{=}CR_2 \xrightarrow{R^1Li} R^3\text{-}\overset{\overset{Li}{|}}{N}\text{-}\underset{\underset{R^1}{|}}{C}{=}CR_2 \xrightarrow{H_2O} R^1COCHR_2$$

$$\downarrow R^2X$$

$$R^3\text{-}N{=}\underset{\underset{R^1}{|}}{C}\text{-}\underset{\underset{R^2}{|}}{C}R_2 \xrightarrow{H_2O} R^1CO\underset{\underset{R^2}{|}}{C}R_2$$

(25) <u>From allyl β-ketocarboxylates</u> by alkylation and Pd(II) catalysed deallylcarboxylation (J. Tsuji, M. Nisar and I. Shimizu, <u>ibid</u>, 1985, <u>50</u>, 3416):

$$RCOCH_2COOCH_2CH{=}CH_2 \xrightarrow[\text{ii. Pd (II)}]{\text{i. Base}/R^1X} RCOCH_2R^1$$

(26) δ,ε-unsaturated ketones are available by a Cope rearrangement of suitable allylic alcohols using bis(benzonitrile)palladium (II) chloride as catalyst (N. Bluthe, M. Malacria and J. Gore, <u>Tetrahedron Lett.</u>, 1983, <u>24</u>, 1157):

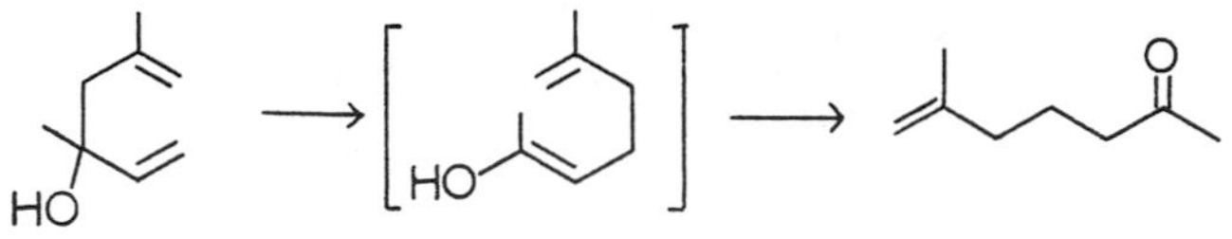

(27) From α-nitroketones using Pd(O) mediated allylation followed by denitration (N. Ono, T. Hamamoto and A. Kaji, J. Org. Chem., 1986, 51, 2832.

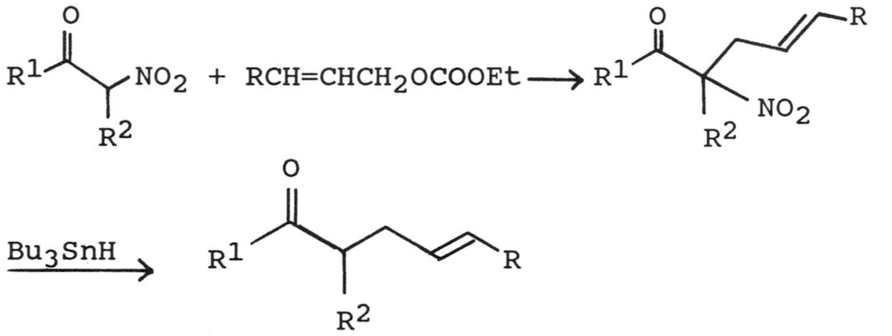

$R^1COCH(R^2)NO_2$ + $RCH{=}CHCH_2OCOOEt$ → $R^1COC(R^2)(NO_2)CH_2CH{=}CHR$

$\xrightarrow{Bu_3SnH}$ $R^1COCH(R^2)CH_2CH{=}CHR$

(28) From alkanes by oxidation using the Gif system (D.H.R. Barton et al, J. Chem. Soc., Perkin Trans. 1, 1986, 947).

(b) Reactions of the ketones

(1) Reduction to secondary alcohols

(i) Ketones are reduced to the corresponding secondary alcohols using several achiral reducing agents as follows (cf. First supplement, p.72):

$R_2CO \longrightarrow R_2CHOH$

(a) by lithium aluminium hydride in non-polar solvents with phase-transfer or crown ether catalysis (G. Gevorgyan and E. Lukevics, J. Chem. Soc., Chem. Commun., 1985, 1234); or (b) by 2-propanol in a Meerwein-Pondorf-Verlay type reduction promoted by either alumina (G.H. Posner and A.W. Runquist, J. Org. Chem., 1977, 42, 1202), or activated nickel (G.P. Boldrini et al, ibid, 1985, 50, 3082), or zirconium oxide (H. Matsushita et al, Chem. Lett., 1985, 731); or (c) using sodium borohydride either supported on alumina (E. Santaniello, F. Ponti and A. Manzocchi, Synthesis, 1978, 891), or in aqueous ethanol with cerium (III) chloride (J-L. Luche and A.L Gemal, J. Amer. Chem. Soc., 1979, 101, 5848), or in methanolic dichloromethane at -78°C (D.E. Ward, C.K. Rhee and W.M. Zoghaib, Tetrahedron Lett., 1988, 29, 517), or using sodium triacetoxyborohydride (A.K. Saksena and P. Mangiaracina, ibid, 1983, 24, 273), or the complex zirconium borohydride, $ClCp_2ZrBH_4$ (T.N. Sorrell, ibid, 1978, 4985); or (d) using trialkylstannanes promoted by Lewis acids or radical initiators, or under high pressure (M. Degueil-Castaing and A. Rahm, J. Org. Chem., 1986, 51, 1672), or tri-n-butyltin hydride on silica gel (N.Y.M. Fung et al, ibid, 1978, 43, 3977); or (e) using sodium dithionite in aqueous dioxan (J.G. Devries, T.J. Vanbergen and R.M. Kellogg, Synthesis, 1977, 246); or (f) using formamidinesulphinic acid in alkaline solution (J.E. Herz and L.A. de Marquez, J. Chem. Soc., Perkin Trans. 1, 1973, 2633); or (g) using diphenylstibine (Y.Z. Huang, Y. Shen and C. Chen, Tetrahedron Lett., 1985, 26, 5171); or (h) using chromium or tungsten carbonyl hydride (P.L. Gaus et al, J. Amer. Chem. Soc., 1985, 207, 2428); or (i) using hydrogen absorbed on the alloy $LaNi_5$ (T. Imamoto, T. Mita and M. Yokoyama, J. Chem. Soc., Chem. Commun., 1984, 164); or (j) using alkali metal reduction in protic solvents (review: S.K. Pradhan, Tetrahedron, 1986, 42, 6351), or (k) by hydrogenation in methanolic sodium hydroxide using a rhodium catalyst (G. Mestroni et al, J. Organomet. Chem., 1978, 157, 345).

(ii) Prochiral ketones can be reduced to secondary alcohols with some degree of asymmetric induction as follows:

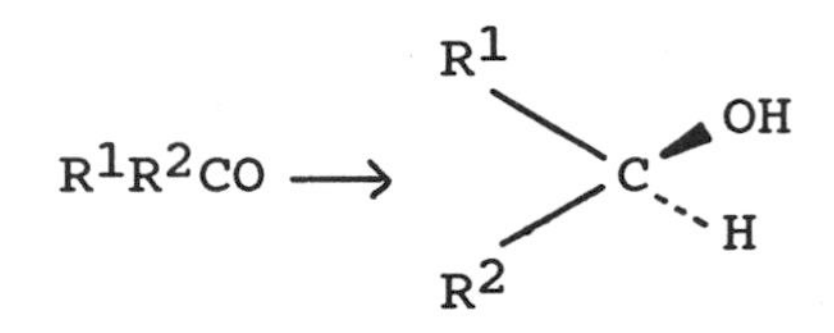

(a) Using chiral auxiliaries with achiral reducing agents such as lithium aluminium hydride (K. Kabuto et al, J. Org. Chem., 1985, 50, 3013), or diisobutylaluminium hydride (T. Mukaiyama, K. Tomimori and T. Oriyama, Chem. Lett., 1985, 813), or sodium borohydride (R. Fornasier et al, J. Org. Chem., 1985, 50, 3209; R. Noyori et al, J. Amer. Chem. Soc., 1988, 110, 629), or lithium borohydride (K. Soai et al, J. Chem. Soc., Chem. Commun., 1985, 138), or borane (S. Itsuno et al, J. Chem. Soc., Perkin Trans. 1, 1985, 2039, or Raney-nickel (T. Osawa, Chem. Lett., 1985, 1609), or in catalytic transfer hydrogenation (P. Kvintovies, B.R. James and B. Neil, J. Chem. Soc., Chem. Commun., 1986, 1810); or using a chiral, rhenium template in hydride reductions (J.M. Fernandez et al, ibid, 1988, 37).

(b) Using chiral reducing agents, often those based on borane chemistry (review: M. Srebnik and P.V. Ramachandran, Aldrichimica Acta, 1987, 20, 9) although other reagents have been advocated (eg. M. Falorni, L. Lardicci and G. Biacomelli, Tetrahedron Lett., 1985, 26, 4949; A.I. Meyer and J.D. Brown, ibid, 1988, 29, 5617). The most useful borane reagents, $R^1R^2R^3B$, are (-)-isopinocampheylborane, (R^1 = (1), R^2 = R^3 = H)

(H.C. Brown and A.K. Mandal, J. Org. Chem., 1984, 49, 2558), or (-)-diisopinocampheylborane, (R^1 = R^2 = (1), R^3 = H) (idem, ibid, 1977, 42, 2996), or 2,5-dimethylborolane, (R^1 = R^2 = -CH(Me)CH_2CH_2CH(Me)-, R^3 = H) (T. Iami et al, J. Amer. Chem. Soc., 1986, 108, 7402), or (R)-Alpine-Borane, (R^1 = (1), R^2 = R^3 = (2)) (H.C. Brown and G.G. Pai, J. Org. Chem., 1985, 50, 1384), or, especially, (-)-diisopinocampheylchloroborane, (R^1 = R^2 = (1), R^3 = Cl) (H.C. Brown, J. Chandrasekharan and P.V. Ramachandran, ibid, 1986, 51, 3394).

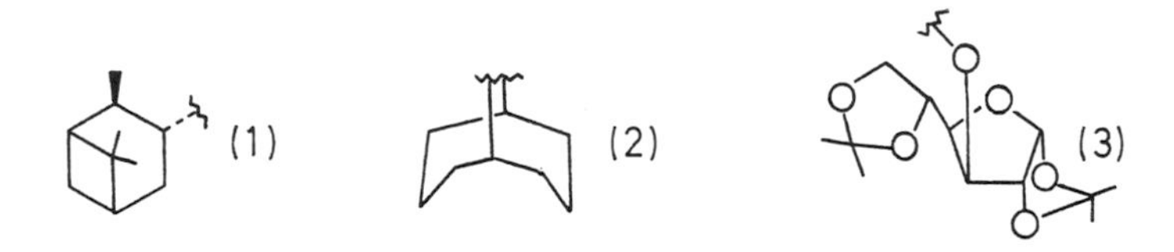

Of the available boron hydride reagents, $R^1R^2R^3BH$, Alpine-Hydride, (R^1 = (1), R^2 = R^3 = (2)), is of limited use with simple ketones (S. Krishnamurthy, F. Vogel and H.C. Brown, ibid, 1977, 42, 2534), but the sugar-modified potassium hydride, R^1 = (3), R^2 = R^3 = (2), affords enantiomeric excesses of up to 70% (H.C. Brown, W.S. Park and B.T. Cho, ibid, 1986, 51, 1934). Potassium triphenylborohydride, (R^1 = R^2 = R^3 = Ph), is a diastereoselective (syn) reductant for α-branched ketones (N.M. Yoon, K.E. Kim and J. Kang, ibid, 1986, 51, 226). The efficacy of several of the foregoing reagents, in the asymmetric reduction of a variety of ketones, has been compared (H.C. Brown et al, ibid, 1987, 52, 5406).

(c) Using natural reducing systems (review: C.J. Sih and C.S. Chen, Angew. Chem. Int. Ed. Engl., 1984, 23, 570). The simplest is baker's yeast (reviews: S. Servi, Synthesis, 1990, 1;

S. Butt and S.M. Roberts, Chem. Brit., 1987, 127) which is effective for difunctional ketones (eg. D. Seebach et al, Helv. Chim. Acta, 1987 70, 1605) but less so for simple ketones (eg. J.B. Jones, Tetrahedron, 1986, 42, 3351). A convenient isolated enzyme is alcohol dehydrogenase immobilised on a solid support (E. Keinan et al, J. Amer. Chem. Soc., 1986, 108, 162).

Other reductive or oxidative transformations

(2) Reduction to methylene

$$R_2CO \longrightarrow R_2CH_2$$

The Clemmensen reduction has been reviewed (E. Vedejs, Org. React., 1975, 22, 401). Alternative procedures include the use of triethylsilane/BF_3 (J.L. Fry et al, J. Org. Chem., 1978, 43, 374), or of sodium borohydride and acetic acid (on the ketonic tosylhydrazone) (R.O. Hutchins and N.R. Natale, ibid, 1978, 43, 2299), or of high-temperature catalytic hydrogenation (W.F. Maier et al, Tetrahedron Lett., 1981, 22, 4227).

(3) Dimethyltitanium dichloride converts ketones to the gem-dimethyl derivatives (M.T. Reetz, J. Westermann and R. Steinbach, J. Chem. Soc., Chem. Commun., 1981, 237):

$$R_2CO \longrightarrow R_2CMe_2$$

(4) Baeyer-Villiger oxidation

$$R^1COR^2 \longrightarrow R^1COOR^2 \ (+\ R^1OCOR^2)$$

The oxidation can be effected using benzeneseleninic acid/H_2O_2 (P.A. Grieco et al, J. Chem. Soc., Chem. Commun., 1977, 870), or bis(trimethylsilyl) peroxomonosulphate (M. Camporeale et al, J. Org. Chem., 1990, 55, 93), or, especially on a large-scale, permaleic acid generated in situ from maleic anhydride with 30% H_2O_2 (I. Bidd et al, J. Chem. Soc., Perkin Trans. 1, 1983, 1369). With unsymmetrical ketones, the product distribution is determined by the relative migratory aptitudes of R^1 and R^2 (for a review of this and other oxidative rearrangements see G.R. Krow, Tetrahedron, 1981, 37, 2697). Asymmetric Baeyer-Villiger oxidations can be achieved using monooxygenase enzymes (review: C.T. Walsh and Y-C. Chen, Angew. Chem. Int. Ed. Engl., 1988, 27, 333; cf. M.J. Taschner and D.J. Black, J. Amer. Chem. Soc., 1988, 110, 6892). A double oxidation is possible whereby the ketone is converted to a dialkyl carbonate (W.F. Bailey and M. Shih, ibid, 1982, 104, 1769):

$$R_2CO \longrightarrow (RO)_2CO$$

(5) Other oxidative rearrangements (review: G.R. Krow, op. cit.) can be used to convert the ketones into amides or lactams. These include the Schmidt rearrangement using hydrazoic acid, and the Beckmann rearrangement that can be applied directly to ketone substrates using hydroxylamine-O-sulphonic acid in a one-pot process (G.A. Olah and A.P. Fung, Synthesis, 1979, 537):

$$\text{cyclohexanone} \xrightarrow[\text{ii. } \Delta]{\text{i. } H_2NOSO_3H/H.COOH} \text{caprolactam}$$

(6) Oxidative cleavage gives two carboxylic acids from acyclic ketones, eg. using potassium superoxide (M. Lissel and E.V. Dehmlow, Tetrahedron Lett., 1978, 3689), or a dicarboxylic acid from cyclic ketones, eg. using chromium (VI) oxide. With nitrosyl chloride/EtOH an ω-oximinoester is produced (M.M. Rogic, J. Vitrone and M.D. Swerdloff, J. Amer. Chem. Soc., 1977, 99, 1156):

$$\text{cyclohexanone} \xrightarrow[\text{EtOH}]{\text{NOCl}} \text{CO}_2\text{Et}\ldots\text{CH}{=}\overset{+}{\text{N}}\text{HOH}\ \text{Cl}^-$$

(7) Reductive amination, by the Leuckart reaction and related procedures, has been reviewed (M.V. Klynev and M.L. Khidekel, Russ. Chem. Rev., 1980, 49, 14). The same outcome is possible by reduction ($LiAlH_4$) of the adduct of the ketone with trimethylsilyl azide (E. Kyba and A.M. John, Tetrahedron Lett., 1977, 2737):

$$R_2CO \longrightarrow R_2CHNH_2$$

(8) Reduction to ethers or thioethers. Reaction of the ketone with an alcohol and triethylsilane/TFA gives an ether (M.P. Doyle, D.J. De Bruyn and D.A. Kooistra, J. Amer. Chem. Soc., 1972, 94, 3659), while thiols with BF_3 afford thioethers (Y. Kikugawa, Chem. Lett., 1981, 1157):

$$R_2CHOR^1 \xleftarrow[\text{Et}_3\text{SiH/TFA}]{R^1\text{OH}} R_2CO \xrightarrow[\text{BF}_3]{R^1\text{SH}} R_2CHSR^1$$

(9) Reductive coupling

(a) Vicinal diols are produced by the coupling of two molecules of a ketone, either thermally or photochemically (review: B.E. Kahn and R.D. Rieke, Chem. Rev., 1988, 88, 733). The thermal reaction can be performed with a variety of metals in aqueous or non-aqueous media. Unsymmetrical products can be obtained either by coupling an inactive ketone (R = alkyl), used in excess, with an activated ketone (R^1 or R^2 is electron withdrawing) (A. Clerici and O. Porta, J. Org. Chem., 1982, 47, 2852, and Tetrahedron, 1983, 39, 1239), or by coupling a dialkyl ketone and a diaryl ketone using ytterbium metal (Z. Hou et al, J. Chem. Soc., Chem. Commun., 1988, 668):

$$R_2CO + R^1COR^2 \longrightarrow R_2C(OH)C(OH)R^1R^2$$

(b) γ-Lactones result from samarium iodide promoted coupling of ketones with α,β-unsaturated esters (S. Fukuzawa et al, J. Chem. Soc., Perkin Trans. 1, 1988, 1669):

$$R_2CO + R^1CH{=}C(R^2)CO_2Et \xrightarrow{SmI_2} \text{5,5-}R_2\text{-4-}R^1\text{-3-}R^2\text{-tetrahydrofuran-2-one}$$

(c) β-Aminoalcohols (predominant threo isomers) are produced by diastereoselective coupling of ketones with aldimines using niobium (III) chloride (E.J. Roskamp and S.F. Pederson, J. Amer. Chem. Soc., 1987, 109, 6551):

$$R_2CO + R^1CH{=}NR^2 \longrightarrow R_2CH(OH)CH(NHR^2)R^1$$

(d) Alkenes are formed by reductive coupling mediated by $TiCl_3$ or $TiCl_4$ in the presence of reductants that provide low valency titanium species involved in the alkene-forming step (reviews: J.E. McMurry, Acc. Chem. Res., 1983, 16, 405; D. Lenoir, Synthesis, 1989, 883). Unsymmetrical reactions are sometimes feasible if an excess of one ketone is used (J.E. McMurry, J. Org. Chem., 1978, 43, 3255):

$$R^1COR^2 + R^3COR^4 \longrightarrow R^1R^2C{=}CR^3R^4$$

Reactions leading to tertiary carbinols

(10) Grignard reactions can often be performed (in THF solvent) by mixing the ketone, alkyl halide and activated magnesium, ie without prior formation of the organomagnesium species (C. Blomberg and F.A. Hartog, Synthesis, 1977, 18):

$$R_2CO + R^1X \xrightarrow[\text{THF}]{\text{Mg}} R_2C(R^1)OH$$

Conversion of a Grignard reagent into its titanium or zirconium counterpart is useful for sensitive ketones (B. Weidmann and D. Seebach, Angew. Chem., Int. Ed. Engl., 1983, 22, 31), indeed clean alkylation of ketones containing ester, nitrile or amide functionality is possible, eg. (M.T. Reetz et al, Chem. Ber., 1985, 118, 1421):

$$RMgX \xrightarrow[\text{ii. } R^1COR^2]{\text{i. } ClTi(OPr^i)_3} R^1R^2C(R)OH$$

(11) Vinyl alanes are good alternatives to Grignard reagents for the preparation of allylic alcohols (N. Okuda and E. Negishi, Tetrahedron Lett., 1978, 2357):

$$R_2CO \xrightarrow[\text{BuLi}]{R^2(R^1)C{=}CH(AlMe_2)} R^2(R^1)C{=}CH{-}C(OH)R_2$$

(12) Homoallylic alcohols can be prepared by reacting the ketone and an allylic halide with zinc and ammonium chloride (C. Einhorn and J.L. Luche, J. Organomet. Chem., 1987, 322, 177), or with tin (and ultrasound) (idem, ibid., 1987, 322, 177), or with cerium amalgam (T. Imamato et al, J. Org. Chem., 1984, 49, 3904). Alternatively, the ketone can be exposed to allyl-tri-n-butyltin (G. Daude and M. Pereyre, J. Organomet. Chem., 1980, 190, 43):

$$R_2CO \longrightarrow R_2C(OH)CH_2CH{=}CH_2$$

(13) *α*-Hydroxyaldehyde acetals result from treatment of the ketone with 1,3-dioxalan in the presence of samarium iodide (M. Matsukawa, J. Inanaga and M. Yamaguchi, Tetrahedron Lett., 1987, 28, 5877):

$$R_2CO \xrightarrow{\text{1,3-dioxolane},\ SmI_2} R_2C(OH){-}\text{(1,3-dioxolan-2-yl)}$$

(14) *α-Hydroxyketones* are formed by carbonylation of a ketone in the presence of the appropriate alkyllithium (D. Seyferth, R.M. Winstein and W. Wang, *J. Org. Chem.*, 1983, *48*, 1144), or by coupling the ketone and an acyl chloride (cf. reaction 9) using samarium iodide (J. Souppe, J-L. Namy and H.B. Kagan, *Tetrahedron Lett.*, 1984, *25*, 2869):

$$R_2CO \xrightarrow[\text{or } R^1COCl,\ SmI_2]{\text{i. } R^1Li,\ CO \quad \text{ii. } H_3O^+} R_2C(OH)COR^1$$

(15) *Hydroxymethylation* to a vicinal diol is the outcome of treating the ketone with benzyl chloromethyl ether and samarium iodide, after hydrogenolysis of the benzyl moiety (T. Imamoto, T. Takeyama and M. Yokoyama, *ibid*, 1984, *25*, 3225):

$$R_2CO \xrightarrow[SmI_2]{ClCH_2OCH_2Ph} R_2C(OH)CH_2OCH_2Ph$$

$$\xrightarrow[Pd/C]{H_2} R_2C(OH)CH_2OH$$

(16) *Halogenomethylation* results from reaction of the ketone with a dihalogenomethane using samarium iodide, eg. (T. Imamoto, T. Takeyama and H. Koto, *ibid*, 1986, *27*, 3243):

$$R_2CO \xrightarrow{CH_2I_2/SmI_2} R_2C(OH)CH_2I$$

By using a hindered lithium base, the same reactants produce α,α-dihalogenoalcohols (H. Taguchi, H. Yamamoto and H. Nozaki, Bull. Chem. Soc. Jpn., 1977, 50, 1588):

$$R_2CO \xrightarrow{CH_2X_2,\ base} R_2C(OH)CX_2$$

(17) α-Hydroxyamides. In a procedure related to the Passerini reaction (review: I. Ugi, Angew. Chem. Int. Ed. Engl., 1982, 21, 810), titanium (IV) chloride can be used to promote a reaction of ketones with isonitriles (M. Schiess and D. Seebach, Helv. Chim. Acta, 1983, 66, 1618):

$$R^1\text{-}NC \xrightarrow{TiCl_4} R^1\text{-}N{=}C(Cl)TiCl_3 \xrightarrow[ii.\ H_3O^+]{i.\ R_2CO} R_2C(OH)CONHR^1$$

(18) β-Hydroxy-α-ketoesters are produced in aldol-type reactions of ketones with an oxalate (bissilyl)enolate (M.T. Reetz, H. Heinbach and K. Schwellnus, Tetrahedron. Lett., 1984, 25, 511):

$$R_2CO + (EtO)(Me_3SiO)C{=}C(OSiMe_3)(OEt) \xrightarrow[ii.\ H_3O^+]{i.\ ZnCl_2} R_2C(OH)COCO_2Et$$

(19) Chiral resolution of ketone enantiomers is possible via the diastereoisomeric adducts that are formed with lithio-N,S-dimethyl-S-

phenylsulphoxime (C.R. Johnson and J.R. Zeller, Tetrahedron, 1984, 40, 1225):

$$\overset{*}{R}COR^1 + Ph\text{-}\overset{O}{\underset{MeN}{\overset{\|}{\underset{\|}{S}}}}{}^*CH_2Li \longrightarrow Ph\text{-}\overset{O}{\underset{MeN}{\overset{\|}{\underset{\|}{S}}}}{}^*\text{-}CH_2\text{-}C(\overset{*}{R})(R^1)\text{-}OH$$

$$\xrightarrow[\text{ii. }\Delta\text{(arrows)}]{\text{i. resolve}} \overset{*}{R}COR^1$$

Alkene-forming reactions

(20) Phosphorus ylides convert the ketones into alkenes or functional alkenes by either (a) the Wittig procedure (using phosphonium salts) (review: H.J. Bestmann and O. Vostrowsky, Top. Curr. Chem., 1983, 109, 85) or (b) the Horner-Wadsworth-Emmons procedure (using phosphonates) (reviews: W.S. Wadsworth, Org. React., 1977, 25 73; W.J. Stec, Acc. Chem. Res., 1983, 16, 411). These procedures are of broad scope and are amenable to stereocontrol with respect to the new double bond (review: B.E. Maryanoff and A.B. Reitz, Chem. Rev., 1989, 89, 863).

$$R^1COR^2 \xrightarrow[\text{or (b) } R^3R^4CHPO(OEt)_2\text{/base}]{\text{(a) } R^3R^4CHPPh_3\text{/base}} (R^1)(R^2)C{=}C(R^3)(R^4)$$

With hindered ketones, reluctant olefinations are assisted by high pressure conditions (N.S. Isaacs and G.N. El-Din, Tetrahedron Lett., 1987, 28, 2191), or by gas phase transfer catalysis (E. Angeletti, P. Tundo and P. Venturello, J. Chem. Soc., Perkin Trans. 1, 1987, 713), or by promotion with ultrasound (J.V. Sinisterra, A. Fuentes and J.M. Marinas, J. Org. Chem., 1987, 52, 3875).

By using α-functional phosphonates, $(EtO)_2POCH_2X$ or $(EtO)_2POCHXY$, the Horner-Wadsworth-Emmons procedure can be used to convert ketones R_2CO, via the reactive vinyl intermediates $R_2C{=}CHX$ or $R_2C{=}CXY$, into homologous aldehydes R_2CHCHO (M. Ceruti, I. Degani and R. Fochi, Synthesis, 1987, 79), or into carboxylic acids, esters or amides ($R_2CHCOOH$, $R_2CHCOOR^1$ or $R_2CHCONR^1{}_2$) (S.E. Dinizo et al, J. Amer. Chem. Soc., 1977, 99, 182), or into condensed vinylcyclopropanes (R.T. Lewis and W.B. Motherwell, J. Chem. Soc., Chem. Commun., 1988, 751):

$$R_2CO \xrightarrow{(EtO)_2POCH{=}\overset{+}{N}{=}\overset{-}{N}} [R_2C{=}C:] \xrightarrow{\text{cyclohexene}} \text{bicyclo[4.1.0]heptane}{=}CR_2$$

Reactive cumulenes can be prepared using an appropriate enamine-phosphonate (T. Rein, B.A. Kermark and P. Helquist, Acta Chem. Scand., Ser. B, 1988, 42, 569):

$$R_2CO \xrightarrow{(EtO)_2POCH_2(CH{=}CH)_nN\text{(piperidino)}} R_2C{=}CH(CH{=}CH)_nN\text{(piperidino)}$$

(21) The Peterson reaction

Ketones are converted into alkenes via β-oxysilanes in the Peterson olefination procedure (reviews: D.J. Peterson, Organomet. Chem. Rev. A, 1972, 7, 295; D.J. Ager, Org. React., 1990, 38, 1; "Silicon Reagents in Organic Synthesis", E.W. Colvin, Academic Press, San Diego, 1988, p.63):

$$R_2CO \longrightarrow [R_2C(OH)CHR^1SiMe_3] \longrightarrow R_2C{=}CHR^1$$

The usual reagents are α-trimethylsilylalkylmagnesium halides but the corresponding organolithiums in the presence of cerium (III) chloride provide less basic but more nucleophilic species (C.R. Johnson and B.D. Tait, J. Org. Chem., 1987, 52, 281):

$$\text{eg. } R_2CO \xrightarrow[CeCl_3]{Me_3SiCH_2Li} R_2C(OH)CH_2SiMe_3 \longrightarrow R_2C{=}CH_2$$

In the final, alkene-forming step acidic and basic elimination tend to have opposite E/Z specificities in appropriate cases, allowing stereospecific alkene formation (P.F. Hudrlik and D. Peterson, J. Amer. Chem. Soc., 1975, 97, 1464). By using α-functional silanes $R^1CH(X)SiMe_3$, with LDA to provide the nucleophile, the Peterson reaction can be applied to the preparation of eg. alkylidine-β-lactams (S. Kano et al, Synthesis, 1978, 746), or α,β-unsaturated esters (cf. reaction 25) (H. Taguchi et al, Bull. Chem. Soc. Jpn., 1974, 47, 2529), or vinyl nitriles (I. Ojima and M. Kumagai, Tetrahedron Lett., 1974, 4006), or enol ethers (P.D. Magnus and G. Roy, J. Chem. Soc., Chem. Commun., 1979, 822), or vinyl silanes (J. Barluenga et al, Synthesis, 1988, 234).

(22) *Methylene formation* (cf. reactions 21 and 22).

$$R_2CO \longrightarrow R_2C{=}CH_2$$

(a) Triphenylstannylmethyl lithium, Ph_3SnCH_2Li, brings about the tin (IV) analogue of a Peterson reaction, giving the alkene *via* elimination from a β-hydroxystannane intermediate (E. Muruyama *et al*, *Chem. Lett.*, 1984, 1897; review: T. Sato, *Synthesis*, 1990, 259);

(b) Direct methylenation is brought about using the Tebbe reagent (4, M=Ti, $X=H_2$) (F.N. Tebbe, G.W. Parshall and G.S. Reddy, *J. Amer. Chem. Soc.*, 1978, *100*, 3611).

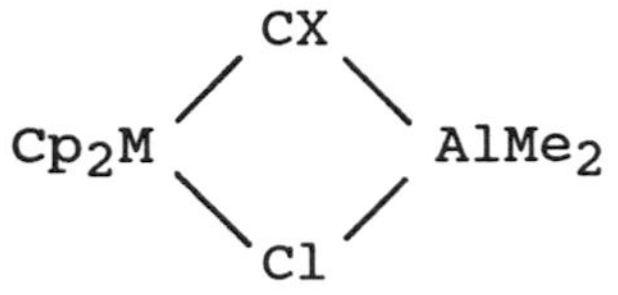

```
          CX
        /    \
Cp2M            AlMe2      (4)
        \    /
          Cl
```

or the zirconium analogue (4, M=Zr, $X=H_2$) (S.M. Clift and J. Schwartz, *ibid*, 1984, *106*, 8300), or the complex produced by aluminium and dichloromethane in ether (A.M. Piotrowski *et al*, *J. Org. Chem.*, 1988, *53*, 2829).

(c) A sulphur-analogous Peterson procedure involves the treatment of β-hydroxysulphoximes (cf. reaction 19) (C.R. Johnson and R.A. Kirchoff, *J. Amer. Chem. Soc.*, 1979, *101*, 3602), with aluminium amalgam and acid to give the alkene by elimination (M.F. Ansell *et al*, *J. Chem. Soc., Perkin Trans. 1*, 1984, 1061; cf. D.R. Morton and F.C. Brokaw, *J. Org. Chem.*, 1979, *44*, 2880):

$$R_2CO \longrightarrow PhS(O)(=NMe)CH_2C(OH)R_2 \xrightarrow[H^+]{Al\text{-}Hg} CH_2{=}CR_2$$

(23) Allenes can be prepared from ketones using a modified Tebbe reagent (4,M=Ti, X= $=CR^1R^2$) (T. Yoshida and E. Negishi, J. Amer. Chem. Soc., 1981, 103, 1276). Alternatively, terminal allenes result from the palladium catalysed decomposition of propargylic carbonate esters produced from ketones using ethynylmagnesium bromide and methyl chloroformate (J. Tsuji, T. Suguira and I. Minami, Synthesis, 1987, 603):

$$R_2CO \xrightarrow{(4,M=Ti,\ X=\ =CR^1R^2)} R_2C{=}C{=}CR^1R^2$$

R_2CO → i. ≡-MgBr; ii. $ClCO_2Me$ →

$$R_2C(OCO_2Me)C{\equiv}CH \xrightarrow[[Pd],Bu_3P]{HCO_2NH_4} R_2C{=}C{=}CH_2$$

(24) Vinylsulphones can be prepared from ketones and metallated sulphones. Reduction of the sulphone products affords alkenes (V. Pascali et al, J. Chem. Soc., Perkin Trans. 1, 1973, 1166):

$$R_2CO \xrightarrow{PhSO_2C(R^1)Li_2} R_2C{=}C(R^1)SO_2Ph \xrightarrow{Al\text{-}Hg} R_2C{=}CHR^1$$

(25) *α,β-Unsaturated esters* can be produced from ketones in the following ways:

(a) By Peterson olefination (reaction 21 above), or by condensation with a ketene silyl acetal (section 2a, part viii).

(b) By Reformatsky reaction with an α-haloester using activated zinc (review: M.W. Rathke, Org. React., 1975, 22, 423), or using commercial grade indium at room temperature (S. Araki, H. Ito and Y. Butsugan, Synth. Commun., 1988, 18, 453):

$$R_2CO \xrightarrow[\text{Zn or In}]{R^1CH(X)CO_2Me} [R_2C(OH)CH(R^1)CO_2Me]$$

$$\longrightarrow R_2C{=}C(R^1)CO_2Me$$

(c) By condensation with the lithium enolate of an ester (M.W. Rathke and D.F. Sullivan, J. Amer. Chem. Soc., 1973, 98, 3050):

$$R^1CH_2CO_2Me \xrightarrow[\text{ii. } R_2CO]{\text{i. LDA}} R_2C(OLi)CH(R^1)CO_2Me$$

$$\xrightarrow{H_3O^+} R_2C{=}C(R^1)CO_2Me$$

(d) By treatment with an alkoxyacetylide followed by acidic work-up (G.A. Olah *et al*, Synthesis, 1988, 537):

$$R_2CO \xrightarrow[\text{ii. } H_3O^+]{\text{i. } \ominus{\equiv}{-}OR^1} R_2C{=}CHCO_2R^1$$

Epoxide and episulphide formation

(26) Sulphonium ylides convert ketones into epoxides ("Sulphur Ylides", B.M. Trost and L.S. Melvin, Academic Press, New York, 1975; cf. First Supplement, p. 82). The reaction can be performed under solid-phase conditions using a dimethylsulphonium-polystyrene resin (M.J. Farrall, T. Durst and J.M.J. Frechet, Tetrahedron Lett., 1979, 203), or using trimethylsulphonium methylsulphate with phase-transfer catalysis (P. Mosset and R. Gree, Synth. Commun., 1985, **15**, 749).

(27) Lithium methylene halides convert ketones into epoxides (cf. First supplement p. 82). By using homologous reagents, $LiC(X)R^1R^2$, tri- or tetra-substituted epoxides can be produced (G. Cainelli, N. Tangari and A. Umani-Ronchi, Tetrahedron, 1972, **28**, 3009). Alternatively, disubstituted epoxides, R^1=R^2=H, are formed in good yield by the application of ultrasound to a mixture containing the ketone, lithium metal and bromochloromethane (C. Einhorn, C. Allaveno and J-L. Luche, J. Chem. Soc., Chem. Commun., 1988, 333):

$$R_2CO \xrightarrow{R^1R^2CHal_2} \text{epoxide } (R, R, R^1, R^2 \text{ on oxirane ring})$$

(28) N-p-tolylsulphonylsulphoximes convert the ketones into epoxides (C.R. Johnson et al, J. Amer. Chem. Soc., 1973, **95**, 4287; cf. First Supplement, p. 82), but use of excess of the reagent can result in the isolation of an oxetane as final product (S.C. Welch et al, ibid, 1983, **105**, 252):

$$R_2CO + Me\text{-}\overset{O}{\overset{\|}{\underset{\underset{TsN}{\|}}{S}}}\text{-}CR^1_2Na \longrightarrow R_2C\overset{O}{\underset{CR^1_2}{\triangle}} \xrightarrow[\text{excess reagent}]{R^1=H} R_2C\square O$$

(29) *α,β*-Epoxysilanes are produced by modification of the Peterson olefination protocol as described above for the preparation of ketones (see Section 2a, entry 4b) (cf. P.D. Magnus *et al*, *ibid*, 1977, *99*, 4536, and *Tetrahedron*, 1983, *39*, 867; R.F. Cunico, *Tetrahedron. Lett.*, 1986, *27*, 4269).

(30) Episulphides can be prepared using the anion of a 2-thiomethyldihydro-1,3-oxazine (A.I. Meyers, *J. Org. Chem.*, 1976, *41*, 1735):

$$R_2CO + LiCH_2S\text{-(4,4,6-trimethyl-5,6-dihydro-4H-1,3-oxazin-2-yl)} \longrightarrow R_2C\overset{S}{\underset{CH_2}{\triangle}}$$

α-Substitution reactions

α-Substitution is often performed *via* stoichiometric conversion of the ketone into a reactive derivative (cf. Sections 2d - 2h below) either as a separate reaction or as part of a one-pot protocol. The following procedures begin with the ketone.

(31) *β*-Ketoesters can be prepared using Mander's reagent, methyl cyanoformate (L.N. Mander and P. Sethi, *Tetrahedron Lett.*, 1983, *24*, 5425):

$$RCOMe \xrightarrow[\text{ii. } NCCO_2Me]{\text{i. Base}} RCOCH_2CO_2Me$$

(32) *β*-Ketonitriles are available from ketones (but not methyl ketones) using p-tosyl cyanide at low temperature (D. Kahne and D.B. Collum, ibid, 1981, 22, 5011):

$$\mathrm{RCOCHR^1_2} \xrightarrow[\text{ii. TSCN, -78}^\circ\text{C}]{\text{i. LDA}} \mathrm{RCOC(CN)R^1_2}$$

(33) *β*-Ketosulphoxides are useful chiral derivatives of the ketones for subsequent asymmetric alkylation procedures. They are prepared, using eg. (S)-(-)-menthyl p-tosylsulphinate, and revert to the ketone by reduction (eg. Al-Hg, or Ni) (M.C. Carreno et al, J. Chem. Soc., Perkin Trans. 1, 1989, 1335):

$$\mathrm{R^1COCH_2R^2} \xrightarrow[\text{ii. (S)-p-tol.SO}_2\text{menthyl}]{\text{i. i-Pr}_2\text{NMgBr}} \mathrm{R^1COCH(R^2)S(O)\text{-}p\text{-}tol}$$

(34) *α*-Thiocyanation of the ketones is possible using copper (II) thiocyanate (S.M. Ali et al, J. Chem. Res. (S), 1981, 234):

$$\mathrm{RCOCHR^1_2} \xrightarrow{\mathrm{Cu(SCN)_2}} \mathrm{RCOC(SCN)R^1_2}$$

(35) *α*-Hydroxyketones and their esters

$$\mathrm{RCOCHR^1_2} \longrightarrow \mathrm{RCOC(OR^2)R^1_2}$$

Direct hydroxylation of the ketones is achievable using molybdenum peroxide/pyridine/HMPA at -70 °C (E. Vedejs, J. Amer. Chem. Soc., 1974, 96 5944) or, for methyl ketones (R^1=H) using o-iodosylbenzoic acid (R.M. Moriarty and K-C. Hou, Tetrahedron Lett., 1984, 25, 691).

α-Acetoxylation (R^2=MeCO) can be brought about using mercuric acetate (D.J. Rawlinson and G. Sosnovsky, Synthesis, 1973, 567), while iodosobenzene methanesulphonate, PhI(OH)(OMs), affords the corresponding methanesulphonate (R^2=$MeSO_2$) (J.S. Lodaya and G.F. Koser, J. Org. Chem., 1988, 53, 210).

(36) α-Allylketones can be prepared from ketones using O-allyl-iso-dicyclohexylurea with a palladium catalyst (Y. Inoue, M. Toyofuko and H. Hashimoto, Chem. Lett., 1984, 1227):

$$RCOCHR^1_2 + CH_2{=}CHCH_2O{-}C({=}NC_6H_{11})NHC_6H_{11} \xrightarrow{PdL_4} RCOC(R^1_2)CH_2CH{=}CH_2$$

(37) Aldol-type reactions. Crossed-aldol reactions involving ketonic substrates are considered elsewhere [cf. Sections 2d-2h, 3a(viii), 3b(iii), 3c]. The bimolecular condensation of a ketone to an α,β-enone can be carried out in the presence of N,O-bis(trimethylsilyl)-β-alanine (G. Schulz and W. Steglich, Angew. Chem. Int. Ed. Engl., 1977, 16, 251), or of basic alumina (J. Muzart, Synthesis, 1982, 60):

$$2MeCOR \xrightarrow[\text{or basic alumina}]{Me_3SiNHCH_2CH_2CO_2SiR^1_3}$$

R O R

Miscellaneous reactions

(38) Vinyl halides

$$RCOCHR^1_2 \longrightarrow RC(X)=CR^1_2$$

Vinyl chlorides (X=Cl) result from treatment of the ketones with PCl_5 followed by aqueous alkali (P.F. Hudrlick and A.K. Kulkarni, Tetrahedron, 1985, 41, 1179), while sequential treatment of the ketones with triflic anhydride then magnesium iodide and base affords the corresponding iodide (X=I) (A.G. Martinez et al, Synthesis, 1986, 220). For the use of phosphorus ylides see Reaction (20) and the reviews cited therein.

(39) Gem-difluorides are available in good yield from the reaction of ketones with diethylaminosulphur trifluoride (W.J. Middleton, J. Org. Chem., 1975, 40, 574):

$$R_2CO \xrightarrow{Et_2NSF_3} R_2CF_2$$

(40) Cycloaddition. Ketones undergo cycloaddition with appropriate dienes. For example, the acetoacetic ester bis-enolate, 1,3-dimethoxy-1-trimethylsilyloxy-1,3-butadiene, allows access to lactones (M.M. Midland and R.S. Graham, J. Amer. Chem. Soc., 1984, 106, 4294):

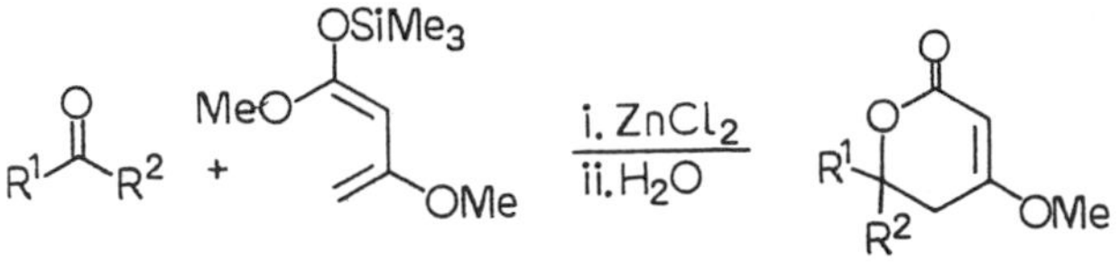

$$R^1COR^2 + \text{MeO(OSiMe}_3\text{)C=CH-C(OMe)=CH}_2 \xrightarrow[\text{ii. } H_2O]{\text{i. } ZnCl_2} \text{2,3-dihydro-4-pyranone}$$

(41) Thioketones, R_2CS (cf. First Supplement, p.91), are conveniently prepared from the parent (oxo) ketones using Lawesson's reagent, 2,4-bis(4-methoxyphenyl)-1,3-dithia-2,4-diphosphetane-2,4-disulphide (B.S. Pedersen et al, Bull. Soc. Chim. Belges, 1978, 87, 229; cf. M.P. Cava and M.I. Levinson, Tetrahedron, 1985, 41, 5061). The chemistry of thioketones has been reviewed (A. Ohno in "Organic Chemistry of Sulphur", S. Oae (Ed.), Plenum, New York, 1977, p.189):

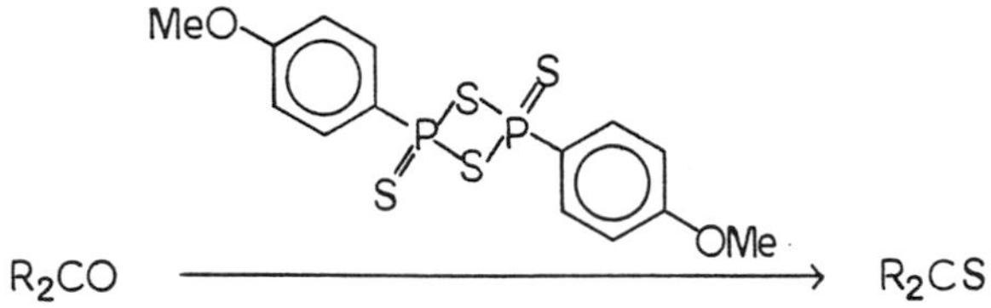

$$R_2CO \longrightarrow R_2CS$$

(42) Carbonyl transposition involves the isomerisation of a ketone by rearrangement of the carbonyl group to another site (generally a neighbouring carbon atom). Many strategies have evolved to effect the transposition, and useful reviews are available (D.G. Morris, Chem. Soc. Rev., 1982, 11, 397; V.V. Kane et al, Tetrahedron, 1983, 39, 345):

$$RCOCH_2R^1 \longrightarrow RCH_2COR^1$$

(c) Cycloalkanones

Medium- and large-ring ketones are comparable to their acyclic counterparts in terms of functional group transformations. However, specific preparations and reactions often apply to small-ring ketones.

(i) Cyclopropanones can be obtained (a) from ketenes using diazoalkanes (E.F. Rothgery, R.J. Holt and H.A. McGee, J. Amer. Chem. Soc., 1975, 97, 4971), or (b) from cyclopropene acetals via conjugate addition and hydrolysis (E. Nakamura, M. Isaka and S. Matsuzawa, ibid, 1988, 110, 1297):

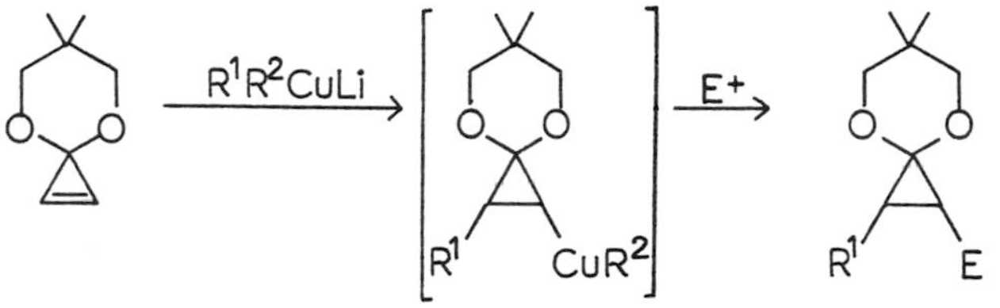

Despite their lability, several cyclopropanones have been isolated and investigated (H.H. Wasserman et al, Top. Curr. Chem., 1974, 47, 73).

(ii) Cyclobutanones are available (a) by [2+2]-cycloaddition of ketenes with alkenes (D. Bellus and B. Ernst, Angew. Chem. Int. Ed. Engl., 1988, 27, 797; cf. L.A. Paquette, R.S. Valpey and G.D. Annis, J. Org. Chem., 1984, 49, 1317), or (b) by ring-expansion of cyclopropane derivatives (J. Salvan and B. Karkour, Tetrahedron Lett., 1988, 29, 1537), or (c) by cyclisation/oxidation of a 4-bromoalkyl-phosphonium salt (L. Fitjer and U. Quabeck, Synthesis, 1987, 299).

The considerable synthetic utility of cyclobutanones has been reviewed (D. Bellus and B. Ernst, op. cit.).

(iii) Cyclopentanones derive from (a) cyclobutanones by ring-expansion or (b) pent-4-enols by rhodium catalysed cyclisation (K. Sakai et al, Tetrahedron, 1984, 25, 961), or (c) 1-vinylcyclopropyl silyl ethers by thermal rearrangement (J. Ollivier and J. Salan, ibid, 1984, 25, 1269), or (d) 5-silylpentanoyl chlorides via intramolecular Friedel-Crafts reaction (H. Urabe and I. Kuwajima, J. Org. Chem., 1984, 49, 1140), or (e) from a suitable acyclic ketone by sequential alkylation/Wittig reactions based on 1-phenylthiovinyl-phosphonium reagents (A.G. Cameron and A.T. Hewson, J. Chem. Soc., Perkin Trans. I, 1983, 2979), or (f) from an acyclic α-diazoketone using rhodium catalysed carbenoid cyclisation (M. Kennedy et al, ibid, 1990, 1047).

(iv) Ring-enlargement reactions contribute greatly to the synthetic potential of cyclic ketones. They are excellent substrates for Baeyer-Villiger oxidation to lactones and Beckmann rearrangement to lactams (see Section 2b, Reactions 4, 5). One-carbon cyclohomologation is possible using diazoalkanes (A.E. Greene, M-J. Luche and A.A. Serra, J. Org. Chem., 1985, 50, 3957), or by converting the cycloalkanone to β-hydroxyselenium or sulphur derivatives (J. Labourer and A. Krief, Tetrahedron Lett., 1984, 25, 2713; B.M. Trost and G.K. Michael, J. Amer. Chem. Soc., 1987, 109, 4124, eg:

CH_2N_2 SeR R^1 R^2 OH R^1 R^2 O

(v) Annulation of cyclic ketones

The classical Robinson annulation (First supplement, p.103) was largely superseded by the Stork silyl modification (G. Stork and J. Singh, J. Amer. Chem. Soc., 1974, 96, 6152 and 6181; pertinent reviews: E.W. Colvin, Chem. Soc. Rev., 1978, 7, 15 and I. Fleming, ibid, 1981, 10, 83),

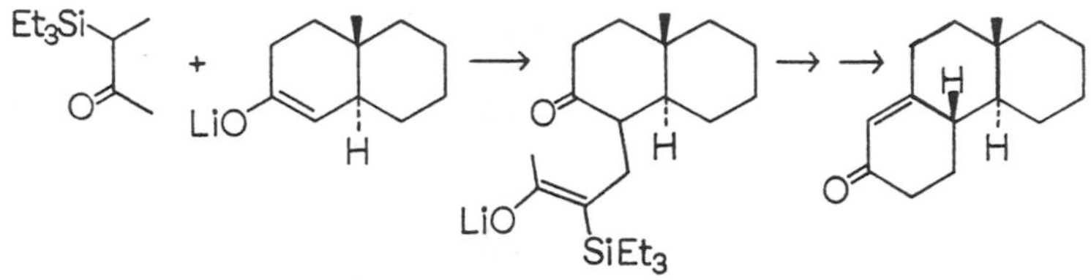

The stereochemical integrity indicated above extends to more complex ketone substrates, eg. those derived from carbohydrates (R.V. Bonnert and P.R. Jenkins, J. Chem. Soc., Chem. Commun., 1987, 6). It is possible to start with silyl enol ethers rather than enolates (J.W. Huffman, S.M. Potnis and A.V. Satish, J. Org. Chem., 1985, 50, 4266), and methods have been developed for annulation α and β to the carbonyl group of a cyclohexanone substrate (W.L. Meyer, ibid, 1985, 50, 438). Methods for 5-membered ring annulations have been reviewed (M. Ramaiah, Synthesis, 1984, 529).

When desired, the angular hydroxyl group at the new ring-junction can be retained using cyclopropane chemistry (J.T. Carey and P. Helquist, Tetrahedron Lett., 1988, 29, 1243):

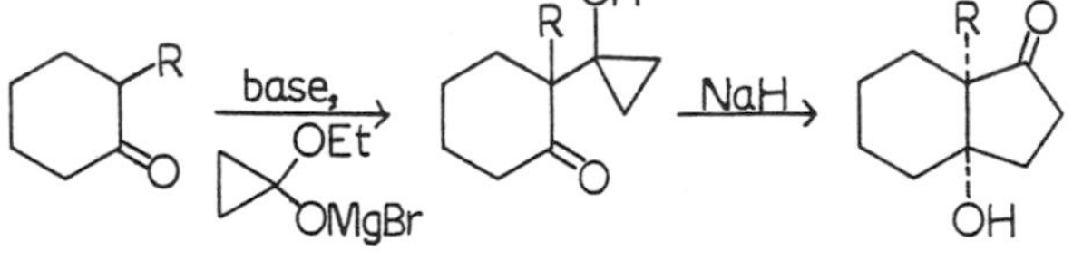

Radical cyclisation of suitable alkynyl ketones leads to exomethylene products with survival of the angular carbinol (G. Pattenden and S. Teague, J. Chem. Soc., Perkin Trans. I, 1988, 1077):

Annulation protocols have been developed for the construction of heterocycles such as 2-pyridines (A.P. Kozikowski, E.R. Reddy and C.P. Miller, ibid, 1990, 195), pyridines and pyrroles (K. Kanomata and M. Nitta, ibid, 1990, 1119).

Spiro-annulation can be carried out using 1-alkoxycyclopropyllithium reagents to give spirocyclobutanones (R.C. Gadwood et al, J. Org. Chem., 1985, 50, 3255) which are cleanly enlarged to spirocyclopentanones if required (B.M. Trost and L.H. Latimer, ibid, 1978, 43, 1031), eg:

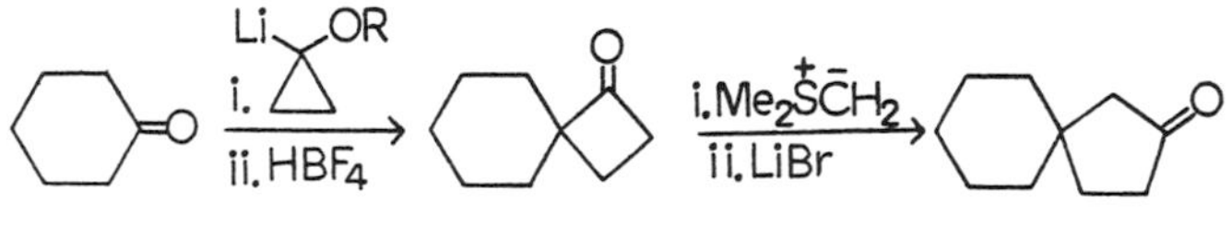

i. Li, OR (1-alkoxycyclopropyllithium); ii. HBF_4

i. $Me_2\overset{+}{S}\overset{-}{C}H_2$; ii. LiBr

(d) Ketone acetals (ketals)

The chemistry of ketals and their sulphur analogues, thioketals, is discussed in detail by R.G. Bergstrom ("The Chemistry of Ethers, Crown Ethers, Hydroxyl Groups and their Sulphur Analogues", S. Patai (Ed.), John Wiley and Sons, New York, 1980, p.881). The ketals are most commonly encountered as protecting groups: oxygen ketals survive most basic and redox media, but thioketals (or selenoketals) are necessary for protection during acidic reactions (monograph: T.W. Greene, "Protective Groups in Organic Synthesis", John Wiley and Sons, New York, 1981).

(i) Preparation of ketals

$$R_2C(OR^1)_2 \longleftarrow R_2CO \longrightarrow R_2C\langle\begin{smallmatrix}O\\O\end{smallmatrix}\rangle(\ )_n$$

Ketals are usually prepared from the ketones using an alcohol or diol with acidic catalysts, eg. *p*-toluenesulphonic acid, camphorsulphonic acid or pyridinium tosylate. Water must be removed for efficient conversion and reagent systems that include an orthoester as water trap are commonly used (for a review of acetalisation see F.A.J. Meskens, *Synthesis*, 1981, 501; cf. T.H. Chan, M.A. Brook and T. Chaly, *ibid*, 1983, 203). It is often convenient to use an acidic resin as catalyst (eg. R. Caputo, C. Ferrari and G. Palumbo, *ibid*, 1987, 386) or to preabsorb the reagents onto a solid support (I. Bidd *et al*, *J. Chem. Soc., Perkin Trans. 1*, 1983, 1369). Reluctant preparations are amenable to high pressure acetalisation (W.G. Dauben, J.M. Gerdes and G.C. Look, *J. Org. Chem.*, 1986, *51*, 4964). Acidic conditions can be avoided by using a tetra-alkoxysilane with iodotrimethylsilane (H. Sakurai *et al*, *ibid*, 1984, *49*, 2808) or 1,2-bis(trimethylsiloxy)ethane with trimethylsilyl triflate (J.R. Hwu and J.M. Wetzel, *ibid*, 1985, *50*, 3946). For some purposes, conventional ketals are inappropriate; 1,3-dioxalan-4-ones (W.H. Pearson and M-C. Cheng, *ibid*, 1987, *52*, 1353), photo-labile acetals (D. Gravel, S. Murray and G. Ladoucer, *J. Chem. Soc., Chem. Commun.*, 1985, 1828) and chiral cyclic acetals (see below) are advantageous in some situations.

(ii) Deprotection of oxygen ketals usually involves acid catalysed hydrolysis or transetherification (T.W. Greene, op. cit.). Silica gel is a useful solid support for a variety of hydrolytic systems (F. Huet et al, Synthesis, 1978, 63), even for quite large-scale deprotections (I. Bidd et al, op. cit.). Milder conditions can be employed if necessary (see A. Fadel, R. Yefsah and J. Salaun, Synthesis, 1987, 37 and references cited therein).

(iii) Reactions of the oxygen ketals $R_2C(OR^1)_2$ include their participation as acceptors in Lewis acid mediated aldol reactions (see Section 2e(iii)). They are reduced to the corresponding mono-ethers (eg R_2CHOR^1) by triethylsilane/trimethylsilyl triflate and converted to allyl monoethers by allyltrimethylsilane/trimethylsilyl triflate (T. Tsunoda, M. Suzuki and R. Noyori, Tetrahedron Lett., 1979, 4679 and 1980, 71). Trimethylsilyl cyanide gives O-trimethylsilyl cyanohydrins, $R_2C(OSiMe_3)CN$ (J.D. Elliott, V.M.F. Choi and W.S. Johnson, J. Org. Chem., 1983, 48, 2294), that can be converted into β-aminoalcohols, $R_2C(OH)CH_2NH_2$ (D.A. Evans, G.L. Carroll and L.K. Truesdale, ibid, 1974, 39, 914), or α-aminonitriles, $R_2C(NR_2{}^1)CN$ (K. Mai and G. Patil, Tetrahedron Lett., 1984, 25, 4583). Partial cleavage of diastereoisomeric acetals provides a convenient resolution of racemic ketones via separation of the cleavage products and hydrolysis to the ketones (A. Mori and H. Yamamoto, J. Org. Chem., 1985, 50, 5444):

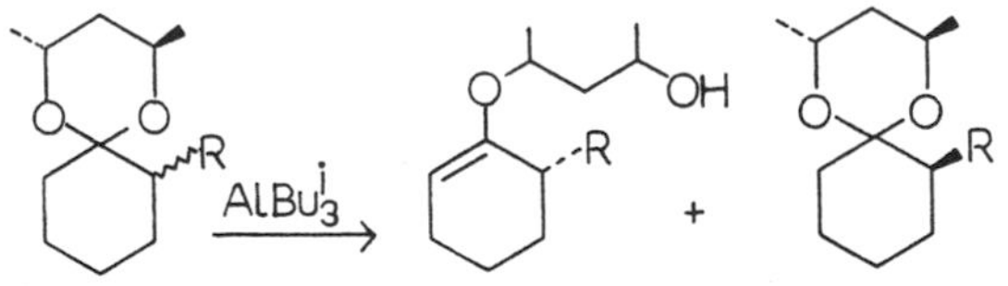

$$\xrightarrow{AlBu^i_3}$$

(iv) Thioketals are formed from ketones using (a) thiols or dithiols with a Lewis acid (see Y. Kamitori et al, J. Org. Chem., 1986, 51, 1427), or with zinc or magnesium triflate (E.J. Corey and K. Shimoji, Tetrahedron Lett., 1983, 24, 169), or (b) using an alkylthio-trimethylsilane (D.A. Evans, K.G. Grimm and L.K. Truesdale, J. Amer. Chem. Soc., 1975, 97, 3229). They also arise during ketone syntheses involving dithiane acyl anion equivalents (D. Seebach and E.J. Corey, J. Org. Chem., 1975, 40, 231) so that the transformation thioketal to ketone may correspond to a step in ketone synthesis as well as to deprotection (cf. Section 2a, preparation 8a):

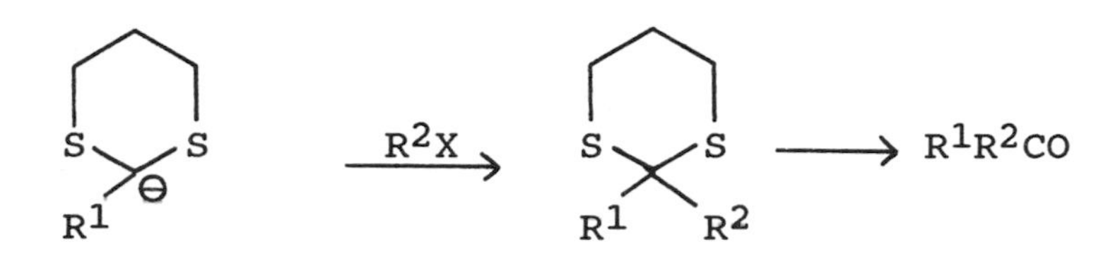

The ease of deprotection depends upon the particular thioketal. Some recommended reagents are mercury (II) compounds (E.J. Corey and B.W. Erickson, ibid, 1971, 36, 3553), DMSO with a halide (G.A. Olah, A.K. Mehrota and S.C. Narang, Synthesis, 1982, 151), lead (IV) oxide/BF_3 (D. Ghiringhelli, ibid, 1982, 580), benzeneseleninic anhydride (N.J. Cussans, S.V. Ley and D.H.R. Barton, J. Chem. Soc., Perkin Trans. 1, 1980, 1654), and clay-supported Fe(III) or Cu(II) chloride (M. Balogh, A. Cornelis and P. Laszlo, Tetrahedron Lett., 1984, 25, 3313); chemical or electrochemical oxidation is an alternative (G.A. Olah, S.C. Narang and G.F. Salem, Synthesis, 1980, 657; M. Platen and E. Stecken, Chem. Ber., 1984, 117, 1679.

(v) Selenoketals, like thioketals, are resistant to acid hydrolysis. Oxidative conditions are required for deprotection; clay supported Fe(III) or Cu(II) nitrate is effective

(P. Laszlo, P. Pennetrou and A. Krief, Tetrahedron Lett., 1986, 27, 3153).

(e) **Enol derivatives**

Enolates, enol ethers and enol esters are important in synthetic applications of ketones. Although the enols are not often isolated, stoichiometric enolisation is a common in-situ procedure during crossed-aldol (review: T. Mukaiyama, Org. React., 1982, 28, 203), α-substitution and other protocols involving ketones. Silicon chemistry underpins many of the reactions (monograph: E.W. Colvin, "Silicon Reagents in Organic Synthesis", Academic Press, San Diego, 1988). With unsymmetrical ketones, regiospecific enol formation is important; unsymmetrical enols can also be prepared indirectly.

(i) Preparation from ketones

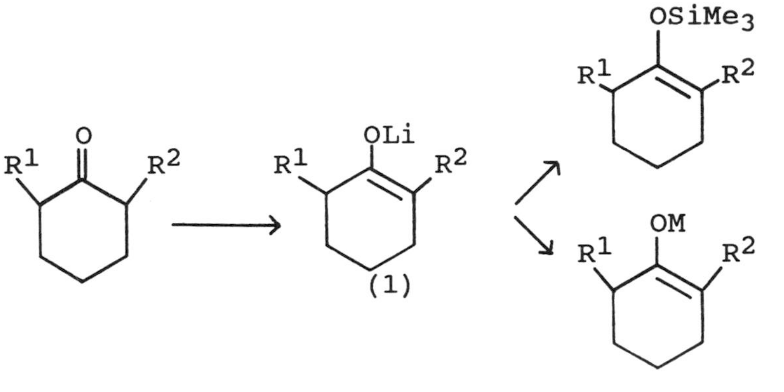

(2)

(3)

Deprotonation (eg. LDA) at low temperature gives a kinetic enolate (1,R^1 = alkyl, R^2 = H) with ca. 99% regiospecificity. The silyl enol ether (2,R^1 =alkyl, R^2=H) is formed if a trapping agent, eg. Me_3SiCl, is present (I. Fleming and I. Paterson, Synthesis, 1979, 736; E.J. Corey and A.W. Gross,

Tetrahedron Lett., 1984, 25, 491 and 495). Chiral lithium bases induce enantioselective enolisation (R. Shirai, M. Tanaka and K. Koga, J. Amer. Chem. Soc., 1986, 108, 543). Trimethylsilyl triflate with an amine base is useful for hindered ketone substrates (L.N. Mander and S.P. Sethi, Tetrahedron Lett., 1984, 25, 5099).

Enolisation in equilibrating conditions, eg Et_3N/Me_3SiCl for 90h at 130 ^oC, gives the thermodynamic enol ether (2, R^1=H, R^2 alkyl) but with regiospecificity of only ca. 80-90% (E.W. Colvin, op. cit.); methyl lithium then gives the enolate (1, R^1=H, R^2=alkyl) at room temperature (G. Stork and P.F. Hudrlik, J. Amer. Chem. Soc., 1968, 90, 4462 and 4464).

Transmetallation converts lithium enolates (1) into others (3), eg. the Mg or Zn analogues (T. Mukaiyama, op. cit.), and also into enolates of cerium (T. Imamoto, T. Kusamato and M. Yokoyama, Tetrahedron Lett., 1983, 24, 5233), or of zirconium (D.A. Evans and L.R. McGee, ibid, 1980, 21, 3975), or of germanium (Y. Yamamoto and Y. Yamada, J. Chem. Soc., Chem. Commun., 1988, 802), or of tin (IV) (B.M. Trost and C.R. Self, J. Org. Chem., 1984, 49, 468).

Silyl enol ethers are precursors for several reactive enolates (3) (review: I. Kuwajima and E. Nakamura, Acc. Chem. Res., 1985, 18, 181) including the so-called "naked enolates" formed with tris(dimethylamino)sulphur (trimethylsilyl)-difluoride (R. Noyori, I. Ishida and J. Sakata, J. Amer. Chem. Soc., 1983, 105, 1598).

The identity of the oxygen substituent or counterion (M) affects the basicity and reactivity of the enol derivatives. Boron enolates (review: I. Paterson, Chem. Ind. (London), 1988, 390; cf. T. Mukaiyama, op. cit.), and tin (II) enolates (review: T. Mukaiyama and N. Iwasawa, Nippon Kagaku Kaishi, 1987, 1099; cf. T. Mukaiyama et al, Tetrahedron, 1984, 40, 1381) combine low basicity with useful reactivity in several applications.

(ii) Indirect preparation of enol derivatives

Unsymmetrical lithium enolates can be obtained (a) from ketenes by the addition of alkyl lithiums to C=O (T.T. Tidwell *et al*, *J. Org. Chem.*, 1985, **50**, 2105 and *J. Amer. Chem. Soc.*, 1985, **107**, 5391), or (b) from carboxylate esters with 2 equivalents of trimethylstannylmethyl lithium (T. Sato *et al*, *Tetrahedron Lett.*, 1986, **27**, 4339), or (c) from α,β-enones by conjugate addition or reduction (K.K. Heng and R.A.J. Smith, *Tetrahedron*, 1979, **35**, 425; A.R. Chamberlein and S.H. Reich, *J. Amer. Chem. Soc.*, 1985, **107**, 1440). The corresponding silyl enol ethers are obtained (d) by the above methods (a-c) performed in the presence of Me_3SiCl, or (e) from α,β-enones by conjugate addition of hydridosilanes (I. Ojima and T. Kogure, *Organometallics*, 1982, **1**, 1390), or (f) by rearrangement of α,β-epoxysilanes using BF_3 (I. Fleming and T.W. Newton, *J. Chem. Soc. Perkin Trans. 1*, 1984, 119), or (g) from α-chloroketones by dehalogenation using Zn/Me_3SiCl (G.M. Rubottom, R.C Mott and D.S. Krueger, *Synth. Commun.*, 1977, **7**, 327), or (h) from acyl silanes using organolithium reagents (H.J. Reich, R.C. Holtan and S.L. Borkowsky, *J. Org. Chem.*, 1987, **52**, 312).

(iii) Crossed aldol reactions

Stoichiometric, regiospecific enol formation allows clean, diastereospecific conversion of ketones into β-hydroxyketones (ie ketonic aldols) *via*, 6-membered metal (M) chelate formation (T. Mukaiyama, *op. cit.*):

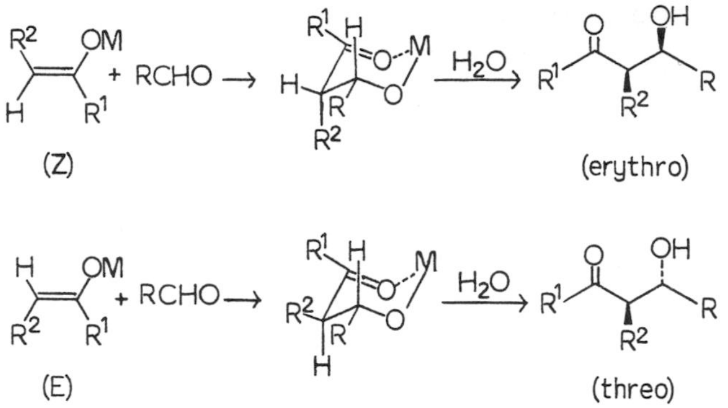

Kinetic coupling (low temperature, short reaction time) favours the stereochemical outcome illustrated, whereas under thermodynamic control equilibration of the chelate intermediate favours _threo_ aldol formation (maximum equatorial substitution in the chelate) (cf. C.H. Heathcock in "Asymmetric Synthesis", vol. 3, J.D. Morrison (Ed.), Academic Press, London, 1984, p.111). Less basic conditions are involved when boron enolates are employed rather than lithium, aluminium, tin, magnesium, zinc or other metal enolates, and boron mediation has been reviewed (D.A. Evans, J.V. Nelson and T.R. Taber, _Top. Stereochem._, 1982, _13_, 1; cf. I. Paterson, _op. cit._, and T. Mukaiyama, _op. cit._, p.234).

Silyl enol ethers are alternatives to enolates as donor species, catalysed by titanium (IV) chloride (or other Lewis acids), trityl salts, iodotrimethylsilane or trimethylsilyl triflate (C.H. Heathcock _et al_, _J. Org. Chem._, 1986, _51_, 3027; H. Sakurai, K. Sasaki and A. Hosomi, _Bull. Chem. Soc. Jpn._, 1983, _56_, 3195; S. Murata, M. Suzuki and R. Noyori, _Tetrahedron_, 1988, _44_, 4259). When the donor is a silyl enol ether and the acceptor is an acetal or ketal (Lewis acid catalysis), the product is a β-alkoxyketone and the procedure is called a Mukaiyama reaction (review: T. Mukaiyama and M. Murakami, Synthesis, 1987, 1043):

$$R^1C(OSiMe_3)=CHR^2 + R^3CH(OR)_2 \xrightarrow{TiCl_4} R^1C(=O)CH(R^2)CH(OR)R^3$$

Enantioselective aldol reactions have been performed using chiral enolates (M. Braun, Angew. Chem. Int. Ed. Engl., 1987, 26, 24), or enolates bearing removable chiral auxiliaries (F. Schneider and R. Simon, Synthesis, 1986, 582), or chiral base catalysts (M. Muraoka, H. Kawasaki and K. Koga, Tetrahedron Lett., 1988, 29, 337; C. Agami, Bull. Soc. Chim. Fr., 1988, 499). Enantioselective Mukaiyama reactions are achieved using chiral acetal acceptors (I.R. Silverman et al, J. Org. Chem., 1987, 52, 180).

Attractive catalyst-regeneration procedures have been developed that obviate the need for stoichiometric enol formation in aldol reactions involving tin (II) enolates (T. Yura, N. Iwasawa and T. Mukaiyama, Chem. Lett., 1986, 187), and in Mukaiyama reactions mediated by trityl perchlorate, a promoter that can be polymer bound for solid-phase reactions (see T. Mukaiyama and M. Muraikami, op. cit.).

(iv) Michael addition

Ketone enolates add in a Michael fashion to the olefinic bond of α,β-enones (D.A. Oare and C.H. Heathcock, J. Org. Chem., 1990, 55, 157) and of nitroalkenes (R.W. Stevens and T. Mukaiyama, Chem. Lett., 1985, 855), anti-addition being favoured. Ketone silyl enol ethers give Lewis acid catalysed Michael reactions with α,β-unsaturated acetals (including ketals) (K.

Narasaka et al, Bull. Chem. Soc. Jpn., 1976, 49, 779) and with nitroalkenes (D. Seebach and M.A. Brook, Helv. Chim. Acta, 1985, 68, 319).

Tin (II) enolates give Michael reactions with greater selectivity than other enolates; conjugated tin (II) enolates react exclusively via the γ-position giving enone products (R.W. Stevens and T. Mukaiyama, Chem. Lett., 1985, 851) eg:

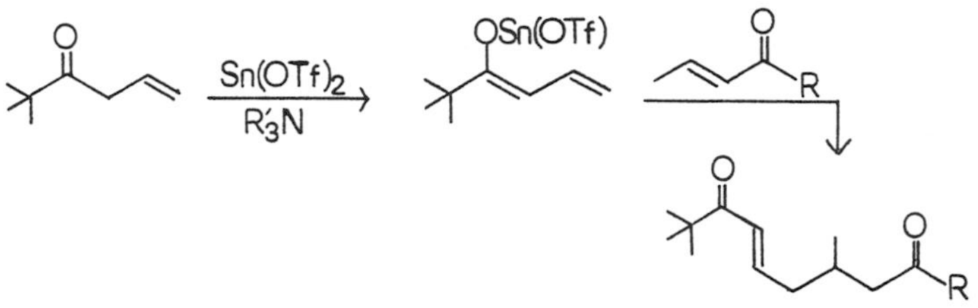

$Sn(OTf)_2$, R'_3N

Michael addition of enolates to aryl vinylselenoxides produces α-cyclopropylketones (R. Ando et al, Bull. Chem. Soc. Jpn., 1984, 57, 2897):

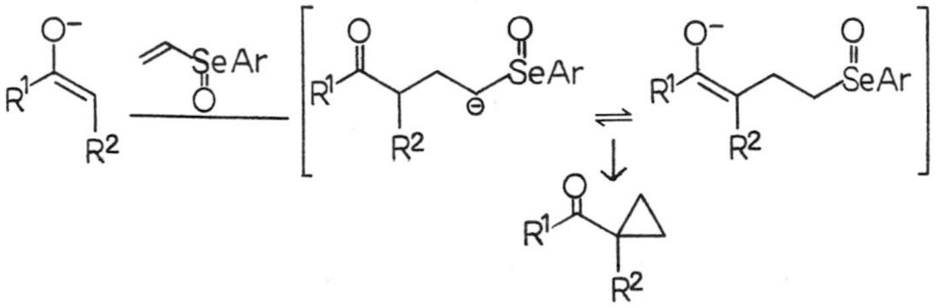

(v) Other α-alkylation reactions

Ketone silyl enol ethers are monoalkylated by alkylhalides (M.T. Reetz et al, Tetrahedron Lett., 1979, 427, I. Paterson, ibid., 1979, 1519). They give α-arylketones with diazonium salts (T. Sakakura, M. Hara and M. Tanaka, J. Chem. Soc., Chem. Commun., 1985, 1545), and α-allylketones with allyl ethers (T. Mukaiyama et al, Chem. Lett., 1986, 1009; cf. R. Hunter and

R.D. Simon, <u>Tetrahedron Lett.</u>, 1986, <u>27</u>, 1385). α-Alkoxymethylation and α-mercaptomethylation occur with chloroalkyl ethers and thioethers (A. Hosomi, Y. Sakata and H. Sakurai, <u>Chem. Lett.</u>, 1983, 405); the latter have been applied to the conversion of cyclic silyl enol ethers into exo-methylene cycloalkanone derivatives (I. Paterson, <u>Tetrahedron</u>, 1988, <u>44</u>, 4207):

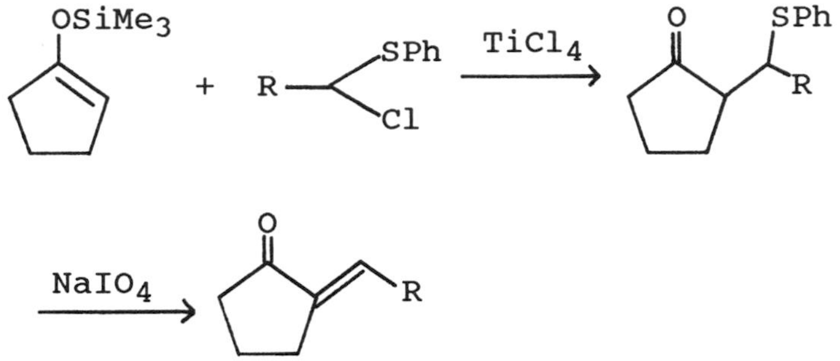

$OSiMe_3$ + R–CH(SPh)Cl $\xrightarrow{TiCl_4}$ 2-[CH(SPh)R]cyclopentanone $\xrightarrow{NaIO_4}$ 2-(RCH=)cyclopentanone

(vi) <u>Miscellaneous reactions</u>

Enolates give α-hydroxyketones after oxygenation in the presence of triethyl phosphite and a phase transfer catalyst (M. Masui, A Ando and T. Shiori, <u>Tetrahedron Lett.</u>, 1988, <u>29</u>, 2835). Lithium enolates give either (E) or (Z) γ,δ-enones with dicarbonyl-iron π complexes depending on the conditions employed (M. Marsi and M. Rosenblum, <u>J. Amer. Chem. Soc.</u>, 1984, <u>106</u>, 7264). Silyl enol ethers are converted to methyl β-ketoesters by Mander's reagent (methyl cyanoformate) (L.N. Mander and P. Sethi, <u>Tetrahedron Lett.</u>, 1983, <u>24</u>, 5425). Thermal cyclisation of suitable alkynyl enols can be used to prepare useful functionalised cyclopentanes (J. Drouin, M-A. Boaventura and J-M. Conia, <u>J. Amer. Chem. Soc.</u>, 1985, <u>107</u>, 1726):

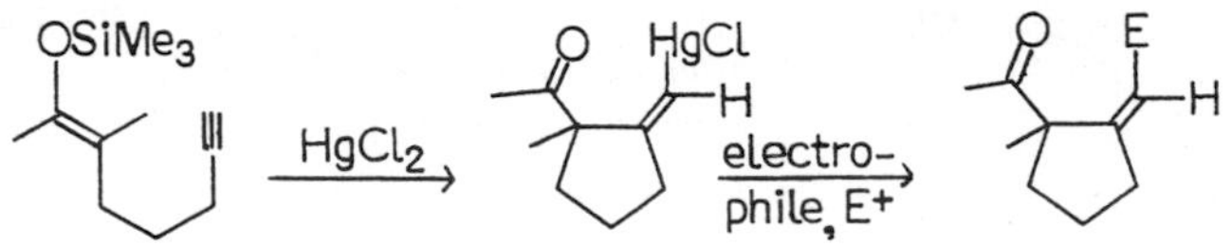

A furan annulation of cyclic enol ethers has been developed (E.J. Corey and A.K. Ghosh, Tetrahedron Lett., 1987, 28, 175) and the Mukaiyama reaction has been modified to give a general synthesis of furans (T. Mukaiyama, T. Ishihara and K. Inomata, Chem. Lett., 1975, 527):

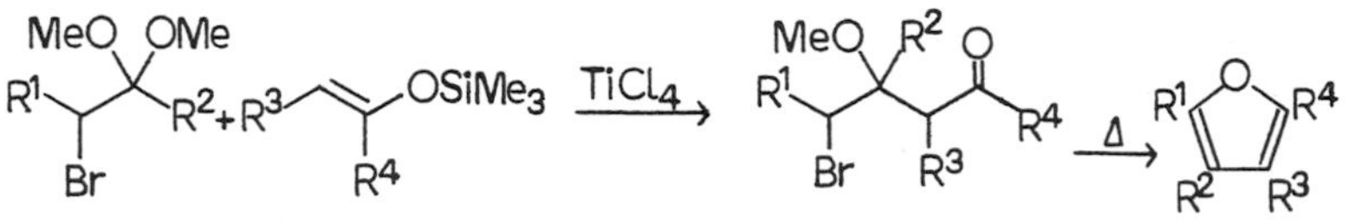

(vii) Enol esters

Enol triflates are formed from ketones using LDA and N-phenyltrifluoromethanesulphonimide (J.E. McMurry and W.J. Scott, Tetrahedron Lett., 1983, 24, 979). They give alkanes by hydrogenation (V.B. Jigajinni and R.H. Wightman, ibid, 1982, 23, 117), and couple with Michael esters in the presence of a palladium catalyst to give dienoates (W.J. Scott et al, J. Org. Chem., 1985, 50, 2302):

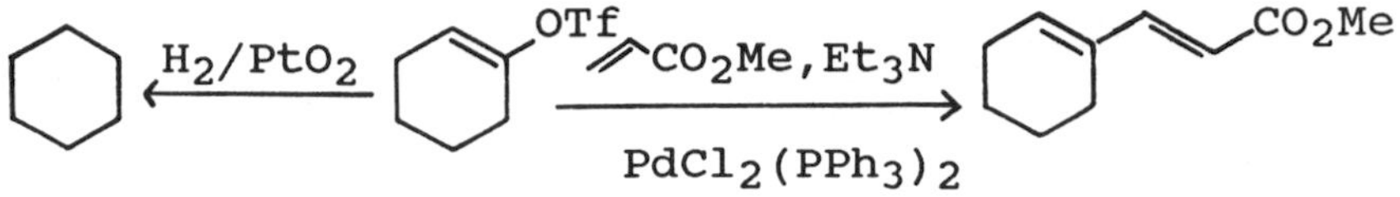

Enol phosphates, from enolates with diethyl phosphochloridate, are reduced to alkenes using activated titanium (S.C. Welch and M.E. Walters, ibid, 1978, 43, 2715):

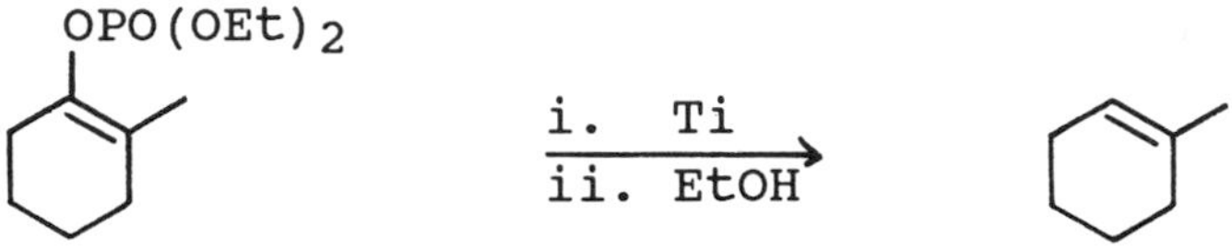

(viii) Enols as Diels-Alder dienes

2-(Trimethylsilyloxy)-1,3-butadiene (an enol of methyl vinyl ketone) undergoes cycloaddition with suitable alkenes giving cyclohexanones (M.E. Jung and C.A. McCombs, Tetrahedron Lett., 1976, 2935). Similarly, the bis-enol, trans-1-methoxy-3-(trimethylsilyoxy)-1,3-butadiene (Danishefsky's diene, review: S. Danishefsky, Acc. Chem. Res., 1981, 14, 400) is an important synthon of β-methoxycyclohexanones and of cyclohexenones (cf. R.C. Gupta, P.A. Harland and R.T. Stoodley, Tetrahedron, 1984, 40, 4657). Trimethylsilyl triflate is a useful reagent for the elimination step (P.E. Vorndam, J. Org. Chem., 1990, 55, 3693):

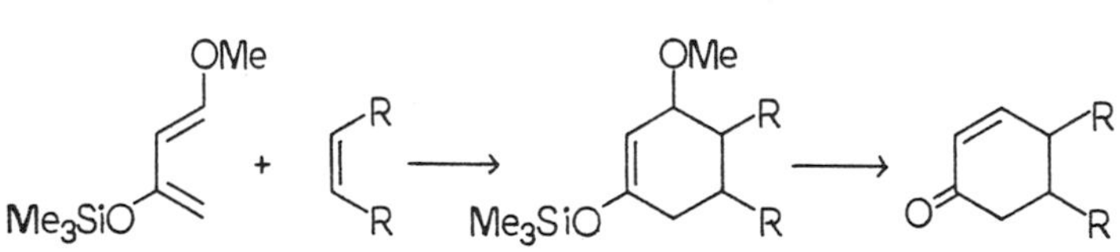

Despite potential steric hindrance, use has also been made of the ketene acetal, 1,1-dimethoxy-3-(trimethylsilyloxy)-1,3-butadiene, in β-diketone synthesis (S. Danishefsky et al, J. Amer. Chem. Soc., 1979, 101, 7020.

(f) **Nitrogen derivatives**

Imines, enamines, oximes and hydrazones are discussed in "The Chemistry of the Carbon-Nitrogen Double Bond", S. Patai (Ed.), Interscience, New York, 1970. Like ketals they sometimes serve as protecting groups (T.W. Greene, op. cit.), in which case exposure to acidic resins can be a convenient method of ketone regeneration (R. Ballini and M. Petrine, J. Chem. Soc., Perkin Trans. I, 1988, 2563; B.C. Rann and D.C. Sarkar, J. Org. Chem., 1988, 53, 878). Other methods are listed in "Advanced Organic Chemistry", 3rd Ed'n, J. March, John Wiley and Sons, New York, 1985, and in "Compendium of Organic Synthetic Methods", Vol. 4, L.G. Wade, John Wiley and Sons, New York, 1980, p. 287).

(i) Imines, $R_2C{=}NR^1$, can be obtained by treating a ketone with a suspension of alumina containing the amine R^1NH_2 (R^1 = alkyl or aryl) (F. Texier-Boullet, Synthesis, 1985, 679), or with bis(dichloroaluminium) phenylimide ($R^1{=}Ph$) (J.J. Eisch and R. Sanchez, J. Org. Chem., 1986, 51, 1848). Classical procedures are given by J. March (op. cit.). Swern oxidation of amines, R_2CHNHR^1 gives imines (D. Keirs and K. Overton, J. Chem. Soc., Chem. Commun., 1987, 1660). Silicon-containing imines ($R^1{=}SiMe_3$, CH_2SiMe_3) can be obtained directly from ketones (T. Morimoto and M. Sekiya, Chem. Lett., 1985, 1371; O. Tsuge, S. Kanemasa and K. Matsuda, J. Org. Chem., 1984, 49, 2688).

Imines can be metallated (eg LDA) and used in a variety of C-C bond forming reactions including alkylation (N. Ikoto et al, Chem. Pharm. Bull.

1986, 34, 1050; M.C. Carreno et al, J. Chem. Soc., Perkin Trans. I, 1989, 1335), aldol reactions (T. Mukaiyama, Org. React., 1982, 28, 203), and Michael reactions (M. Pfau et al, J. Amer. Chem. Soc., 1985, 107, 273).

$R^1C(=NR)CH(H)R^2 \xrightarrow{LDA} R^1C(NRLi)=CHR^2 \xrightarrow[\text{ii. } H_2O]{\text{i. electrophile, } E^+} RC(=O)CH(E)R^2$

They can undergo an aza-Wittig reaction to give alkenes (H. Takeuchi et al, J. Org. Chem., 1989, 54, 431, and references cited therein). They are converted into aziridines by N-tosyl-S-methyl-S-phenylsulphoxime/NaH (C.R. Johnson, ibid, 1973, 95, 4287), and into oxaziridines by m-chloroperoxybenzoic acid (T.A. Hamor et al, J. Chem. Soc., Perkin Trans. 2, 1990, 25):

$R_2C{=}NR^1 \longrightarrow R_2C\langle X, NR^1 \rangle$ (three-membered ring) $\quad (X = CH_2,\ O)$

(ii) Enamines, $R^1{}_2C{=}CR^2{-}NR_2$, are usually prepared from ketones with secondary amines (R_2NH). Their preparations and properties have been reviewed (V.G. Granik, Russ. Chem. Rev., 1984, 53, 383; P.W. Hickmott, Tetrahedron, 1982, 38, 1975 and 3363). The valuable Stork enamine reaction leads to C-alkylation by electrophiles, E^+, and thence, after hydrolysis, to homologous ketones, $R^1_2C(E)COR^2$ (review: J.K. Whitesell and M.A. Whitesell, Synthesis, 1983, 517).

Alkyl methylenamines, $CH_2=CR^1NR_2$, can be obtained from either alkynes, $R^1C\equiv CH$ (J. Barluenga, F. Aznar and R. Liz, J. Chem. Soc., Chem. Commun., 1986, 1465), or amides, R^1CONR_2 (S.H. Pine et al, J. Org. Chem., 1985, 50, 1212) so that these are indirect synthons for ketones, CH_3COR^1 and ECH_2COR^1, via Stork reactions. Reaction of a ketone, R_2CO, with α-(N-methylanilino)methyldiphenylphosphine oxide gives the homologous aldenamine, $R_2C=CHN(Me)Ph$, and hence aldehydes R_2CHCHO or $R_2C(E)CHO$ (N.L.J.M. Broekhof and A. van der Gen, Recl. J.R. Neth. Chem. Soc., 1984, 103, 305).

Enamines derived from chiral amines undergo stereoselective alkylation reactions (J.K. Whitesell and S.W. Felman, J. Org. Chem., 1977, 42, 1663), aldol reactions (O. Takazawe, K. Kogami and K. Hayashi, Bull. Chem. Soc. Jpn., 1984, 57, 1876 and 1985, 58, 2427), C-halogenation reactions (F.M. Laskovics and E.M. Schulman, J. Amer. Chem. Soc., 1977, 99, 6672), and Michael reactions (D. Seebach et al, Helv. Chim. Acta, 1985, 68, 162; C. Stetin, B. De Jesu and J-C. Pommier, J. Org. Chem., 1985, 50, 3863; for a ketone annulation sequence using enamine chemistry see D.L.J. Clive, P.L. Beaulieu and L. Set, ibid, 1984, 49, 1313).

(iii) Oximes, $R_2C=NOH$, are important as substrates for the Beckmann rearrangement to amides and lactams (see C.G. McCarty in S. Patai, op. cit.), and are also widely used as protected ketones. Attention has been given to the discovery of mild methods for ketone regeneration (eg. P. Vankar et al, J. Org. Chem., 1986, 51, 3063; M. Salmon et al, Synth. Commun., 1986, 16, 1827; D.H.R. Barton et al, J. Chem. Soc., Perkin Trans. I, 1986, 2243 and references cited in these papers). Such procedures complete ketone syntheses involving oximation of non-ketonic

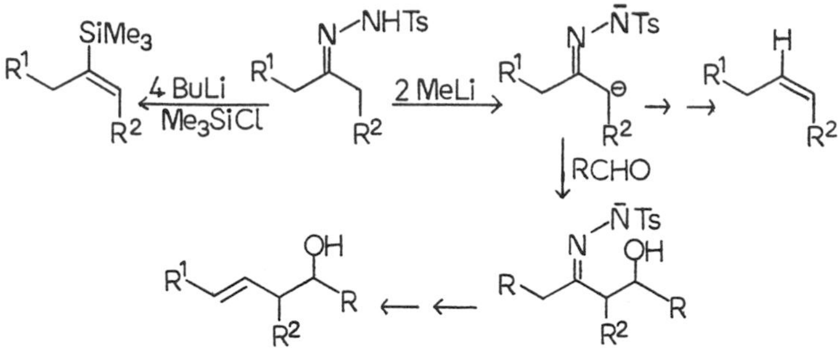

Tosyl hydrazones react with methanolic KCN giving nitriles R^2CHCN (J. Jiricny, D.M. Orere and C.B. Reese, J. Chem. Soc., Perkin Trans. 1, 1980, 1487).

Chiral SAMP or RAMP hydrazones are prepared from ketones and either (S)- or (R)-1-amino-2-methoxymethylpyrrolidine. They can be metallated giving nucleophiles for asymmetric alkylations (D. Enders et al, Tetrahedron, 1984, 40, 1345), aldol reactions (D. Enders et al, Chem. Ber., 1979, 112, 3703 and Angew. Chem. Int. Ed. Engl., 1978, 17, 206), Michael reactions (D. Enders and B.E.M. Rendenbach, Tetrahedron, 1986, 42, 2235), and acetoxylation (and hydroxylation) reactions (D. Enders and V. Bhushan, Tetrahedron Lett., 1988, 29, 2437). The hydrazone products give the corresponding ketones by ozonolysis:

eg:

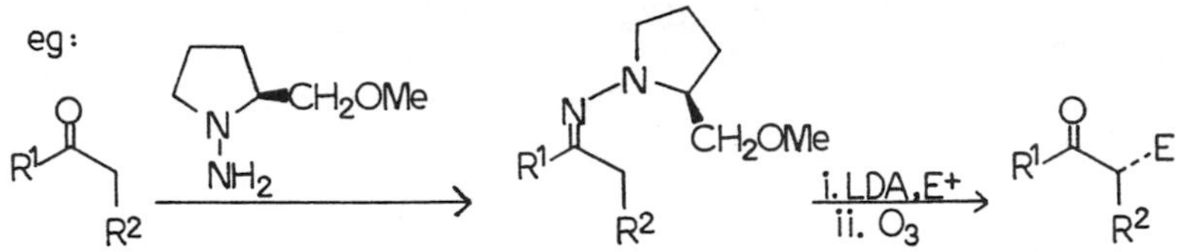

R1COCH2R2 + (S)-1-amino-2-(methoxymethyl)pyrrolidine (CH_2OMe, NH_2) → hydrazone (CH_2OMe) —i. LDA, E+; ii. O_3→ $R^1COCH(E)R^2$

(g) *α*-Haloketones

The chemistry of these compounds has been discussed in detail ("The Chemistry of *α*-Haloketones, *α*-Haloaldehydes and *α*-Haloimines", N. DeKimpe and R. Verhe, John Wiley and Sons, Chichester (UK), 1988).

(i) *α*-Fluoroketones are obtained by fluorination of the parent ketone enolate (S. Rozen and M. Brand, Synthesis, 1985, 665), or enol acetate (S. Rozen and Y. Menahem, J. Fluorine Chem., 1980, 16, 19), or silyl enol ether (S.T. Purrington, N.V. Lazaridis and C.L. Bumgardner, Tetrahedron Lett., 1986, 27, 2715). Racemic *α*-fluoroesters give chiral *α*-fluoroketones after treatment with a chiral lithio-methyl sulphoxide (P. Bravo and G. Resnati, J. Chem. Soc., Chem. Commun., 1988, 218):

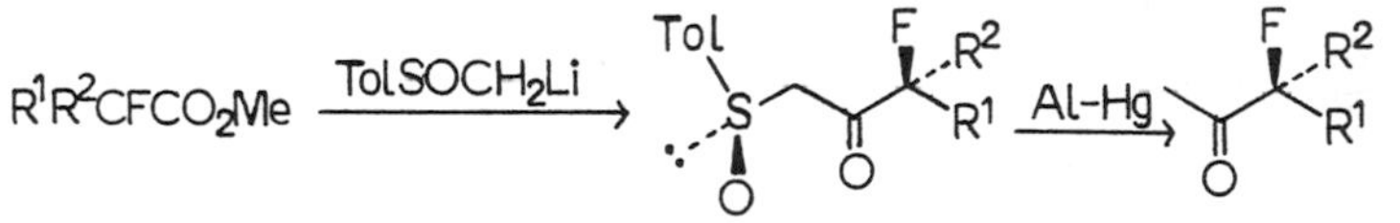

$R^1R^2CFCO_2Me$ —TolSOCH2Li→ Tol–S(O)–CH2–CO–CF(R1)(R2) —Al-Hg→ CH3–CO–CF(R1)(R2)

(ii) *α-Chloroketones* can be obtained (a) from aldehydes using chloromethyl phenylsulphoxide/LDA as a chloromethylating agent (V. Rentrakal and W. Kanghae, Tetrahedron Lett., 1977, 1225), (b) from a ketone enol derivative using either N-chlorosuccinimide or a metal chloride with lead (IV) acetate (S.E. Denmark and M.S. Dappen, J. Org. Chem., 1984, 49, 798; S. Motohashi et al, Synthesis, 1982, 1021) or from the silyl enol ether and sulphuryl chloride (G.A. Olah et al, J. Org. Chem., 1984, 49, 2032), (c) from diazoketones with HCl (G.A. Olah and J. Welch, Synthesis, 1974, 896), or (d) from α-chloroacyl chlorides using an organomanganese reagent or a Grignard reagent at low temperature (G. Friour, G. Cahiez and J.F. Normant, ibid, 1984, 37; B. Frolisch and R. Flogans, ibid, 1984, 734).

(iii) *α-Bromoketones* are available using appropriate modifications of the methods used for α-chloroketones (see the references cited above). Alternatively, specific methods include (a) treatment of the parent ketone with t-butyl bromide/DMSO (E. Armani et al, Tetrahedron, 1984, 40, 2035) or with 4-dimethylaminopyridinium bromide perbromide (A. Arrieta, I. Ganboa and C. Palomo, Synth. Commun., 1984, 14, 939; but cf. S. Cacchi, L Cagliota and E. Cernia, Synthesis, 1979, 64), (b) exposure of esters to dibromomethane/LDA and subsequent elimination and hydrolysis (C.J. Kowalski and M.S. Hague, J. Org. Chem., 1985, 50, 5140; cf. R. Tarhouni et al, Tetrahedron Lett., 1984, 25, 835):

$$RCO_2Et \xrightarrow{LiCHBr_2} RC(OLi)(OEt)CHBr_2 \xrightarrow{BuLi} RC(OLi){=}CHBr \xrightarrow{H^+} RCOCH_2Br$$

(c) Bromination of α-acylanion equivalents, eg. (M. Ashwell and R.F.W. Jackson, J. Chem. Soc., Chem. Commun., 1988, 645):

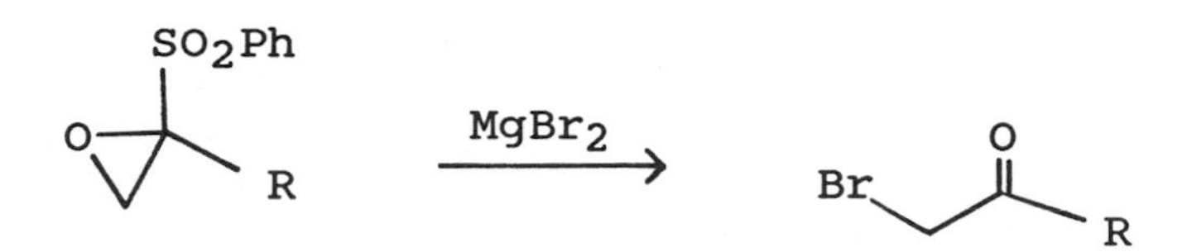

(iv) α-Iodoketones, despite their lability, can be obtained (a) from the parent ketone and iodine/cerium ammonium nitrate (C.A. Horuichi and S. Kiji, Chem. Lett., 1988, 31; but cf. J. Barluenga et al, Synthesis, 1986, 678), (b) from a ketone silyl enol ether with either iodine/silver acetate or N-iodosuccinimide (G.M. Rubottom and R.C. Mott, J. Org. Chem., 1979, 144, 1731; Y.D. Vankar and G. Kumaravel, Tetrahedron Lett., 1984, 25, 233; cf. R.C. Cambie et al, J. Chem. Soc., Perkin Trans. 1, 1978, 126), or (c) by oxidation of alkenes using either iodine/silver chromate (G. Cardillo and M. Shimizu, J. Org. Chem., 1977, 42, 4268) or bis(sym-collidine)iodine (I) tetrafluoroborate /DMSO (R.D. Evans and J.H. Schauble, Synthesis, 1986, 727).

(v) α,α'-Dihaloketones are readily available from appropriate α-haloesters via chloromethyl-dealkoxy substitution (J. Barluenga, J. Chem. Soc., Perkin Trans. 1, 1990, 417):

$$RCHXCO_2Et \xrightarrow[\text{ii. MeLi iii. } H^+]{\text{i. LiBr/CRCH}_2\text{Cl}} RCHXCOCH_2Cl \quad (X=Cl,Br)$$

(vi) Reactions of α-haloketones

(a) Aldol reactions result when the enolate obtained by dehalogenation of an α-haloketone is intercepted by a carbonyl substrate. Most effective are magnesium enolates (obtained with Mg metal) (T. Makata and Y. Kishi, Tetrahedron Lett., 1978, 2745), or aluminium enolates obtained using either zinc/diethylaluminium chloride (K. Maruoka et al, J. Amer. Chem. Soc., 1977, 99, 7705) or tri-n-butylstannyllithium/diethylaluminium chloride (S. Matsubara et al, Bull. Chem. Soc. Jpn., 1984, 57, 3242 and Tetrahedron Lett., 1984, 25, 2569). Cerium (III) enolates can be prepared and used in one of two ways (S. Fukuzawa et al, J. Chem. Soc., Perkin Trans. 1, 1987, 1473):

CeI_3 → [$OCeI_2$ enolate] → RCHO

NaI, $CeCl_3$ → [$OCeCl_2$ enolate] → R_2CO

(b) Reduction. α-Haloketones are reduced to the parent ketones by sodium dithionite/viologen in a two-phase system (K.K. Park et al, ibid, 1990, 2356). Asymmetric reduction to α-haloalcohols can be achieved using Alpine-borane (H.C. Brown and G.G. Pai, J. Org. Chem., 1983, 48, 1784), or bakers yeast (M. Utaka, S. Konishi and A. Takeda, Tetrahedron Lett., 1986, 27,

4737), or a Noyori ruthenium-BINAP system (M. Kitamura et al, J. Amer. Chem. Soc., 1988, 110, 629). Other reagents that reduce α-haloketones to the parent ketone are samarium diiodide (G.A. Molander and G. Hahn, J. Org. Chem., 1986, 51, 1135) and several metal chlorides in the presence of sodium iodide (A. Ono et al, Synthesis, 1986, 570).

(c) Favorskii rearrangement, (cf. First supplement, p. 85), is an important reaction of α-haloketones (review: P.J. Chenier, J. Chem. Educ., 1978, 55, 286) that allows stereospecific ring-contraction of cyclic substrates, eg (R.B. Mitra and A.S. Khanra, Synth. Commun., 1977, 7, 245):

O Br OH i. NaOMe ii. $POCl_3$ CO_2Me

When the substrate is an α,α′-dihaloketone, α,β-unsaturated carboxyl compounds can be obtained (T. Sakai et al, Bull. Chem. Soc. Jpn., 1987, 60, 2295):

X O X ^{-}OH CO_2H X CO_2H

(d) Miscellaneous. Halogen displacement reactions can be used to convert α-haloketones into cyclopropanols (T. Imamota, T. Takeyama and H. Koto, Tetrahedron Lett., 1986, 27, 3243) and into α-allylketones (G.E. Heck and J.B. Yates, J. Amer. Chem. Soc., 1982, 104, 5829):

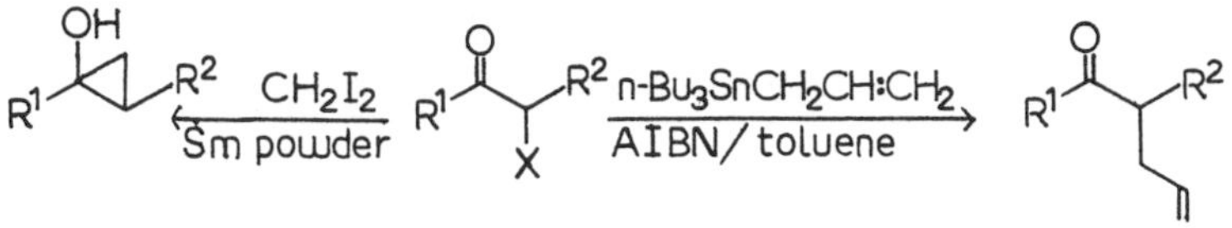

Allylic alcohols are the eventual products when α-chloroketones are treated with trimethylstannylmethyl lithium at -78 ^{o}C, and the mixture is warmed to 0 ^{o}C (E. Murayama et al, Chem. Lett., 1984, 1897; cf. T. Sato, Synthesis, 1990, 259):

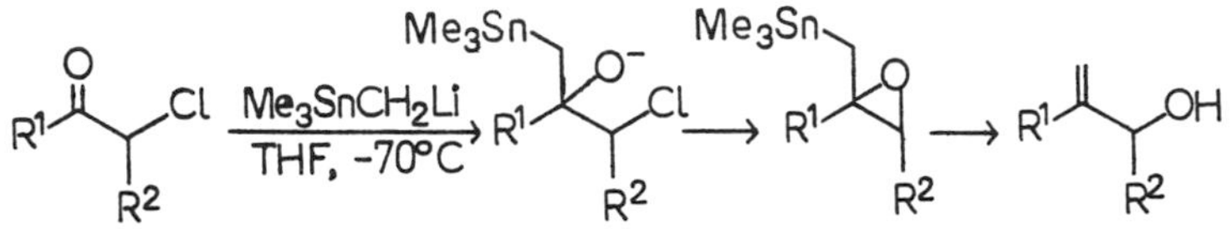

(h) Silicon-containing ketones

(i) Acylsilanes, R^1COSiR_3, have points in common with ordinary ketones. They can be prepared by oxidation (Swern) of α-hydroxysilanes (R.E. Ireland and D.W. Norbeck, J. Org. Chem., 1985, 50, 2198) to which they revert on reduction (eg. J.D. Buynak, J. Chem. Soc., Chem. Commun., 1989, 89). They give enolates with LDA from which aldol products can be obtained (D. Schinzer, Synthesis, 1989, 179), undergo Grignard reactions to tertiary alcohols (M. Nakada et al, J. Amer. Chem. Soc., 1988, 110, 4826) and Wittig reactions to vinylsilanes (J.A. Soderquist and C.L. Anderson, Tetrahedron Lett., 1988, 29, 2425 and 2777). However, in the presence of oxygen nucleophiles the Brook rearrangement occurs, involving silicon migration from carbon to oxygen; aldehydes can be formed by subsequent elimination of a silanol (review: A.G. Brook, Acc. Chem. Res., 1974, 7, 77):

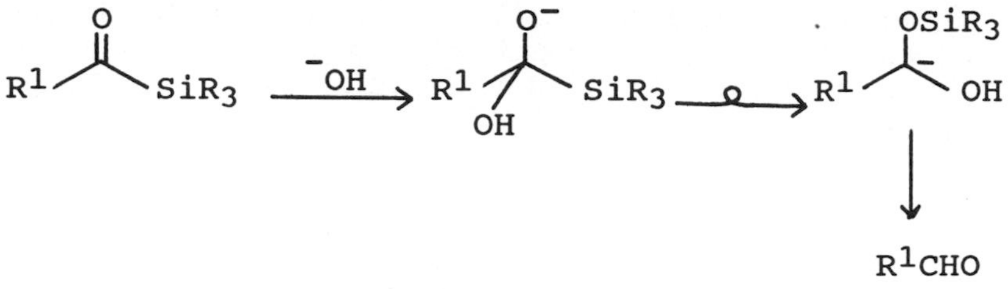

The chemistry of acyl silanes has been reviewed (P.C.B. Page, S.S. Klair and S. Rosenthal, Chem. Soc. Rev., 1990, 19, 147).

(ii) *α-Silylketones* (β-ketosilanes) can be obtained as follows:

(a) from the corresponding α-bromoketone by formation of the silyl enol ether and lithiation (P. Sampson, G.B. Hammond and D.F. Weimer, J. Org. Chem., 1986, 51, 4342):

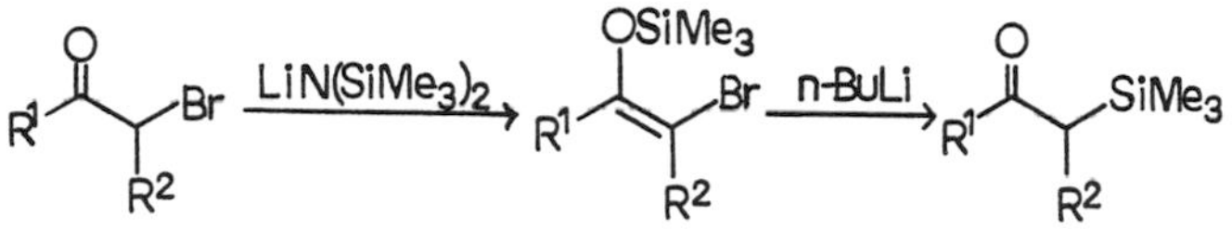

(b) from an aldehyde using a trimethylsilylalkyl- lithium or magnesium reagent, followed by oxidation of the carbinol product (R.A. Ruden and B.L. Gaffrey, Synth. Commun., 1975, 5, 1975; P.A. Brown et al, Tetrahedron Lett., 1987, 28, 693), eg. [E.W. Colvin, op. cit., (see Section 2e)]:

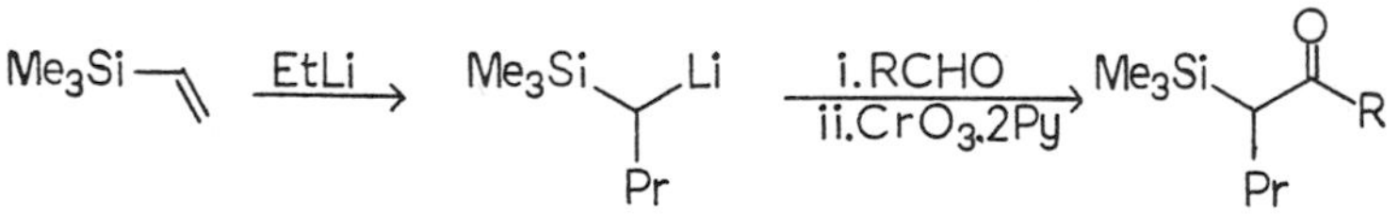

(c) from α-haloacylsilanes (see P.C.B. Page, S.S. Klair and S. Rosenthal, op. cit.), using Grignard reagents (I. Kuwajima et al, Bull. Chem. Soc. Jpn., 1984, 57, 2167):

$$\mathrm{ClCH(R^1)C(O)SiMe_3} \xrightarrow{\mathrm{RMgX}} \mathrm{ClCH(R^1)C(OMgX)(R)SiMe_3} \longrightarrow \mathrm{R^1CH(SiMe_3)C(O)R}$$

(d) from α,β-dihydroxysilanes by sila-pinacol rearrangement (R.F. Cunico, Tetrahedron Lett., 1986, 27, 4269):

$$\mathrm{R^1C(OH)(SiR_3)CH(OH)R^2} \xrightarrow{\mathrm{TFA}} \mathrm{R^1C(O)CH(R^2)SiR_3}$$

(e) from α-silyl allylic alcohols by rhodium catalysed isomerisation (S. Sato et al, ibid, 1984, 25, 769):

$$\mathrm{R^1CH(SiR_3)CH(OH)C(R^2){=}CHR^3} \xrightarrow{[(Ph_3P)_4RhH]} \mathrm{R^1CH(SiR_3)C(O)CH(R^2)CH_2R^3}$$

(f) from α-silyl esters with suitable Grignard reagents (G.L. Larson, I.M. de Lopez-Cepero and L.E. Torres, ibid, 1984, 25, 1673):

$$R^1CH(SiMePh_2)CO_2Et \xrightarrow{RCH_2MgX} R^1CH(SiMePh_2)C(OMgX)=CHR \xrightarrow{H_2O} R^1CH(SiMePh_2)COCH_2R$$

(g) from α,β-epoxysilanes with magnesium iodide (M. Obayashi, H. Ultimoto and H. Nozaki, ibid, 1977, 1807 and 1978, 1383).

(iii) Regiospecific aldol reactions can be achieved using α-silylketones. LDA (and an aldehyde) gives alkylation at the α' position, whereas Lewis acids (eg $TiCl_4$, $SnCl_4$, BF_3) lead to α-alkylation (T. Inoue, T. Sato and I. Kuwajima, ibid, 1984, 49, 4671):

$$R^1CH_2COCH(R^2)CH(OH)R \xleftarrow{\text{i. LDA; ii. RCHO, iii. } H^+} R^1CH(SiMe_3)COCH_2R^2 \xrightarrow[\text{RCHO}]{\text{Lewis acid}} RCH(OH)CH(R^1)COCH_2R^2$$

Dibutylborane triflate gives the same regiospecificity as LDA but via a boron enolate. Good stereospecificity can be obtained using this approach (D. Enders and B.B. Lohray, Angew. Chem. Int. Ed. Engl., 1988, 27, 581).

3. UNSATURATED KETONES

(a) Ketenes, $R^1R^2C=C=O$

Ketenes are often prepared and used *in situ*, but their acetals can be isolated and stored. The chemistry of these compounds has been reviewed (see "The Chemistry of Ketenes, Allenes and Related Compounds", S. Patai, Wiley Interscience, New York, 1980). Some important families of higher order ketenes (beyond the scope of this chapter) are α-oxoketenes, RCOCH=C=O (R.K. Dieter, *Tetrahedron*, 1986, *42*, 3029), α-cyanoketenes, RC(CN)=C=O (H.W. Moore and M.D. Gheorgiu, *Chem. Soc. Rev.*, 1981, *10*, 289), and alkoxyketenes, $RC(OR^1)=C=O$ (H. Meier *et al*, *Chem. Ber.*, 1988, *121*, 1643). Conjugated ketenes have been reviewed elsewhere (H.W. Moore and O.H.W. Decker, *Chem. Rev.*, 1986, *86*, 821).

(i) *Ketene*, ($R^1=R^2=H$), (First supplement, p.96) is conveniently stored as the cyclic dimer, diketene (review: R.J. Clemens, *Chem. Rev.*, 1986, *86*, 241). *Aldoketenes* (R^1=alkyl, R^2=H), and *ketoketenes* ($R^1=R^2$=alkyl) can be obtained in the following ways (see B.B. Snider, *Chem. Rev.*, 1988, *88*, 793 and references cited therein); (a) by base induced dehydrochlorination of acyl chlorides, $R^1R^2CHCOCl$; (b) by dehalogenation of α-haloacyl halides, R^1R^2CXCOX (cf. L.M. Baigrie *et al*, *J. Org. Chem.*, 1985, *50*, 2105); (c) by Wolff rearrangement of α-diazoketones, $R^1COC(N_2)R^2$.

(ii) *Acetal derivatives*

(a) *Ketene acetals*, $R^1R^2C=C(OR)_2$, are available by elimination from suitable α,α-dialkoxyphosphine oxides (T.A.M. van Schaik, A.V. Henzen and A. van der Gen, *Tetrahedron Lett.*, 1983, *24*, 1303):

$$Ph_2POCH(OEt)_2 \xrightarrow[\text{ii. RCHO}]{\text{i. LDA}} Ph_2POC(OH)(R)(OEt)_2$$

$$\xrightarrow{KOBu^t} RCH{=}C(OEt)_2$$

(b) Ketene O,S-acetals ($R^1R^2C{=}C(OR)(SR^3)$), can be obtained from ketones by Petersen olefination, eg. (A. de Groot and B.J.M. Jansen, Synth. Commun., 1983, 13, 985):

$$HC(OMe)(SPh)(SiMe_3) \xrightarrow[\text{ii. } R^1R^2CO]{\text{i. BuLi}} R^1R^2C{=}C(OMe)(SPh)$$

(c) Ketene dithioacetals, $R^1R^2C{=}C(SR)_2$, can be prepared from aldehydes, ketones, esters, thioesters, dithioacetals and gem-dihaloalkenes. The preparations and reactions have been reviewed (M. Kolb, Synthesis, 1990, 171).

(d) Ketene silyl acetals, $R^1R^2C{=}C(OR)(OSiMe_3)$, are best obtained from esters $R^1R^2CHCO_2R$, by enolisation (LDA) and interception with chlorotrimethylsilane (C. Ainsworth and Y-N. Kuo, J. Organomet. Chem., 1972, 46, 73; but cf. C.S. Wilcox and R.E. Babston, Tetrahedron Lett., 1984, 25, 699). However, esters of acetic acid do not react cleanly apart from phenyl acetate (giving R=Ph) (N. Slougi and G. Rousseau, Synth. Commun., 1987, 17, 1), unless t-butyldimethylsilyl chloride in HMPT is used as trapping reagent (M.W. Rathke and D.F. Sullivan, ibid, 1973 3, 67).

(e) Ketene silyl thioacetals, $R^1R^2C{=}C(SR)(OSiMe_3)$ can be prepared as above but from the thioesters, $R^1R^2CHCOSR$ (C. Gennnari and P.G. Cozzi, Tetrahedron, 1988, 44, 5965), or from the corresponding ketene O,S-acetals (when R=Me) by silyl-demethyl substitution using iodotrimethylsilane (S. Hackett and T. Livinghouse, Tetrahedron Lett., 1984, 25, 3539).

(f) Ketene bis(trimethylsilyl)acetals, $R^1R^2C{=}C(OSiMe_3)_2$, are obtained by trapping (Me_3SiCl) either the enolate of a trimethylsilyl ester, or the dianion of a carboxylic acid (using LDA in either case) (C. Ainsworth et al, J. Organomet. Chem., 1972, 46, 59 and 73).

(iii) Nitrogen derivatives

(a) Keteneimines are unstable although the phenylimine, $Me_2C{=}C{=}NPh$, is known (U. Behrens and T. Sielisch, ibid, 1986, 310, 179). Keteneimine metal complexes, $[(R_2C{=}C{=}NR^1)(MLn)]$, can be prepared from isonitriles, R^1NC, with metal carbene complexes, $(LnM{=}CR_2)$ (R. Aumann, Angew. Int. Ed. Engl., 1988, 27, 1456).

(b) Keteniminium salts are useful synthetic intermediates (see below) and can be prepared in situ from suitable amides using triflic anhydride/collidine, eg. (L. Ghosez et al, J. Amer. Chem. Soc., 1985, 107, 2192):

CON → $=C=\overset{+}{N}$

(iv) Haloketenes, $RC(X){=}C{=}O$ and $CX_2{=}C{=}O$, by analogy with ordinary ketenes, are prepared in situ either using dehydrohalogenation of haloacyl halides, $RC(X)HCOX$ or X_2CHCOX, or using dehalogenation (Zn-Cu couple) of trihaloacetyl halides. The preparations and reactions have been reviewed (W.T. Brady, Tetrahedron, 1981, 37, 2949).

(v) Trimethylsilylketene, $Me_3SiCH=C=O$, is fairly stable and can be prepared by pyrolysis of the appropriate t-butyl acetylenic ether (E. Valenti, M.A. Pericas and F. Serrantosa, J. Org. Chem., 1990, 55, 395):

$$Me_3Si{-}{\equiv}{-}OCMe_3 \xrightarrow{\Delta} CH_2{=}CMe_2 + Me_3SiCH{=}C{=}O$$

It gives rise to a bis(trimethylsilyl)acetal that is useful in the preparation of unsaturated carboxylic acids (see Section (viii) below).

(vi) Cycloaddition reactions and related processes. The [2+2]-cycloaddition of ketenes with alkenes produces cyclobutanones but is practicable only as an intramolecular reaction unless haloketenes are used. However, the intramolecular process, involving any of the ketenes described above or keteniminium salts, is of wide scope and has been reviewed (B.B. Snider, Chem. Rev., 1988, 88, 793).

Intermolecular reactions with haloketenes (see W.T. Brady, op. cit.) are useful for the preparation of cyclobutanones and hence (see Section 2c) cyclopropane and cyclopentanone derivatives:

$Cl_2C{=}C{=}O$ + RCH=CHR

Stepwise replacement of chlorine by hydrogen can be achieved using zinc/acetic acid or tri-n-butyltin hydride (A. Amman, M. Rey and A.S. Drieding, Helv. Chim. Acta, 1987, 70, 321; R.L. Danheiser and S. Savariar, Tetrahedron Lett., 1987, 28, 3299). [2+2]-Cycloaddition involving haloketones has also been used to prepare cyclobutenones (with alkynes), spirocyclobutanones (with alkylidenecycloalkanes), vinylcyclobutanones (with allenes), tropolones (with cyclopentadienes), β-lactones (with carbonyl compounds) (for examples and references, see W.T. Brady, op. cit.), and β-lactams (with imines) (cf. J-E. Dubois and G. Axiotis, Tetrahedron Lett., 1984, 25, 2143).

Ketene allyl acetals are suitable substrates for Claisen rearrangement. In the Ireland-Claisen reaction, a silyl allyl acetal rearranges into a γ,δ-unsaturated acid (R.E. Ireland et al, J. Amer. Chem. Soc., 1976, 98, 2868 and 1984, 106, 3668):

$$\text{R-CH(O-C(=CH_2)OSiBu^tMe_2)-CH=CH_2} \longrightarrow \text{RCH=CH-CH_2-CH_2-COOH}$$

(vii) Addition reactions

Ketenes, their acetals and thioacetals exhibit many of the addition reactions expected for alkenes. Electrophilic and nucleophilic processes have been identified although concerted or radical mechanisms, rather than ionic, have been proposed in some cases (reviews: H.R. Seikaly and T.T. Tidwell, Tetrahedron, 1986, 42, 2587; D.P.N. Satchell and R.S. Satchell, Chem. Soc. Rev., 1975, 4, 231).

Electrophilic addition normally proceeds *via* oxygen- or sulphur-stabilised carbocations, often requires acid catalysis, and leads to carboxylic acid derivatives, eg. hydrogen halides (H-X) give acyl halides:

$$R_2C{=}C{=}O \xrightarrow{H^+} R_2CH{-}\overset{+}{C}{=}O \longrightarrow R_2CHCOX$$

Similarly, carboxylic acids ($H{-}OCOR^1$) give anhydrides while aqueous mineral acids give carboxylic acids. α-Thioacyl chlorides are given by arylsulphenyl chlorides (ArS-Cl), and chloromethyl methyl ether ($CH_3OCH_2{-}Cl$) produces β-methoxyacyl chlorides using $ZnCl_2$ as catalyst. Electrophilic ring-closure reactions involving ketene dithioacetal moieties are useful for the construction of di- and poly-cyclic systems, eg. (R.S. Brinkmeyer, Tetrahedron Lett., 1979, 207):

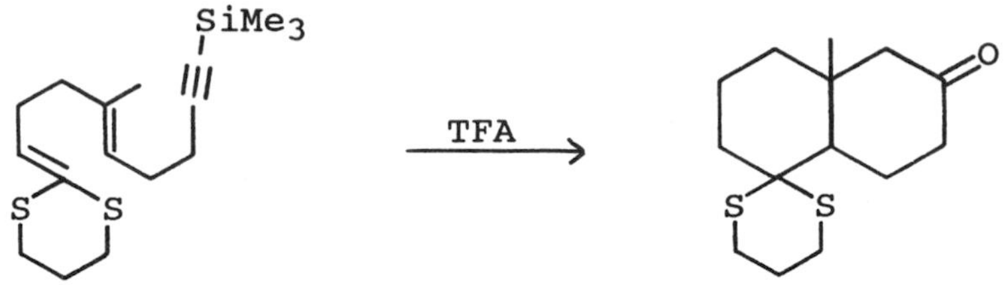

$$\xrightarrow{\text{TFA}}$$

<u>Nucleophilic addition</u> involves attack at the carbonyl carbon producing an enolate. The procedure can be used to produce hindered enolates not easily accessible by the usual methods, eg. (L.M. Baigrie <u>et al</u>, <u>J. Org. Chem.</u>, 1985, <u>50</u>, 2105):

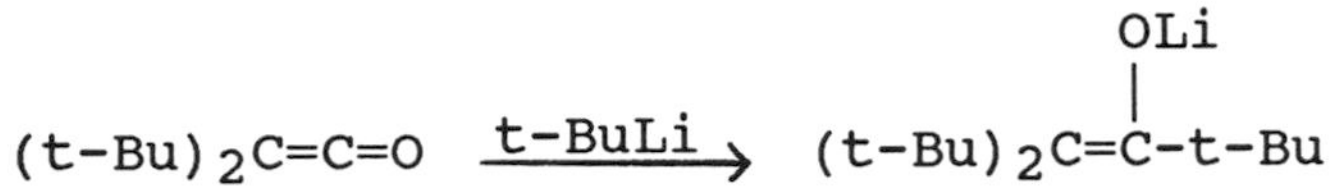

$$(t\text{-}Bu)_2C{=}C{=}O \xrightarrow{t\text{-}BuLi} (t\text{-}Bu)_2C{=}C(OLi)\text{-}t\text{-}Bu$$

The enolates can be used <u>in situ</u>, eg. for aldol reactions (R. Haner, T. Laube and D. Seebach, <u>J. Amer. Chem. Soc.</u> 1985, <u>107</u>, 5396), or converted into enol ethers (eg. with Me_3SiCl) or esters (eg with RCOCl). Lithium aluminium hydride (source of H^-) followed by acetyl chloride gives the enol acetate of the corresponding aldehyde ($R_2C{=}CHOAc$). Protic nucleophiles lead eventually to carboxylic acid derivatives; alcohols give esters, thiols give thioesters and amines give amides (see H.R. Seikaly and T.T. Tidwell, <u>op. cit.</u>, for examples of all these reactions):

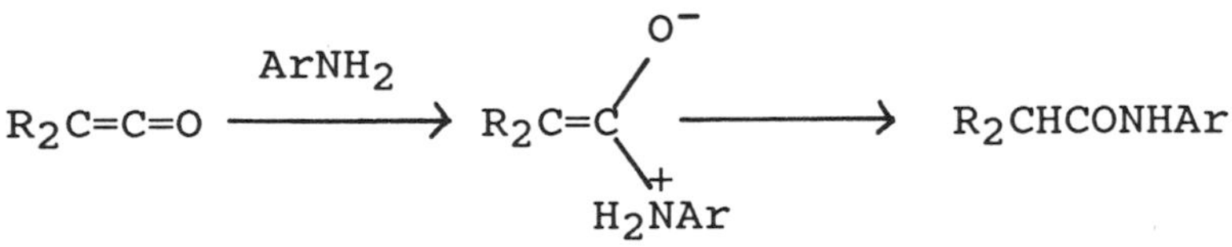

$$R_2C{=}C{=}O \xrightarrow{ArNH_2} R_2C{=}C(O^-)(\overset{+}{H_2}NAr) \longrightarrow R_2CHCONHAr$$

(viii) C-C bond-forming reactions of ketene acetals. Ketene silyl acetals, silyl thioacetals and bis-silyl acetals undergo aldol-type reactions with aldehydes or ketones, eg. (C. Gennari and P.G. Cozzi, Tetrahedron, 1988, 44, 5965):

$$\text{OBu}^t\text{-CH(CH}_3\text{)-CHO} + \text{CH}_2\text{=C(OSiBu}^t\text{Me}_2\text{)(SBu}^t\text{)} \xrightarrow{SnCl_4} \text{product (OBu}^t\text{, OH, O, SBu}^t\text{)}$$

With O-acetals the product is an ester, while bis-silylacetals generally give carboxylic acids due to the lability of the Si-O bond in silyl esters. Under Mukaiyama conditions it is possible to chlorinate the carbinol product *in situ* (M. Bellassoved, J-E. Dubois and E. Bertounesque, Tetrahedron Lett., 1988, 29, 1275):

$$\text{PhCHO} + \text{RCH=C(OSiMe}_3\text{)}_2 \xrightarrow[\text{ii. } H_2O]{\text{i. } TiCl_4} \text{PhCH(Cl)CH(R)CO}_2\text{H}$$

When trimethylsilylketene acetals are used, the product is an α,β-unsaturated acid derivative (M. Bellassoved and M. Gaudemar, ibid, 1988, 29, 4551; B.Y. Chung, M.S. Kim and S.D. Lee, Bull. Korean Chem. Soc., 1988, 9, 67):

$$\text{RCHO} + \text{Me}_3\text{SiCH=C(OSiMe}_3\text{)}_2 \xrightarrow[\text{ii. } H_2O]{\text{i. } ZnBr_2} \text{RCH=CHCO}_2\text{H}$$

Michael reactions are also feasible; even nitrobenzene can be used as a substrate (T.V. Rajan Babu and T. Fukunaga, J. Org. Chem., 1984, 49, 4571). Addition to α,β-unsaturated esters gives 1,5-dicarboxylates (M. Kawai, M. Onaka and Y. Izumi, Bull. Chem. Soc. Jpn., 1988, 61, 2157), while α,β-enones give 5-oxocarboxylates (T.V. Rajan Babu, J. Org. Chem., 1984, 49, 2083):

$$R^4COCH{=}CHR^3 + R^1R^2C{=}C(OSiMe_3)(OMe) \longrightarrow R^4COCH_2CHR^3C(R^1R^2)CO_2Me$$

Ketene silyl acetals are alkylated by aldehyde dioxalans (to yield β-ketoester acetals) (Y. Hayashi, K. Wariishi and T. Mukaiyama, Chem. Lett., 1987, 1243), and by dienyl-iron complexes, eg. (A.J. Pearson and M.K. O'Brien, Tetrahedron Lett., 1988, 29, 869):

OMe, Fe(CO)$_3$ + OSiMe$_3$, OSiMe$_3$ $\xrightarrow{\Delta}$ OMe, Fe(CO)$_3$, CO$_2$Me

Ketene phenylthioacetals can be lithiated directly and the metalloderivatives are alkylated by a wide range of electrophiles, eg (T. Cohen and R.B. Weisenfeld, J. Org. Chem., 1979, 44, 3601; cf. M. Kolb, op. cit.):

$$R_2C{=}C(SPh)_2 \xrightarrow{\text{Li naphthalide}} R_2C{=}C(SPh)Li \xrightarrow{R^1X} R_2C{=}C(SPh)R^1$$

(b) α,β-Enones

Most aspects of the chemistry of α,β-enones have been comprehensively reviewed ("The Chemistry of Enones", Parts 1 and 2, S. Patai and Z. Rappoport (Eds.), John Wiley and Sons, Chichester, 1989).

(i) Preparation, cf. Section (iii) - Homologation reactions.

(a) From ketones by oxidation and elimination of their α-seleno or α-thio derivatives (H.J. Reich, J.M. Renga and I.L. Reich, J. Amer. Chem. Soc., 1975, 97, 5434; B.M. Trost, Acc. Chem. Res., 1978, 11, 453), or of the same derivatives after rearrangement (D. Liotta, ibid, 1984, 17, 28) eg:

SePh PhSe NaIO$_4$ NaIO$_4$

Alternatively a ketone can be converted to a homologous enone using TOSMIC (J. Moskal and A.M. van Leusen, Tetrahedron Lett., 1984, 25, 2585):

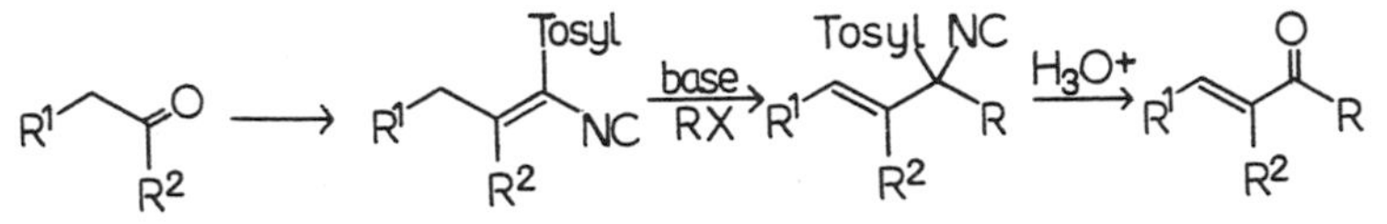

(b) From ketone silyl enol ethers using oxidation by diallyl carbonate and a Pd catalyst (I. Minani et al, Tetrahedron, 1986, 42, 2971; review, J. Tsuji and I. Minami, Acc. Chem. Res., 1987, 20, 140). α-Methylene ketones can be prepared via either cyclopropanation (I. Ryu, S. Murai and N. Sonada, J. Org. Chem., 1986, 51, 2389) or sulphonation (E. Block et al, ibid, 1984, 49, 3664):

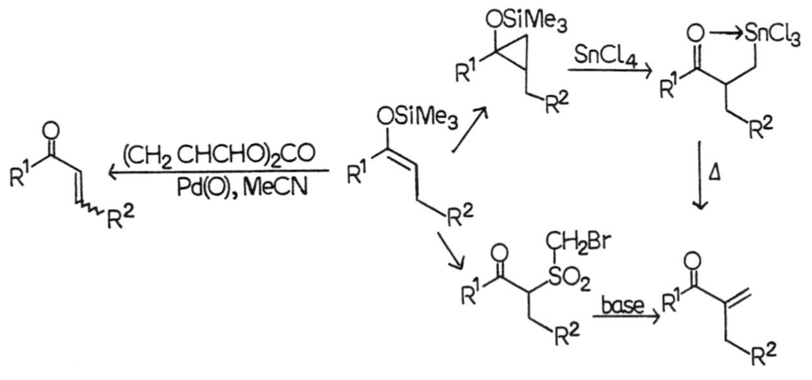

(c) From ketone enamines using acylation followed by hydrogenation and elimination (P. Zhang and L. Li, Synth. Commun., 1986, 16, 957):

$$R^1C(NR_2)=CHR^2 \xrightarrow{R^3COCl} R^1C(NR_2)=C(R^2)COR^3 \xrightarrow[\text{ii. } \Delta]{\text{i. } H_2/PtO_2} R^1CH=C(R^2)COR^3$$

(d) From acyl halides by halide displacement using a vinyl cuprate and Pd(O) (N. Jabri, A. Alexakis and J.F. Normant, Tetrahedron, 1986, 42, 1369), or a vinyl allane with $ZnCl_2$ and Pd(O) (E. Negishi et al, Tetrahedron Lett., 1983, 24, 5181), or a vinyl stannane and Pd(O) (J.B. Verlhac, J-P. Quintard and M. Pereyre, J. Chem. Soc., Chem. Commun., 1988, 503), or a vinyl silane (I. Fleming and E. Pearce, J. Chem. Soc., Perkin Trans. 1, 1980, 2485). Alternatively, suitable alkenes will couple to acylcobalt (II) Schiff's base complexes formed from acyl halides (review: G. Pattenden, Chem. Soc. Rev., 1988, 17, 361):

$$RCOCl \xrightarrow{[M-CH=CHR^1]} RCOCH=CHR^1$$

$$RC(=O)-Co(Salophen)Py \xrightarrow[h\nu]{CH_2=CHPh} RCOCH=CHPh$$

(e) From aldehydes using crossed-aldol reactions and elimination of water from the hydroxyketone product (see Section 2e(iii)) (cf. I. Matsuda et al, Tetrahedron Lett., 1984, 25, 3879; T. Hirao et al, Chem. Lett., 1984, 367).

(f) From alkyl halides and α,β-unsaturated acyl anion equivalents (eg. U. Hertenstein, S. Hunig and M. Oller, Chem. Ber., 1980, 113, 3783). A particularly versatile synthesis involves elaboration of an allylic sulphone with three selected alkyl halides (K. Ogura et al, J. Org. Chem., 1986, 51, 700):

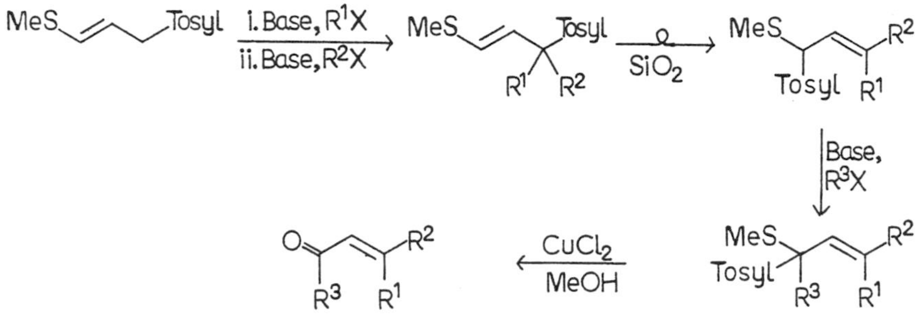

A similar approach uses the methoxyallyl sulphide, $CH_2{=}C(R^1)CH(OMe)SPh$, to give the ketone, $R^3CH{=}CR^1COR^2$, using R^2X and R^3X (T. Mandai et al, Tetrahedron Lett., 1984, 25, 5913). Allylic selenides can also be used (R.S. Brown, S.C. Eyley and P.J. Parsons, J. Chem. Soc., Chem. Commun., 1984, 438).

(g) Alkynes, as their trialkylboronate derivatives, provide a flexible preparation involving either one or two alkyl migration steps (A. Pelter and M.E. Colclough, Tetrahedron Lett., 1986, 27, 1935):

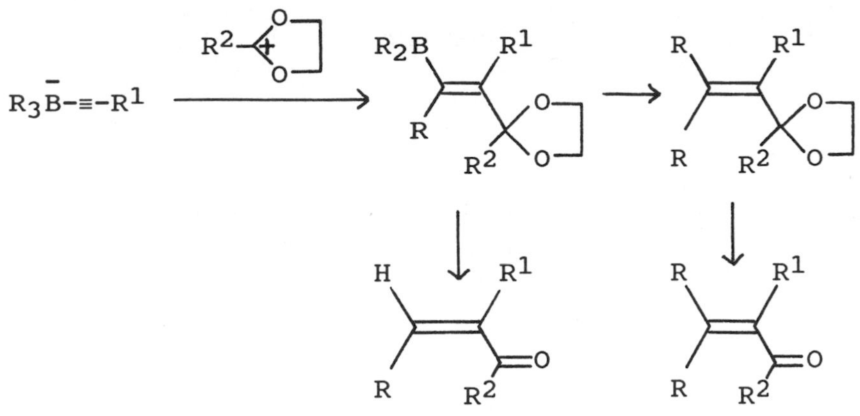

α,β-Ynones are reduced to the corresponding enones by chromium (II) salts (A.B. Smith, P.A. Levenberg and J.Z. Suits, Synthesis, 1986, 184), or the acetals of α,β-ynones undergo conjugate addition by alkylcuprates, eg. R_2CuLi, to give β-alkyl-α,β-enones (A. Alexakis et al, Tetrahedron, 1984, 40, 715).

(h) From alkenes using allylic oxidation by SeO_2/t-butylhydroperoxide (M.A. Umbreit and K.B. Sharpless, J. Amer. Chem. Soc., 1977, 99, 5526), or from allylic alcohols using, eg. imidazolium dichromate (S. Kim and D.C. Lhim, Bull. Chem. Soc. Jpn., 1986, 59, 3297).

(ii) Cyclopentenones are important in the context of natural products but can be awkward to prepare. They may be obtained from acyclic precursors in the following ways:

(a) from α,β-unsaturated acyl chlorides by rhodium catalysed carbenoid cyclisation (E. Wenkert et al, J. Org. Chem., 1990, 55, 311):

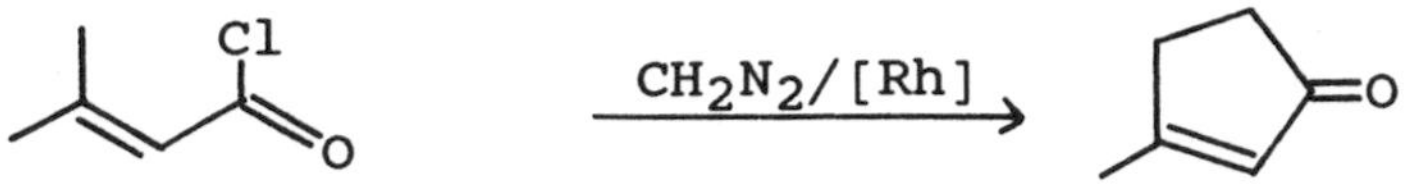

$\xrightarrow{CH_2N_2/[Rh]}$

(b) from divinyl ketones by Lewis acid catalysed Nazarov cyclisation, especially when a β-silyl group is present to direct the ring-closure (S.E. Denmark, K.L. Habermas and G.A. Hite, Helv. Chim. Acta, 1988, 71, 168; cf. P. Magnus and D. Quagliato, J. Org. Chem., 1985, 50, 1621):

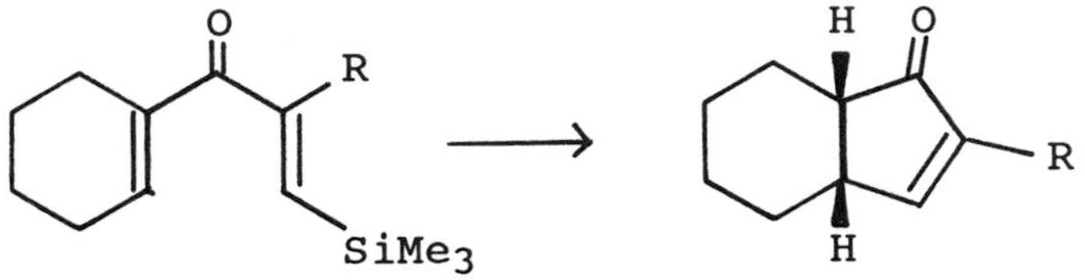

(c) by Pauson-Khand ring closure of appropriate enynes (P.L. Pauson, Tetrahedron, 1985, 41, 5855) with cobalt carbonyl (for a similar approach using organozirconium chemistry, see E. Negishi et al, J. Amer. Chem. Soc., 1985, 107, 2568). Quite complex ketones can be obtained in a single step, eg. (P. Magnus and D.P. Becher, ibid, 1987, 109, 7495):

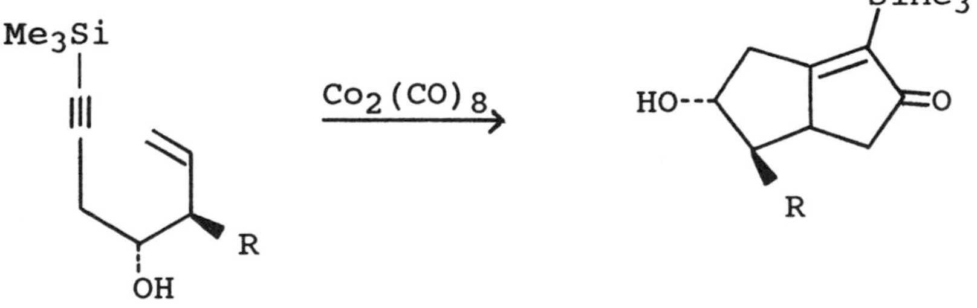

$\xrightarrow{Co_2(CO)_8}$

(d) from 1-ethynylprop-2-enyl acetates by Pd catalysed cyclisation and hydrolysis of the resulting enol acetate (V. Rautenstrauch, J. Org. Chem., 1984, 49, 950):

i. $[(MeCN)_2PdCl_2]$
ii. H_2O

(e) from cyclopropenone acetals via cycloaddition to electron deficient alkenes (D.L. Boger and C.E. Brotherton, J. Amer. Chem. Soc., 1984, 106, 805).

(iii) Homologation of α,β-enones

Homologation at the α position is best achieved by aldol reactions using an aldehyde with B-phenylselenenyl-9-borabicyclo[3.3.1]nonane (W.R. Leonard and T. Livinghouse, J. Org. Chem., 1985, 50, 730), or with a rhodium or ruthenium complex catalyst (S. Sato et al, Chem. Lett., 1985, 1875, and J. Organomet. Chem., 1988, 340, C5), or with DABCO (N. Daude, U. Eggert and H.M.R. Hoffman, J. Chem. Soc. Chem. Commun., 1988, 206):

RCHO

For homologation at the β-position, the triphenylphosphonium homoenolate derivative is useful, eg (1):

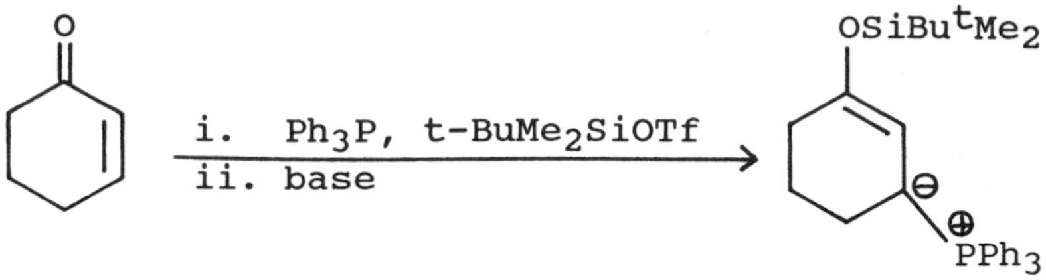

$$\text{cyclohex-2-enone} \xrightarrow[\text{ii. base}]{\text{i. } Ph_3P,\ t\text{-}BuMe_2SiOTf} \text{1-}(OSiBu^tMe_2)\text{cyclohexenyl-3-ylide } (\ominus / \oplus PPh_3) \qquad (1)$$

The homoenolate gives a Michael adduct, or an O-silyl aldol derivative (2) (S. Kim and P.H. Lee, Tetrahedron Lett., 1988, 29, 5413), or the Wittig product (3) that can be elaborated by a vinylogous aldol reaction (A.P. Kozikowski and S.H. Jung, J. Org. Chem., 1986, 51, 3400):

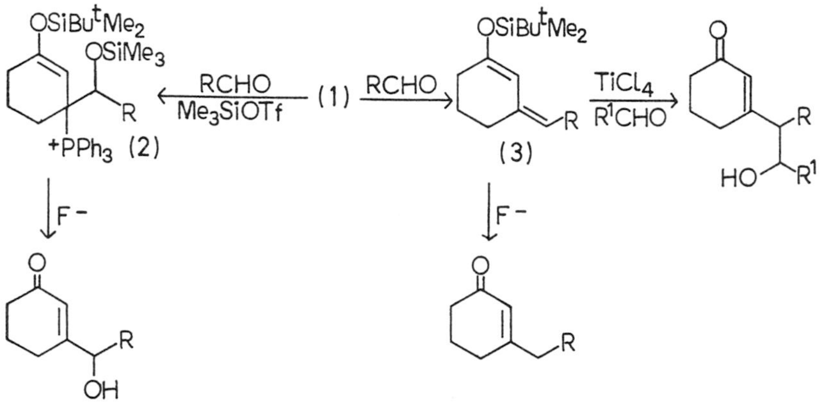

OSiBu^{t}Me$_2$ OSiMe$_3$ R $^+$PPh$_3$ (2) $\xleftarrow[Me_3SiOTf]{RCHO}$ (1) $\xrightarrow{RCHO}$ OSiBu^{t}Me$_2$ R (3) $\xrightarrow[R^1CHO]{TiCl_4}$ O R HO R^1

(2) $\xrightarrow{F^-}$ O R OH

(3) $\xrightarrow{F^-}$ O R

An alternative strategy employs conjugate addition to the α,β-enone in the presence of trimethylchlorosilane (Section v) followed by oxidation (cf. Y. Hatanka and I. Kuwajima, ibid, 1986, 51, 1932).

Suitable α,β-enones couple directly with alkylcobalt (II) Schiff's base complexes (V. Patel and G. Pattenden, J. Chem. Soc. Chem. Commun., 1987, 871):

Co(salen)Py + R, O → hν → R, O

(iv) Reduction

Carbonyl reduction, giving allylic alcohols, requires careful application of common reducing agents to suppress conjugate reduction. Successful reductions have been reported using hydrogen (E. Farnetti et al, ibid., 1986, 746), isopropanol (T. Nakano et al, J. Org. Chem., 1988, 53, 3752), lithium aluminium hydride (R. Noyori et al, J. Amer. Chem. Soc., 1984, 106, 6709 and 6717; S. Fukuzawa et al, J. Chem. Soc. Perkin Trans. 1, 1986, 1929), boranes (S. Krishnamurthy and H.C. Brown, J. Org. Chem., 1977, 42, 1197), or sodium borohydride (R. Fornasier et al, ibid., 1986, 51, 1769).

Conjugate reduction, giving the saturated ketones, is possible using dihydropyridine reagents (Y. Inoue et al, Bull. Chem. Soc. Jpn., 1988, 61, 3020; K. Nakamura et al, Chem. Lett., 1984, 925), or 2-phenylthiazoline/$AlCl_3$ (H. Chikashita, M. Miyazaki and K. Itoh, Synthesis, 1984, 308; cf. H. Chikashita and K. Itoh, Bull. Chem. Soc. Jpn., 1986, 59, 1747), or a complex copper hydride (W.S. Mahoney, D.M. Breslensky and J.M. Stryker, J. Amer. Chem. Soc., 1988, 110, 291), or diphenylsilane catalysed by Pd(O)/$ZnCl_2$ (E. Keinan and N. Greenspoon, ibid., 1986, 108, 7314), or sodium dithionite with phase-transfer catalysis (F. Camps, J. Coll and J. Guitart, Tetrahedron, 1986, 42, 4603).

(v) Conjugate addition

Attack by a nucleophile, Nu^-, at the β-position of α,β-enones can be used to prepare (a) saturated ketones by quenching (eg. H_2O), or (b) silyl enol ethers (by interception with eg. Me_3SiCl) that can be further elaborated (Section 2e), or (c) α-alkylated derivatives by exposure to an electrophile (E^+), eg:

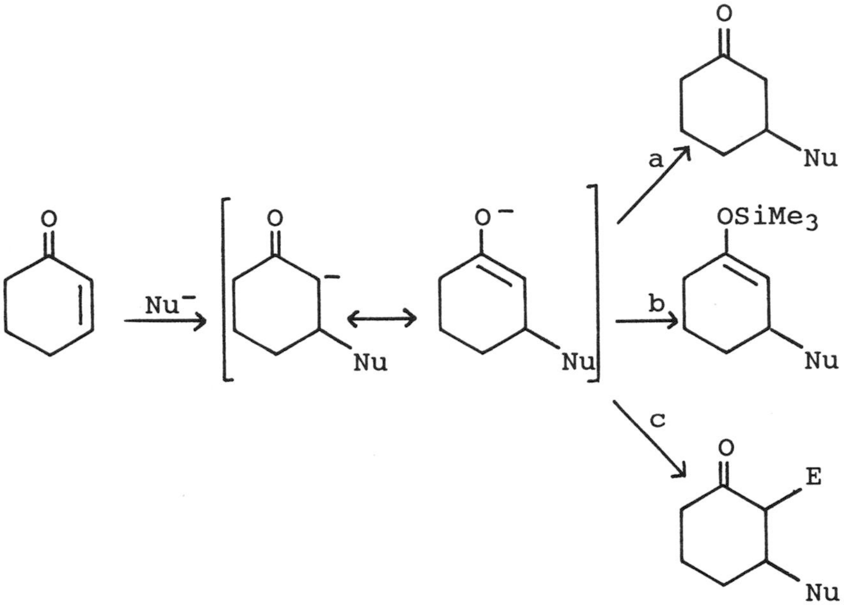

The usual reagents for alkylation (ie Nu=R) are lithium dialkylcuprates, R^2CuLi, from which only one R group is transferred (reviews: G.H. Posner, Org. React., 1972, 19, 1; H.O. House, Acc. Chem. Res., 1976, 9, 59; cf. "An Introduction to Synthesis using Organocopper Reagents", G.H. Posner, John Wiley and Sons, New York, 1980), and higher order cuprates, eg.

R(CN)CuLi, (review: B.H. Lipshutz, Synthesis, 1987, 325), although organozinc reagents can also be used (K. Soai, J. Org. Chem., 1988, 53, 4148). Reviews are available covering conjugate additions in the presence of trapping agents (R.J.K. Taylor, Synthesis, 1985, 364), and conjugate addition/ α-alkylation procedures involving cyclopentenones (R. Noyori and M. Suzuki, Angew. Chem. Int. Ed. Engl., 1984, 23, 847).

Vinyl or aryl groups (Nu) can be introduced from appropriate cuprates (G.H. Posner, op. cit., but see G.A. Kraus and P. Gottschalk, Tetrahedron Lett., 1983, 24, 2727, for an alternative arylation procedure). Vinyl groups can also be transferred from vinyl boranes (P. Jacob and H.C. Brown, J. Amer. Chem. Soc., 1976, 98, 7832), or vinyl allanes (N. Okukado and E. Negishi, Tetrahedron Lett., 1978, 2357). Alkynyl cuprates do not transfer the alkyne moiety but alkynyl boranes can be used instead (H.C. Brown, U.S. Racherla and D. Basavaiah, Synthesis, 1984, 303). Bis(trialkylsilyl) cuprates can be used to reversibly protect the olefinic bond during multistage syntheses (I. Fleming and T.W. Newton, J. Chem. Soc., Perkin Trans. 1, 1984, 1805):

$$R^1COCH{=}CHR^2 \underset{CuBr_2}{\overset{(R_3Si)_2CuLi}{\rightleftharpoons}} R^1COCH_2CH(R^2)SiR^3$$

For asymmetric additions either the reagent or the substrate needs to be chiral. For use with prochiral substrates, chiral auxiliaries have been incorporated into higher order cuprate reagents (eg. E.J. Corey, R. Naef and F.J. Hannon, J. Amer. Chem. Soc., 1986, 108, 7114). Alternatively, removable chiral sulphinyl auxiliaries on the substrates are of proven value (review: G.H. Posner, Acc. Chem. Res., 1987, 20, 72):

Br, Tol, S, O, R_2CuLi, R, Al-Hg

Conjugate addition is a reaction of great scope, applicable to a wide range of carbon and other nucleophiles. Acyl cuprates produce diketones by conjugate acylation (D. Seyferth and R.C. Hui, Tetrahedron Lett., 1986, 27, 1473; cf. D.J. Ager and M.B. East, J. Org. Chem., 1986, 51, 3983), and even enantioselective hydroxylation can be realised, *via* conjugate silylation using a chiral binaphthyl palladium catalyst (T. Hayashi, Y. Matsumoto and Y. Ito, J. Amer. Chem. Soc., 1988, 110, 5579):

R^1, R^2, O, $PhCl_2SiSiMe_3$, BINAP.$PdCl_2$, $PhCl_2Si$, $OSiMe_3$, MeLi, OLi, RX, i HBF_4, ii H_2O_2, R, OH

The more classical, Michael-type additions involving enol nucleophiles have been extensively studied. Detailed investigations have been made of conjugate additions to α,β-enones involving silyl enol ethers (C.H. Heathcock, M.H. Norman and D.W. Uehling, ibid., 1985, 107, 2797), amide and thioamide enolates (C.H. Heathcock et al, J. Org. Chem., 1990, 55, 132), and ester and ketone enolates (D.A. Oare and C.H. Heathcock, ibid., 1990, 55, 157).

(vi) Miscellaneous reactions

α,β-Enones are readily converted into their ketal derivatives, eg. with pyridinium or collidinium tosylate as catalyst (T.J. Nitz and L.A. Paquette, Tetrahedron Lett., 1984, 25, 3047). With a diol and chlorotrimethylsilane (or tetrachlorosilane) the enones give β-chloroketals of the corresponding saturated ketones (G. Gil, ibid., 1984, 25, 3805). Ketals prepared using chiral diols undergo asymmetric addition reactions to the olefinic bonds, eg alkylation (A. Ghribi, A. Alexakis and J.F. Normant, ibid, 1984, 25, 3079 and 3083) or cyclopropanation (E.A. Mash and K.A. Nelson, Tetrahedron, 1987, 43, 679).

Grignard reagents, RMgX, do not add cleanly to α,β-enones but regiospecific addition to the carbonyl group is possible using either organocadmium or organoytterbium reagents (S. Araki, H. Ito and Y. Butsugan, J. Organomet. Chem., 1988, 347, 5; K. Yokoo et al, Chem. Lett., 1983, 1301). The Grignard reagents will, however, react cleanly with α,β-enone iron tricarbonyl complexes, but the product results from conjugate addition with carbonyl insertion (T.N. Danks, D. Rakshit and S.E. Thomas, J. Chem. Soc., Perkin Trans. 1, 1988, 2091):

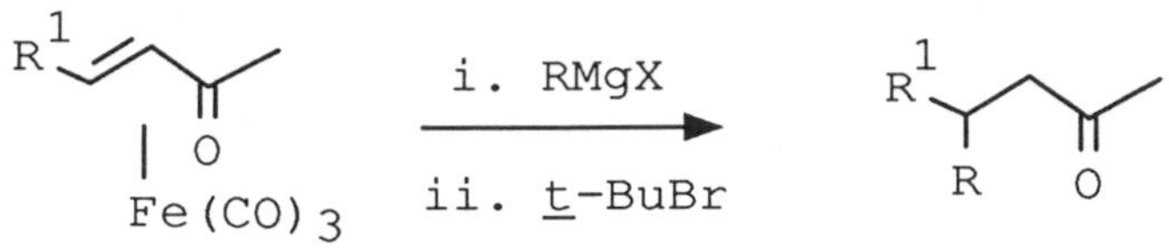

Several methods have been developed for the conversion of α,β-enones into aldehyde derivatives ((a) J-L. Parrain, A. Duchene and J-P. Quintard, ibid., 1990, 187; (b) D.H. Hua et al, J. Amer. Chem. Soc., 1987, 109, 5026; (c) T. Hosokawa, T. Ohta and S-I. Muraheshi, J. Chem. Soc., Chem. Commun., 1983, 848; (d) T.K. Jones and S.E. Denmark, J. Org. Chem., 1985, 50, 4037):

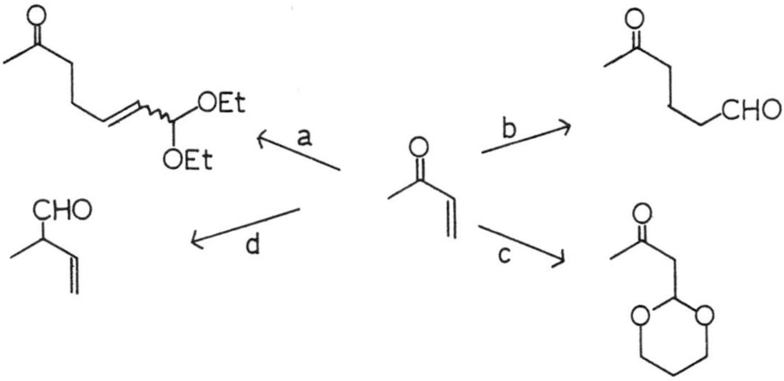

The olefinic bond of α,β-enones can be epoxidised (L.A. Arias et al, ibid, 1983, 48, 888) or converted to the corresponding cis-vicinal diol (C.R. Johnson and M.R. Barbachyn, J. Amer. Chem. Soc., 1984, 106, 2459). Olefination of the carbonyl group affords conjugated dienes (P.R.

Jenkins et al, J. Chem. Soc., Chem. Commun., 1987, 1540, and J. Chem. Soc., Perkin Trans. 1, 1986, 1129). Annulation procedures have been developed for the construction, onto the olefinic bond, of a pyridine ring (K. Kanomata and M. Nitta, ibid, 1990, 1119) or a tetrahydrofuran ring (R.J. Linderman and A. Godfrey, J. Amer. Chem. Soc., 1988, 110, 6249).

(c) α,β-Ynones, RCOC≡CR^1

(i) Preparation. α,β-Ynones can be prepared from acyl chlorides, RCOCl, using either lithium or trimethylsilyl acetylides, M-C≡C-R^1 (M. Yamaguchi et al, Synthesis, 1986, 421; D.R.M. Walton and F. Waugh, J. Organomet. Chem., 1972, 37, 45). Boron reagents are especially useful; lithium alkynyltrifluoroborates give the ynones with carboxylic anhydrides (H.C. Brown, U.S. Racherla and S.M. Singh, Tetrahedron Lett., 1984, 25, 2411), while lithium acetylides are cleanly incorporated into the ynones either using α-iodovinylboranes and subsequent oxidation (M. Hoshi, Y. Masud and A. Arase, Bull. Chem. Soc. Jpn., 1985, 58, 1683), or using (chiral) alkylborinates via decomposition of their lithium alkynylborinate complexes (H.C. Brown et al, J. Org. Chem., 1988, 53, 1391) eg:

(3-tetrahydrofuryl)-B(OEt)$_2$ —Li-≡-R^1→ [(3-tetrahydrofuryl)-B(OEt)$_2$(≡-R^1)] Li —HCl_2COMe→ (3-tetrahydrofuryl)-C(=O)-≡-R^1

Suitable alkynes can be oxygenated to α,β-ynones using t-butylhydroperoxide and catalytic CrO_3/TosOH (J. Muzart and O. Piva, Tetrahedron Lett., 1988, 29, 2321). Aldehydes, RCHO, give the ynones when treated with alkynyl vanadium dichlorides, $R^1C{\equiv}CVCl_2$ (T. Hirao, D. Misu and T. Agawa, Tetrahedron, 1986, 42, 4603).

(ii) Reactions. α,β-Ynones undergo asymmetric reduction to ynols with Alpine-borane (M.M. Midland et al, Tetrahedron, 1984, 40, 1371) or with BINAL-H (R. Noyori et al, J. Amer. Chem. Soc., 1984, 106, 6709 and 6717). Reduction with sodium cyanoborohydride affords the corresponding allene, $RCH{=}C{=}CHR^1$ (G.W. Kabalka et al, J. Chem. Soc., Chem. Commun., 1978, 726).

Conjugate addition by iodide produces allene enolates that give aldol reactions (M. Taniguchi et al, Tetrahedron Lett., 1986, 27, 4763):

R–CO–C≡CH —(Bu_4NI, $TiCl_4$)→ [R–C(O^-)=C=CH–I] —(R^1CHO)→ R–CO–C(=CH–I)–CH(OH)–R^1

Ynones, $RCOC{\equiv}CR^1$ (R^1 = butyl or longer), rearrange to (E,E)-2,4-dienones in the presence of ruthenium or palladium catalysts (D. Ma et al, ibid, 1988, 29, 1045; B.M. Trost and T. Schmidt, J. Amer. Chem. Soc., 1988, 110, 2301).

Second Supplements to the 2nd Edition of Rodd's Chemistry of Carbon Compounds, Vol. 1C, edited by M. Sainsbury
© 1992 Elsevier Science Publishers B.V., Amsterdam

Chapter 9a

MONOBASIC ALIPHATIC SATURATED ACIDS

J. BERGMAN

Introduction

This survey covering the period 1970-1990 deals mainly with syntheses and reactions of monocarboxylic acids and their derivatives. A comprehensive coverage of their physical properties can be found in the 2nd edition of Chem. Carbon Compounds (M.F. Ansell and R.H. Gigg, Vol Ic; pp 92-218) and also in the first supplement C.Y. Hopkins, Vol Ic; pp 109-122). Other useful general references include the following:

"The Chemistry of Carboxylic Acids and Esters," Ed. S. Patai, Wiley, New York (1969)

"The Chemistry of Acyl Halides," Ed. S. Patai,Wiley, New York (1972)

"The Chemistry of Acid Derivatives," Ed. S. Patai,Wiley, New York (1979)

Houben-Weyl, "Methoden der Organischen Chemie," 4th ed, Vol E4 (acid derivatives), G. Thieme, Stuttgart-New York (1983)

Houben-Weyl, "Methoden der Organischen Chemie," 4th ed. Vol E5 (acids and derivatives), Part 1, G. Thieme, Stuttgart-New York (1985)

Houben-Weyl, "Methoden der Organischen Chemie," 4th ed. Vol E5 (acids and derivatives), Part 2, G. Thieme, Stuttgart-New York (1985)

Houben-Weyl, "Methoden der Organischen Chemie," 4th ed. Vol E13 (peracids), Part 1, G. Thieme, Stuttgart-New York (1988)

Finally it might be noted that a second supplement in the Patai series on acid derivatives has been announced for publication in 1993.

(1) SATURATED MONOCARBOXYLIC ACIDS

(a) *Formic acid and its derivatives*

(i + ii) *Formic acid,* HCOOH

At room temperature and normal pressure, 95% of formic acid vapour consists of dimerized acid. The ring-type dimeric structure exists both in the vapour phase and in solution. Liquid formic acid consists of long chains of molecules linked to each other by hydrogen bonds. Solid formic acid can be isolated in two polymorphic forms (α and β). The system formic acid/formate gives rise to strong hydrogen bonds (H. Basch and W.J. Stevens, J. Am. Chem. Soc., 1991, 113, 95):

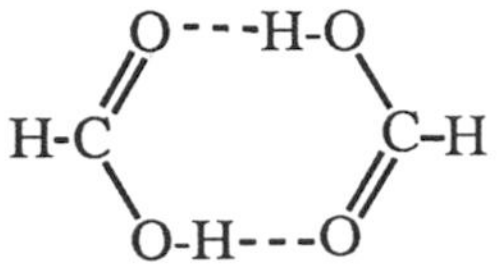

Dimer of formic acid

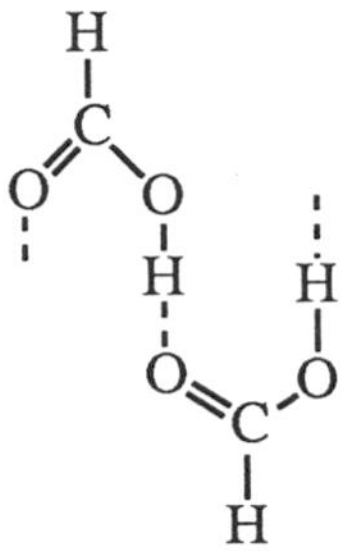

Portion of the chain in an α-type formic acid crystal

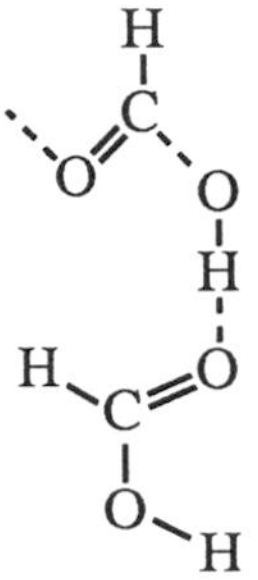

Portion of the chain in a β-type formic acid crystal

The major source of formic acid today is from hydrolysis of methyl formate, which in turn is prepared by base-catalyzed carboxylation of methanol. A direct synthesis of formic acid from carbon monoxide and water cannot be carried out economically because of the unfavourable position of the equilibrium. Other industrial sources include by-product recovery from hydrocarbon oxidation and pentaerythrit production.

Formic acid as a formylating agent has recently been discussed and compared with other formylating agents (G.A. Olah, L. Ohannesian and M. Arvanaghi, Chem. Rev., 1987, 87, 671).

Formic acid is sometimes used as a formylating agent in conjunction with dehydrating agents. Thus N-formylation of 2-pyridone is easily achieved by using formic acid and DCC at 0°C (F. Effenberger et al. Chem. Ber., 1980 113, 2086). Formic acid also acts as a useful C_1 building block (D.J. Schreck, D.C. Busby and R.W. Wegman, J. Mol. Catal., 1988, 47, 117). The utility of ammonium formate in organic synthesis has been reviewed (S. Ram and R.E. Ehrenkaufer, Synthesis, 1988, 91). Ammonium formate is an efficient deoxygenation agent for several classes of compounds, including N- oxides (R. Balichi, Synthesis, 1989, 645).

Ammonium formate, together with Pd on C, selectively reduces the heterocyclic rings of quinolines and isoquinolines (P. Batczewski and J.A. Joule, Synth. Commun., 1990, 20, 2815).

Formic anhydride can be prepared in several ways and spectroscopic data are available for this compound (G.A. Olah, Angew. Chem., 1979, 91, 649):

$$HCOF + HCO_2Na \longrightarrow (HCO)_2O$$

Formic anhydride is known to decompose to carbon monoxide and formic acid at room temperature within about 1h (H. Kühne et al., J. Mol. Spectr., 1979, 77, 251). Consequently this anhydride has, in contrast to acetic formic anhydride, found no use in organic synthesis.

Acetic formic anhydride, is readily prepared from acetyl chloride and sodium formate. Several other methods are available. A comprehensive review dealing with acetic formic anhydride (preparation, physical properties and reactions) has recently been published (P. Strazzolini, A.G. Giumanini and S. Cauci, Tetrahedron, 1990, 46, 1081).

(iii) *Halides and cyanides of formic acid,* HCOX

Formyl fluoride is prepared from benzoyl fluoride and sodium formate, and can be used as a formylating agent (G.A. Olah et al., Chem. Rev., 1987, 87, 671).Ghosez et al. (Chem. Comm., 1979, 1180) have reported an improved preparation of formyl chloride:

$$HCOOH + (CH_3)_2C{=}C(X)NMe_2 \longrightarrow (CH_3)_2CH{-}C^{\oplus}(O{-}CHO)NMe_2\ X^{\ominus} \longrightarrow$$

$$\longrightarrow HCOX + (CH_3)_2CHCONMe_2 \qquad X{=}Cl\ (F,Br,I)$$

Another method for the synthesis of HCOCl, due to H. Takeo and C. Matsumura (J. Chem. Phys., 1976, 64, 4536), involving the passage of formic acid vapour through powdered PCl_5, allowed the recording of the near-UV spectra of this notoriously unstable molecule (H.G. Libuda et al. J. Phys. Chem., 1990, 94, 5860). Formyl chloride is an important intermediate in the atmospheric degradation of several chlorinated hydrocarbons, e.g. CH_3Cl and CH_2Cl_2 (E.C. Tuazon et al., Int. J. Chem. Kinet., 1988, 20, 241).

Formyl cyanide is a highly reactive, and possibly a cosmic molecule which can be generated by retro-ene cleavage using flash vacuum thermolysis of allyloxy-acetonitrile (Y. Vallée, J. Chem. Res. [S], 1990, 40):

O C≡N H ⟶ O C≡N H + Propene.

(iv) *Esters of formic acid (formates)*, HCOOR

The carboxylation of methanol with carbon monoxide, catalyzed by sodium methoxide, has been known since 1925 and is still the most important process for the preparation of formates and formic acid.

$$CH_3OH + CO \longrightarrow HCOOCH_3$$

The process can easily be extended to higher alcohols. Problems with the process, include inactivation of the catalyst by the formation of CH_3OCH_3.

$$NaOCH_3 + HCOOCH_3 \longrightarrow HCOONa + CH_3OCH_3$$

It is also necessary to exclude moisture and to use CO_2-free carbon monoxide (S.P. Tonner et al., J. Mol. Catal., 1983, 18, 215; A. Aguilo, T. Horlenko, Hydrocarbon Proc., 1980, 59, 120; K. Kudo, Chem. Lett., 1987, 577). Problems with the catalyst has triggered the development of several alternative catalysts such as guanidines (M.J. Green, EP 104875, 1984, and ruthenium compounds (G. Jenner and G. Bitsi, J. Mol. Catal., 1988, 45, 255).

Alkyl formates can be isomerized to carboxylic acids (D.J. Schreck, D.L. Busby and R. Weyman, J. Mol. Catal., 1988, 44, 117). In fact, this type of conversion could be used as an alternative synthesis of acetic acid if methyl formate becomes less expensive.

A new laboratory method for formates has been developed (M. Dymicky, Org. Prep. Proced. Int., 1982, 14, 177):

$$HCOOH + ROH \xrightarrow{BF_3 \cdot 2\,MeOH} HCOOR$$

(v) *Formamides,* $HCONR_2$

Formamide, $HCONH_2$, is produced commercially either via the direct synthesis:

$$CO + NH_3 \xrightarrow{NaOCH_3} HCONH_2$$

or the indirect two-step process:

$$CO + CH_3OH \xrightarrow{NaOCH_3} HCOOCH_3$$

$$HCOOCH_3 + NH_3 \longrightarrow HCONH_2 + CH_3OH$$

Formamide is not only a useful solvent but also a valuable component in many synthetic sequences, notably the Bredereck reaction.

Formamide is a useful formylating agent, for example, in the N-formylation of primary and secondary amines. Formamide reacts with acyl chlorides to give acyltriamines. Reaction with NaOMe produces sodium diformylamide (I.C. Gramain and R. Rémuson, Synthesis, 1982, 264).

Dimethylformamide, $HCON(CH_3)_2$, "DMF" is a useful high-boiling (153°C) solvent that can be produced either directly (from CO and dimethylamine) or indirectly as described above for formamide. A detailed spectroscopic study (far-IR) of DMF has recently been published (R. Buchner and J. Yarwood, Mol. Phys., 1990, 71, 65). DMF yields important complexes in combination with reagents like $POCl_3$ and $COCl_2$. These complexes are useful formylating agents (W. Kantlehner in "Advances in Organic Chemistry". H. Böhme and H. Viehe eds, 1979, **9** (part 2), pp 5-64, Wiley, N.Y. cf also the chapter by C. Jutz, **9** (part 1), pp 225-342).

Other secondary formamides yield similar reagents. Particularly interesting is the report that N-formylmorpholine with $POCl_3$ produces a complex that is ten times more reactive than the DMF-$POCl_3$ complex. This deserves further exploration (J.A. de Groot et al., Rec. Trav. Chim., 1982, 101, 35). Another way to potentiate the reagent has recently been reported. Thus trifluoromethanesulphonic anhydride and DMF form a very reactive iminium complex that be can used to substitute even moderately reactive aromatics such as naphthalene and 1,3,5-trimethylbenzene (A. Garcia Martinez et al., Chem. Comm. 1990, 1571; cf also C. Djerassi et al., Synth. Commun. 1984, 14, 383).

Like formamide, DMF can be used as a selective N-formylation agent. Thus primary aliphatic amines give excellent yields while anilines are left untouched (M. Iwata, H. Kuzuhara, Chem. Lett., 1989, 2029).

N-Formylpiperidine formylates efficiently a wide variety of Grignard reagents (G.A. Olah and M. Arvanaghi, Org. Synth., 1986, 64, 114):

N-CHO (piperidine) + RMgX ⟶ R-CHO

A variant of this reaction yields imines (and subsequently, after reduction, amines) (B.L. Feringa and J.F.G.A. Jansen, Synthesis, 1988, 184):

RMgX + H-CO-N(R_1)($SiMe_3$) ⟶ R(H)C=N—R_1

When sodium is reacted with DMF three anionic species result. At 100°C the major product is sodium dimethylamide. This reagent system (Na/DMF) effects condensation between imines and weakly acidic reactants (M. Paventi and A.S. Hay, Synthesis, 1990, 878). An example is given in which phenylpropene is converted into 1,4-diphenyl-1,3-butadiene:

Ph-CH=CH-CH$_3$ + PhN=CH-Ph $\xrightarrow{\text{Na / DMF}}$ Ph-CH=CH-CH=CH-Ph

It is predicted that the combination Na/DMF will prove useful in many other reactions.

(vi) *Hydrogen cyanide,* HCN

Hydrogen cyanide is produced commercially from ammonia and hydrocarbons such as methane:

$$CH_4 + NH_3 \longrightarrow HCN + 3H_2 \quad \Delta H = +252\ KG/mol$$

HCN is mainly used for the commercial production of methyl methacrylate and adiponitrile (by adding 2 eq. of HCN to 1 eq. of butadiene). Hydrogen cyanide reacts with chlorine to form Cl-C≡N which is usually directly trimerized to cyanuric chloride.

As indicated above, catalyzed (either Cu or Ni based) additions of HCN to olefins (hydrocyanation) is of industrial importance. The nickel based process proceeds via nickel hydride intermediates, whereas the copper based process is limited to halide containing systems (E. Puentes et al., J. Catal., 1983, 82, 365). The hydrocyanation of conjugated carbonyl compounds has been reviewed (N. Nagata and M. Yoshioka, Org. React., 1977, 25, 255).

The possible role of HCN and cyano compounds in chemical evolution has been discussed (J.P. Ferris and W.J. Hagan, Tetrahedron, 1984, 40, 1093).

The cyanide ion is a very strong nucleophile. The availability of tetraalkylammonium cyanides allows the reactions to be run homogenously in a wide range of solvents (H. Kobler, K.-H. Schuster and G. Simchen, Ann., 1981, 333).

Lithium cyanide dissolved in THF (commercially available) is an efficient reagent for conversion of halides into nitriles (S. Harusawa et al., Tetrahedron Letters, 1987, 4189). Sodium cyanide in DMSO offers an interesting possibility for the nucleophilic cleavage of esters:

$$RCO_2CH_3 + CN^{\ominus} \longrightarrow RCO_2^{\ominus} + CH_3CN$$

A general review of ester cleavages via S_N2-type dealkylation is available (J. McMurry, Org. React., 1976, 24, 187).

(b) *Saturated carboxylic acids or alkanoic acids,* $C_nH_{2n+1}COOH$

The alkanoic acids include the saturated fatty acids as the most important subgroup. The term "fatty acids" is applied to aliphatic carboxylic acids with carbon chain length in the range of C_6-C_{24}.

Fatty acids obtained by splitting natural fats and oils are quantitatively far more important than those made synthetically. The fats and related compounds are considered in more detail in a later chapter. Several books and reviews on fatty acids are avaiable (E.H. Pryde ed., "Fatty Acids" A.O.C.S. Champaign, Ill. USA, 1979; H. Fineberg, J. Am. Oil. Chem. Soc., 1979, 56, 805A). The field of fatty acid chemistry is also periodically reviewed in the journal Natural Products Reports. The most recent survey gives several entries to books and reviews and more than 700 references to the current research literature (M.S.F. Lie Ken Jie, Nat. Prod. Rep., 1989, 6, 231).

(i) *Chemical properties and reactions of carboxylic acids*

Insertion of carbon monoxide into the O-H bond of methanol has already been discussed. Insertion into the C-O bond is likewise of paramount industrial importance. The Monsanto process for the production of acetic acid involves a rhodium carbonyl catalyst in conjunction with an iodide promoter. Methyl iodide and acetyl iodide are important intermediates in the process. The process has also been extended to other alcohols such as ethanol and propanol (T.W. Deklera and D. Forster, J. Am. Chem. Soc., 1985, 107, 3565) and appears to be quite general:

$$R\text{—}OH + CO \longrightarrow RCOOH$$

Some reactions of related interest have been reported by Kaplan:

$$CO + CO \xrightarrow{HI/\ H_2O} CH_3COOH$$

$$CH_3COOH + CO \xrightarrow{HI/\ H_2O} CH_3CH_2COOH$$

(L. Kaplan, J. Org. Chem., 1982, 47, 5422). Formic acid can be used as a substitute for CO in these reactions. (L. Kaplan, J. Org. Chem., 1985, 50, 5376). Stoichiometric quantities of formic acid, in this type of reaction, can be used to convert alcohols to ^{13}C-labelled carboxylic acids (H. Langhals, I. Mergelsberg and Chr. Rüchardt, Tetrahedron Letters, 1981, 2365). Other illustrations of the Koch-Haaf reaction are cited in reviews. See for example, (H. Bahrmann, in "New Syntheses with Carbon Monoxide", J. Falbe ed., Springer-Verlag, New York, 1980, 372).

For laboratory purposes carboxylation of organometallic derivatives is a useful method.

$$RMgX \xrightarrow[2.\ H^{\oplus}]{1.\ CO_2} RCOOH$$

Alternatively the oxazoline method can be used. When terminal alkynes are readily available the following method offers a third alternative (K Tamao, M. Kumada and K. Maeda, Tetrahedron Letters, 1984, 321):

$$C_6H_{13}C\equiv CH \quad + \quad HSiMe(OEt)_2 \longrightarrow C_6H_{13}CH{=}CHSiMe(OEt)_2$$

$$C_6H_{13}CH{=}CHSiMe(OEt)_2 \xrightarrow[2.\ H_2O_2]{1.\ KHF} C_6H_{13}CH_2\,COOH$$

When the appropriate primary alcohol is readily available oxidation with calcium hypochlorite should be considered (G.W. Kabalka et al, Synth. Commun., 1990, 20, 1617). The new reagent benzotriazole-CrO_3 is, for laboratory purpose, an alternative oxidant. Thus, for example, 1-octadecanol was converted to stearic acid in a yield of 91% (E.J. Parish, H. Honda and D.L. Hileman, Synth. Commun., 1990, 20, 3359).Further oxidative methods can be found in a monograph, A.H. Haines,"Methods for the Oxidation of Organic Compounds", Academic Press, New York, 1988.

(ii) *Transformations based on dianions of carboxylic acids*

Aliphatic acids can be converted to dianions by treatment with two equivalents of lithium diisopropylamide (LDA), (M.W. Rathke et al., J. Am. Chem. Soc., 1973, 95, 3050). Alkylations of the resulting dianion can be effected smoothly with alkyl halides that are not prone to undergo β–elimination:

$$R\text{-}CH_2COOH \xrightarrow{LDA} R\text{-}\overset{\ominus}{C}H\text{-}COO^{\ominus} \xrightarrow[2.\ H_2O,\ H^{\oplus}]{1.\ R_1X} (R)(R_1)CH{-}COOH$$

This methodology, in combination with a resolution step, can be used for the preparation of optically active α-branched carboxylic acids (Ph. E. Sonnet and J. Gazzillo, Org. Prep. Proced Int., 1990, 22, 203 and further refs. given therein). A comprehensive review on dianions is also available (N. Petragnani, Synthesis, 1982, 521). The dianions can also serve as convenient intermediates for a wide range of functionalized acids. One example is their use in the ring opening of epoxides to give γ-hydroxyacids:

$$R\text{-}CH^{\ominus}_2COO^{\ominus} + \text{(2,2-disubstituted oxirane, } R_1, R_1) \xrightarrow[2.\ H_2O,\ H^{\oplus}]{1.\ -20\ ^{o}C} (R_1)_2C(OH)\text{-}CH_2\text{-}CHRCOOH$$

Oxidation of the dianions with O_2 can, depending on the reaction conditions, afford α-hydroperoxy acids. Utilization of bis(trimethylsilyl)-peroxide gives a cleaner product (M.Pohmakitr and C. Winotai, Synth. Commun, 1988, 18, 2141). Coupling of the dianions can be effected by reagents such as $FeCl_3$ or I_2 (J. Bergman and B. Pelcman, Tetrahedron Letters, 1987, 4441; Ph. Rehaud and M.A. Fox, J. Org. Chem., 1988, 53, 3745):

$$R\text{-}CH^{\ominus}\text{-}COO^{\ominus} \xrightarrow[2.\ H_2O,\ H^{\oplus}]{1.\ I_2} \begin{array}{c} R\text{—}CH\text{—}COOH \\ | \\ R\text{—}CH\text{—}COOH \end{array}$$

Dianions can also be generated from α-halocarboxylic acids. Formation of the dianion of chloroacetic acid is fast even at -80°C. These dianions react with ketones to provide glycidic acids in an improved version of Darzens reaction (C.R. Johnson and Th.R. Bade, J. Org. Chem., 1982, 47, 1205; cf. also P. Coutrot and A.E. Gadi, Synthesis, 1984, 115):

$$R(X)CHCOOH \xrightarrow{LDA} R_1(X)C{=}C(OLi)_2 \xrightarrow{R_2COR_2} (R_2)_2C\overset{O}{—}C(R_1)COOH$$

Many of the transformations just discussed can also be performed with carboxylic acids in a masked form:

$R-CH_2COOH$ + (2-amino-2-methylpropanol, NH_2, OH) → (4,4-dimethyloxazoline)-CH_2-R →

$\xrightarrow[2.\ R_1X]{1.\ B^{\ominus}}$ (oxazoline)-CH(R)(R_1) $\xrightarrow[H^{\oplus}]{H_2O}$ R(R_1)CH-COOH

Syntheses of a number of chiral carboxylic acids possessing high optical purity has been realized using chiral amino alcohols. The method often allows the separate preparation of both enantiomers of various α-alkyl-alkanoic acids from a single chiral oxazoline (A.I. Meyers, ACS Symposium Series, 1982, 185, 82):

CH_3CH_2-(oxazoline, Ph, CH_2OMe) $\xrightarrow{LDA}$ (aza-enolate, N–Li - - - - OMe) $\xrightarrow{RI}$

CH_3(H)(R)C-(oxazoline, Ph, CH_2OMe) $\xrightarrow{H_3O^{\oplus}}$ CH_3(H)(R)C-COOH

(iii) *Physical properties of carboxylic acids*

Liquid carboxylic acids consist mainly of dimeric structures [cf.section 1a (i)]. The coordinated motions of the hydrogen atoms in the hydrogen bonds have recently been studied using pulsed NMR spectroscopy and inelastic neutron scattering.

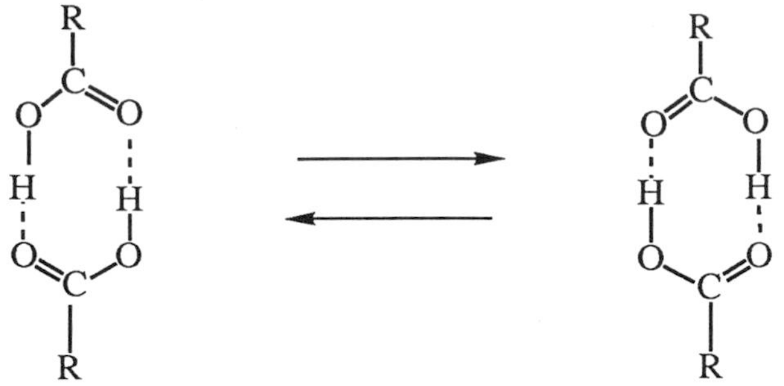

(A.J. Horsewill and A. Aibout, J. Phys. Condens. Matter, 1989, 1, 9609). In protic and dipolar aprotic solvents, at low acid concentration, the dimers are broken up to form 1:1 or 1:2 hydrogen-bonded complexes with the solvent (D. Hadzi and S. Detoni, in "The Chemistry of Acid Derivatives", (Ed., S. Patai) suppl B, Part I, 1979, 213). In a recent collection of reviews, under the general title "Crystallization and Polymorphism of Fats and Fatty acids" [eds. N. Garti and K. Sato, M. Dekker, 1988], individual articles such as "Crystal Structures of Fats and Fatty Acids" by L. Hernquist, and "Crystallization of Fats and Fatty Acids" by K. Sato and "Spectroscopic Aspects of Polymorphism" by M. Kobayashi can be found.

In dilute aqueous solution, generally less than 10^{-4} molar, salts of long-chain carboxylic acids concentrate at the surface of the water, with the charged ends immersed in the bulk phase and the hydrocarbon chains forming a surface layer. This causes a marked reduction in the surface tension of the water, by even minute concentrations of such salts (soaps). If the concentration of the salt is increased, and the surface layer becomes saturated, then at a certain level the salt molecules aggregate forming micelles. This concentration, the critical micelle concentration, occurs over a narrow range, rather than at a precise point, and is recognized by a marked deviation in a range of physical properties of the solution.

Micelles have potent catalytic properties, providing not only non-polar environments in an aqueous system, but also exhibiting localized areas of high charge density at their surfaces. In such areas the rates of many reactions are accelerated (J.H. Fendler and E.J. Fendler, "Catalysis in Micellar and Macromolecular Systems" Academic Press, New York, 1975).

(iv) *Transformations of carboxylic acids*

Several reagents are available for direct or indirect transformations of carboxylic acids into aldehydes and alcohols. In this context alkoxy-aluminium hydrides are of interest. A recent comprehensive review on this area of chemistry is available (J. Malek, Org. React., 1988, 36, 249).

Carboxylic acids are efficiently reduced to alcohols by $BH_3 \cdot THF$ (N.M. Yoon et al., J. Org. Chem., 1973, 38, 2786). Equally facile reductions of carboxylic acids to aldehydes are possible with thexylchloroborane-dimethyl sulphide (H.C. Brown, J. Am. Chem. Soc., 1984, 106, 8001). Fujisawa et al. have proposed a simple and chemo-selective method to convert carboxylic acids to aldehydes using Viehe´s reagents for activation of the carboxylic group (T. Fujisawa, Tetrahedron Letters, 1983, 1543):

$$\text{R-COOH} \xrightarrow{\text{(Cl)(H)C=}\overset{\oplus}{N}Me_2\ Cl^{\ominus}} \xrightarrow[\text{cat. CuI, -78 °C}]{\text{Li (tBuO)}_3\text{AlH}} \xrightarrow{H_2O} \text{R-CHO}$$

Recent developments in the synthesis of aldehydes by the reduction of carboxylic acids and their derivatives have been reviewed (J.S. Sha, Org. Prep. Proced. Int., 1989, 21, 453).

The Schmidt reaction with carboxylic acids has been surveyed (C.I. Koldobskii, V.A. Ostrowskii and B.V. Gidaspov, Russ. Chem. Rev., 1978, 47, 1084):

$$RCOOH + HN_3 \xrightarrow{H^{\oplus}} RNH_2$$

A modern variant of this reaction uses the reagent $N_3PO(OC_6H_5)_2$ (K. Ninomiya, T. Shioiri and S. Yamada, Tetrahedron, 1974, 30, 2151):

$$RCOOH \xrightarrow[R_1OH]{N_3PO(OC_6H_5)_2} RNHCO_2R_1 \longrightarrow RNH_2$$

Individual carboxylic acids

(v) *Acetic acid,* CH_3COOH

The industrially most important process for the synthesis of acetic acid is based on the carbonylation of methanol:

$$CH_3OH + CO \xrightarrow[\text{HI, } H_2O]{\text{Rh-catalyst}} CH_3COOH$$

Mechanistic details about this process are available (R.T. Ely and T.C Singleton, Applied Industrial Catalysis, 1983, 275). Oxidation of acetaldehyde was previously an important process, but due to high prices of ethylene very few plants are now in operation.

Lead tetraacetate. This versatile oxidant is still a favourite among many chemists. A review concerning the oxidative decarboxylation of carboxylic acids induced by $Pb(OAc)_4$ has been published (R.A. Sheldon and J.K Kochi, Org. React., 1972, 19, 279).

Manganese triacetate and thallium triacetate. These two oxidants are more recent addition to chemist´s arsenal of reagents. Both of these reagents and also lead acetate are discussed in recent monographs (W.J. Mijs and C.R.H.I. de Jonge, "Organic Synthesis by Oxidation with Metal Compounds", Plenum, N.Y. 1986; M. Hudlicky, "Oxidations in Organic Chemistry" A.C.S. Monograph 186, Washington D.C. 1990.

2. FUNCTIONAL DERIVATIVES OF CARBOXYLIC ACIDS.

(a) *Esters,* $RCOOR_1$

Various new methods for esterification have been reviewed (E. Haslam, Tetrahedron, 1980, 36, 2469). Other methods for esterifications include the following transformations (see G. Satyanarama and S. Sivaram, Synth. Commun., 1990, 20, 3273; M.A. Brook and T.H. Chan, Synthesis, 1983, 201):

$$RMgBr + (R_1O)_2CO \xrightarrow[\text{2h, 0-10 °C,}]{\text{THF}} RCOOR_1$$

$$R_1C(=O)OH + Me_3SiCl \rightleftharpoons R_1C(=O)OSiMe_3 + HCl \rightleftharpoons$$

$$\xrightleftharpoons{R_2OH} R_1C(=O)OR_2 + [Me_3SiOH] \xrightarrow{Me_3SiCl} Me_3SiOSiMe_3 + HCl$$

A procedure based on N,N-carbonyldiimidazole (CDI) has been extended to t-butyl esters (S. Ohta et al., Synthesis, 1983, 833):

$$R_1C(=O)OH \xrightarrow{CDI} \text{N-acylimidazole (RC(=O)-Im)} \xrightarrow[DBU]{t\text{-}BuOH} R_1C(=O)O\text{-}tBu$$

Reactions of carboxylic acids with equimolar amounts of alkyl chloroformate and triethylamine in the presence of a catalytic amount of 4-dimethylaminopyridine afford the corresponding esters in high yields, without the for-mation of the symmetrical anhydrides in most cases (S. Kim, Y.C. Kim and J.I. Lee, Tetrahedron Letters, 1983, 3365).

Esters of trichloroacetic acid can be used as alkylating agents (J.M. Renga and P.-C. Wang, Synth. Commun., 1984, 14, 77).:

$$Cl_3CCOOMe + RCOOH \xrightarrow[\text{[18-crown-6]}]{K_2CO_3} RCOOMe + CHCl_3 + CO_2$$

Ester interchange of methyl or ethyl carboxylic esters by primary or secondary alcohols is carried out effectively under conditions of solid-liquid phase transfer catalysis (PTC), and can be achieved in the absence of added organic solvents (J. Barry, G. Bram and A. Petit, Tetrahedron Letters, 1988, 4567).

4-Dialkylaminopyridines are particularly useful catalysts for acylation of hindered alcohols. Thus even sensitive alcohols such as linalool can be acetylated using acetic anhydride and 4-pyrrolidinopyridine (W. Steglich and G. Höfle, Synthesis, 1972, 619).

The Mitsunobo reaction can be used for the preparation of esters:

$$RCOOH + (EtO)_3P + EtOOC\text{-}N{=}N\text{-}COOEt \longrightarrow$$

$$\longrightarrow RCOOEt + (EtO)_3PO + EtOOCNHNHCOOEt$$

Enantioselective hydrolysis of prochiral malonates can be effected by pig liver esterases (PLE), followed by decarboxylation of the intermediate product:

$$R_1R_2C(COOR_3)_2 \xrightarrow{PLE} R_1R_2C(COOH)(COOR_3) \longrightarrow R_1R_2CH(COOR_3)$$

Recent reviews on ester hydrolyses catalyzed by PLE are available (M. Ohno and M. Otsuka, Org. React., (1989), 37, 1; L.-M. Zhu and M.C. Tedford, Tetrahedron, 1990, 46, 6587).

Carboxylic acids can be esterfied by oxidation of hydrazones with peracetic acid with iodine as catalyst. This procedure is particularly useful for preparation of diphenylmethyl esters (R. Bywood et al., J. Chem. Soc. Perkin I, 1975, 2019):

$$RCOOH + (C_6H_5)_2C{=}NNH_2 \xrightarrow[I_2]{CH_3C(O)OOH} RCOOCH(C_6H_5)_2$$

An interesting example of a Baeyer-Villiger cleavage is given below:

$$\text{cycloheptanone} \xrightarrow[H_2SO_4,\ EtOH]{K_2SO_5} HO(CH_2)_6CO_2Et$$

(S. Canonica, M. Ferrari and M. Sisti, Org. Prep. Proced. Int., 1989, 21, 253).This oxidative procedure gives better yields (85%) than methods using trifluoroperacetic acid or sodium perborate. In a related approach ketones have been converted to esters by reaction with $Me_3SiOOSiMe_3$ in the presence of $SnCl_4$ or $BF_3 \cdot OEt_2$ (S. Matsubara, K. Takai and H. Nozaki, Bull. Chem. Soc. Japan, 1983, 56, 2029).

The Bouvealt-Blanc reaction, with added trimethylchlorosilane, yields endiols as their silylated forms. This additive, which reacts with basic by-products, prevents side reactions such as ß-eliminations and Claisen condensations, by keeping the reaction mixture essentially neutral (K. Rühlmann, Synthesis, 1971, 236):

$$R_1\text{-}COOR_2 + 4Na + 4\,Me_3SiCl \longrightarrow$$

$$\longrightarrow R_1(Me_3SiO)C{=}C(OSiMe_3)R_1 + 2Me_3SiOR_2 + 4\,NaCl$$

Other aspects of the reductive dimerisation of esters are given in a comprehensive review (J.J. Blomfield, D.C. Owsley and J.M. Nelke, Org. React., (1976), 23, 259).

(b) *Acyl halides*

Methods for the synthesis of acyl halids have been reviewed by M.F. Ansell "The Chemistry of Acyl Halides" S. Patai, ed, pp 35-68, New York, 1972. DMF reacts with $COCl_2$ to produce a salt of Vilsmeier-type, N,N-dimethylchloromethyleneammonium chloride, which is a useful reagent which converts carboxylic acids to acid chlorides (T. Fujisawa and T. Sato, Org. Synth., 1987, 66, 121):

$$HCONMe_2 + (COCl)_2 \longrightarrow Cl(H)C{=}N^{\oplus}Me_2 \quad Cl^{\ominus}$$

$$RCOOH \longrightarrow RCOCl$$

Several other reagents are available for activation of carboxylic acids under mild conditions, for example, 1-chloro-N,N,2-trimethylpropenylamine (T. Fujisawa and T. Sato, Org. Synth., 1987, 66, 116).

Carboxylic acids can be converted into acyl fluorides by treatment with diethylamino sulphur trifluoride (DAST).

$$RCOOH \xrightarrow{DAST} RCOF$$

A review on fluorination reactions using DAST is available (M. Hudlicky, Org. React., 1988, 66, 513).

Aldehydes can be directly converted to acyl bromides by a radical-mediated oxidation (I.E. Marko and A. Mekhalfia, Tetrahedron Letters, 1990), 7237):

$$RCHO \xrightarrow[AIBN,\ CCl_4]{NBS} RCOBr$$

Amines can be dealkylated by acyl chlorides (J.H. Cooley and E.J. Evain, Synthesis, 1989, 1).

Acyl chlorides react with certain tin-organic compounds. The products are useful (cf acyl-silanes) in organic synthesis (T.N. Mitchell and K. Kwetkot, Synthesis, 1990, 1001):

$$RCOCl + R_3SnSnR_3 \xrightarrow[THF]{Pd(0)} R\text{-}C(=O)\text{-}SnR_3$$

Carboxylic acids can, after conversion to acyl chlorides, be decarboxylated using N-hydroxypyridine-2-thione (D.H.R. Barton et al., Tetrahedron, 1987, 43, 2733):

$$\text{N-hydroxypyridine-2-thione (N–OH)} \xrightarrow{RCOCl} \text{N-(OCOR)pyridine-2-thione} \xrightarrow[\text{or } \Delta]{h\nu} R^{\bullet} + CO_2 + \text{2-pyridyl-S}^{\bullet}$$

Aliphatic carboxylic acids react with SF_4, even at low temperatures, to give 1,1,1-trifluoroalkanes as well as $\alpha,\alpha,\alpha^1,\alpha^1$,-tetrafluoroalkyl ethers as unexpected by-products (C.-L. J. Wang, Org. React., 1985, 34, 319):

$$RCOOH \xrightarrow{SF_4} RCOF \xrightarrow{SF_4} RCF_3 + (RCF_2)_2O$$

Treatment of $ClCH_2COF$ with SbF_5 results in the isolation of an acylium salt:

$$ClCH_2COF + SbF_5 \xrightarrow{SO_2\,(l)} ClCH_2C^{\oplus}O\ SbF_6^{\ominus}$$

This salt can be used in aromatic electrophilic substitution reactions (G.A. Olah, H.C. Lin and A. Germain, Synthesis, 1974, 895).

An improved procedure for dehalogenation of α–haloacyl halides has been developed (C.C. McCarney and R.S. Ward, J. Chem.Soc. Perkin I, 1975, 1600):

$$RR_1CX{-}COX \xrightarrow[THF]{Zn} RR_1C{=}C{=}O$$

(c) *Anhydrides,* $(RCO)_2O$

Several new methods for the preparation of carboxylic acid anhydrides have been developed. (J.B. Hendrickson and M.O.S. Hussoin, J. Org. Chem., 1989, 54, 1144). A new convenient method uses a "supported" phosphorus pentoxide reagent in boiling toluene. (S.G. Burton and P.T. Haye, Synth. Commun., 1989, 19, 3331). The following method allows preparations to be carried out at low temperatures (-20°C) (A.Nangia and S. Chandrasekaran, J.Chem. Res. [S], 1984, 100):

$$2\ RCOOH \xrightarrow[THF]{MeSO_2Cl\text{-}Et_3N} (RCO)_2O$$

The synthesis of some fatty anhydrides can be effected from the corresponding acid halides using solid-liquid phase-transfer catalysis (J.-X. Wang, Y.-L. Hu and W.-F. Cui, J. Chem. Res. [S], 1990, 84):

$$RCOCl \xrightarrow[Bu_4NBr\text{-}MeCN]{KHCO_3} (RCO)_2O$$

Acetic anhydride is produced by any of three processes.

The ketene process relies on the following two steps:

(i) $CH_3COOH \xrightarrow{\Delta} H_2C{=}C{=}O + H_2O$

(ii) $H_2C{=}C{=}O + CH_3COOH \longrightarrow (CH_3CO)_2O$

Acetic anhydride can be obtained directly by liquid-phase oxidation of acetaldehyde. The peracetic acid formed reacts with a second molecule of acetaldehyde to form Ac_2O:

(iii) $CH_3CHO + O_2 \longrightarrow H_3C{-}C(=O){-}O{-}OH \xrightarrow{CH_3CHO} (CH_3CO)_2O$.

The most recent method depends on catalytic (Rh or Ni) carbonylation of methyl acetate (S.W. Polichnowski, J. Chem.Ed., 1986, 63, 206):

$$CH_3COOCH_3 + CO \xrightarrow{\text{catalyst}} (CH_3CO)_2O$$

(d) *Peroxy acids*

The most versatile reagent for preparation of percarboxylic acids is hydrogen peroxide.

$$RCOOH + H_2O_2 \underset{}{\overset{H^{\oplus}}{\rightleftarrows}} R{-}C(=O){-}OOH$$

Peracetic acid is a useful co-oxidant for certain oxidations with RuO_4 (D.M. Ketcha and D.S. Swern, Synth. Commun., 1984, 14, 915).

Interactions between acyl fluorides and alkylperoxy silanes provide a mild route to, for example, t-butyl peroxy esters (D. Brandes and A. Blaschette, J. Organometal Chem., 1975, 99, C33):

$$H_3CCOF + \text{t-BuOOSiMe}_3 \longrightarrow H_3C{-}C(=O){-}O{-}O{-}\text{t-Bu}$$

t-Butylperoxyesters can also be obtained in the following way:

$$R{-}C(=O){-}Cl + \text{t-BuOONa} \longrightarrow R{-}C(=O){-}O{-}O{-}\text{t-Bu}$$

(W. Duismann and C. Rüchardt, Ann., 1976, 1834). An updated review on peroxy acids is available in the Patai series (G. Bouillon, C. Lick and K. Schank in "The Chemistry of Peroxides" 1983, pp. 279-309, Wiley, N.Y).

Peroxycarbamidic acids are formed in situ by the addition of H_2O_2 to a nitrile. A trichloromethyl substituent enhances the reactivty of the nitrile.These reagents are also useful for epoxidations (L.A. Arias et al., J. Org. Chem., 1983, 48, 888).

(e) *Amides*

Many condensing agents are available for the synthesis of amides from carboxylic acids and amines. A long list is found in the following reference (J. Cossy and C. Pale-Grosdenange, Tetrahedron Letters, 1989, 2771). It is recommended that molecular sieves are used (140-180°C) to form amides from carboxylic acids and primary amines.

A direct oxidative transformation of aldehydes to amides has been reported by Japanese workers (Y. Tamaru, Y. Yameda and Z. Yoshida, Synthesis, 1983, 474):

$$R_1\text{-CHO} + HNR_2 \xrightarrow[\text{ArBr / base}]{\text{Pd-catalyst}} R_1\text{-C(=O)-}NR_2$$

Nitriles can be hydrated to amides by treatment with MnO_2 on silica gel (K.-T. Liu et al., Synthesis, 1988, 715):

$$R\text{—C}\equiv N + H_2O \xrightarrow[\text{hexane}]{MnO_2/SiO_2} RCONH_2$$

Alternatively, hydrolysis of RCN with sodium percarbonate can be used (G.W. Kabalka et al., Synth. Commun., 1990, 20, 1445).

Aldehydes can be directly converted to amides by a radical-mediated oxidation (I.E. Marko and A. Mekhlfia, Tetrahedron Letters, 1990, 7327):

$$R\text{-CHO} \xrightarrow[HN(R_1)_2]{\text{NBS, AIBN}} R\text{-C(=O)-}N(R_1)_2$$

A recently reported route to amides involves the addition of organolithium, or organomagnesium reagents, to tertiary formamides, followed by oxidation of the intermediate α–aminoalkoxide by an Oppenauer type procedure. In certain cases $HgCl_2$ can be used advantageously in the second step (C.G. Screttas and B.R. Steele, J. Org. Chem., 1988, 53, 5151; Org. Prep. Proced. Int., 1990, 22, 271):

$$BuLi \quad + \quad \xrightarrow[\text{2. } HgCl_2 \text{ (-78°C} \rightarrow \text{RT)}]{\text{1. DMF, MCH-THF/ -78 °C}} \quad BuCONMe_2 \quad 90\%$$

The second paper by Screttas and Steele reviews organometallic carboxamidation:

$$HCON(iPr)_2 \xrightarrow[\text{2. t-BuLi}]{\text{1. LDA, THF}} LiCON(iPr)_2$$

$$LiCON(iPr)_2 + RCOR \longrightarrow R_2C(OH)C(=O)N(iPr)_2$$

The Haller-Bauer reaction can afford a useful synthesis of amides provided that the starting materials (non-enolizable ketones) are readily available:

$$R_1R_2R_3C\text{-}C(=O)R_4 \xrightarrow[\text{benzene}]{NaNH_2} R_1R_2R_3C\text{-}C(=O)NH_2$$

Recent developments have concentrated on the stereochemical outcome and the specific reactivity of the intermediates involved (J.P. Gilday and L.A. Paquette, Org. Prep. Proced. Int., 1990, 22, 131).

The photoamidation procedure to afford amides appears attractive, however, no new entries have been found since the disclosure of the method in 1964. Certainly this reaction deserves further study (D. Elad and J. Rokach, J. Org. Chem., 1964, 29, 1885):

$$R\text{-}CH=CH_2 + HCONHR_1 \xrightarrow{h\nu} RCH_2CH_2CONHR_1$$

N-Chlorosubstituted amides are highly reactive and are useful as synthetic intermediates. A new method involving chlorination of an emulsion of a fatty amide in boiling water has been developed (W.A. Henderson and L.W. Chang, Org. Prep. Proced. Int., 1986, 18, 269):

$$RCONH_2 + Cl_2 \xrightarrow[100^{\circ}C]{H_2O} RCONHCl + HCl$$

Addition reactions of N-halogenoamides to unsaturated compounds have recently been reviewed (N.N. Labeish and A.A. Petrov, Russ. Chem. Rev., 1989, 58, 1048).

Amides hydrolyse in aqueous solutions giving the parent carboxylic acids and amines. The process involves the formation of a tetrahedral intermediate, which subsequently decomposes to products in a rate-limiting step that is subject to general acid-base catalysis. However, amides bearing good leaving groups such as N-nitroso and N-nitro decompose by a nucleophilic catalysed pathway (B.C. Challis and S.P. Jones, J. Chem. Soc. Perkin II, 1979, 703; B.C. Challis et al., J. Chem. Soc. Perkin II, 1989, 1823).

The electron-transfer photochemistry of amides has recently been discussed (J.D. Coyle, Pure Appl. Chem., 1988, 66, 941).

Treatment of long-chain carboxamides with I-hydroxy-I-tosyloxyiodobenzene in hot CH_3CN gives alkylammonium tosylates in excellent yields (A.V. Ganapathy, Diss. Univ. of Akron 1989; A.Vasudevan and G.F. Koser, J. Org. Chem., 1988, 53, 5158):

$$RCONH_2 + PhI(OH)OTs \xrightarrow[\Delta]{CH_3CN} RNH_3^{\oplus}OTs^{\ominus}$$

Similar results can be obtained with $PhI(OCOCF_3)_2$ (G.M. Loudon et al., Org. Synth., 1988, 66, 132).

Amides can be oxygenated with peroxides in the presence of a ruthenium catalyst (S.-I. Murahashi, J. Am. Chem. Soc, 1990, 112, 7820):

$$R_1-CH(R_2)-N(R_3)-C(=O)-R_4 \xrightarrow[\text{Ru cat}]{t\text{-BuOOH}} R_1-C(R_2)(OOtBu)-N(R_3)-C(=O)-R_4$$

Several types of amides, for example, hexaneamide, can be reduced to the corresponding amine by sodium borohydride in acidic DMSO (S.R. Wann, P.T. Thorsen and M.M. Kreevoy, J. Org. Chem.,1981, 46, 2579):

$$RCON(R_1)_2 \xrightarrow[\text{DMSO, } H^{\oplus}]{\text{Na BH}_4} RCH_2N(R_1)_2$$

Even carboxylic acids can be used as substrates.

A review on the physical properties of acetamide including its application as a solvent is available (D.H. Kerridge, Chem. Soc. Rev., 1988, 17, 181).

(f) *Carbohydrazides*

Carbohydrazides are useful intermediates in organic synthesis. They are frequently prepared for the characterization of carboxylic acids. Carbohydrazides can be converted to aldehydes (O.A. Attanasi, F. Serra-Zanetti and G. Tosi, Org. Prep. Proced. Int., 1989, 20, 405):

$$RCONHNH_2 \xrightarrow[\text{NaBH}_4]{\text{CuCl}_2 \cdot 2H_2O} RCHO$$

Under non-reductive conditions acids, esters or amides can be produced (J. Tsuji et al., Tetrahedron, 1980, 36, 1311):

$$RCONHNH_2 \xrightarrow{Cu^{2+}} \begin{cases} \xrightarrow{H_2O} RCOOH \\ \xrightarrow{R_1OH} RCO_2R_1 \\ \xrightarrow{(R)_2NH} RCON(R)_2 \end{cases}$$

A comprehensive review on carbaohydrazides can be found in a monograph (P.A.S. Smith, "Derivatives of Hydrazine and other Hydronitrogens having N-N Bonds" Benjamin, Reading, Mass., 1983).

(g) *Acyl azides*

The chemistry of azides (including acyl azides) has been recently reviewed (F.F. Scriven and K. Turnbull, Chem. Rev., 1988, 88, 297). Several new reagents for the conversion of carboxylic acids into acyl azides have been reported. Thus, carboxylic acids reacts with O,O-diphenylphosphoryl azide (commercially available) to give the corresponding acyl azides (R.J.W. Cremlyn, Aust. J. Chem., 1973, 26, 1591).

Thionyl chloride and DMF yields a complex of the "Vilsmeier-type" which can be used as an activating reagent for the reaction of carboxylic acids with NaN_3 (A. Arrieta, J.M. Aizpurna and C. Palomo, Tetrahedron Letters,

1984, 3365). Similar activation has been achieved with phenyl dichlorophosphate (J.M. Lago, A. Arrieta and C. Palomo, Synth. Commun., 1983, 13, 289). Acyl chlorides reacts smoothly with trimethylsilyl azide (S.S. Washburne and W.R. Peterson, Jr, Synth. Commun., 1972, 2, 227; B.N. Kozhushko, J. Org. Chem. USSR, 1984, 20, 654):

$$RCOCl \xrightarrow{Me_3SiN_3} RCON_3$$

(h) *Nitriles*

Preparative methods

Conversions of alkyl halides to nitriles in homogeneous non-aqueous systems have already been discussed [section 1a (vi)]. Primary alcohols can cleanly be converted to nitriles by first making the corresponding trifluoroacetate, followed by nucleophilic substitiution with sodium cyanide (F. Camps, V. Gasol and A. Guerrero, Synth. Commun., 1988, 18, 445):

$$CH_3(CH_2)_{13}OH \xrightarrow[\text{2. NaCN, THF-HMPA}]{\text{1. TFAA, } CH_2Cl_2} CH_3(CH_2)_{13}C{\equiv}N$$

Amides can be converted to nitriles by treatment with P_2O_5 supported on silica (D. Kaiser, Synth. Commun., 1984, 14, 883; R.B. Perni and G.W. Gribble, Org. Prep. Proced. Int., 1983, 15, 297).

t-Butylamides are cleanly converted into nitriles, via their imidoyl chlorides (R.B. Perni and G.W. Gribble, Org. Prep. Proced. Int., 1983, 15, 297):

$$\text{RCONH-t-Bu} \xrightarrow[\Delta,\ C_6H_6]{POCl_3} \left[\text{R(Cl)C=N-t-Bu} \right] \longrightarrow RC{\equiv}N$$

Multifunctional amides can be chemoselectively converted into nitriles using the Burgess reagent, methyl (carboxysulfamoyl) triethylammonium hydroxide inner salt, as indicated below (D.A. Claremon and B.T. Phillips, Tetrahedron Letters, 1988, 2155):

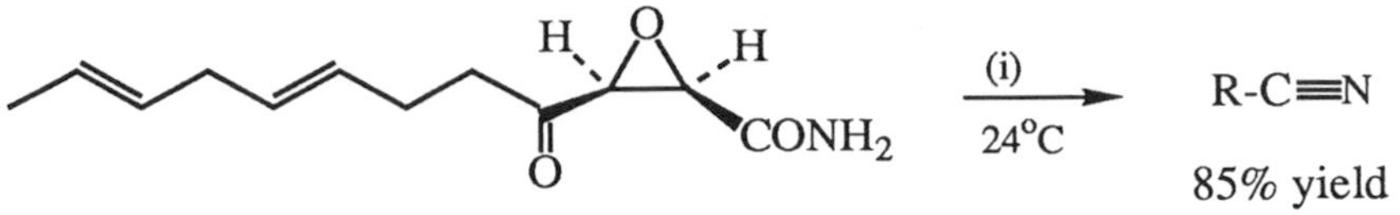

$$\xrightarrow[24^{\circ}C]{(i)} \quad R\text{-}C{\equiv}N \quad 85\%\ \text{yield}$$

(i) $CH_3OCON^{\ominus}\text{-}SO_2\text{-}N^{\oplus}\text{-}Et_3$

Acrylonitrile can be alkylated via radical reactions to yield alkyl- or aralkylnitriles (J. Hershberger et al., Tetrahedron, 1988, 44, 6295):

$$RI + H_2C{=}CHC{\equiv}N \xrightarrow[AIBN]{Bu_3GeH} RCH_2CH_2C{\equiv}N$$

An interesting exchange reaction has been described by Vinokurov et al., (J. Org. Chem. USSR, 1987, 23, 1602):

$$RCONH_2 + MeCN \xrightarrow[80\text{-}100^{\circ}C]{H_2SO_4} MeCONH_2 + RCN$$

Carboxylic acids can be directly converted into nitriles by reactions with ethyl polyphosphate (PPE) (T. Imamoto et al., Synthesis, 1983, 142):

$$R\text{-}COOH + NH_3 \xrightarrow{PPE/CHCl_3} R{-}C{\equiv}N$$

The reversed reaction i.e. hydration of nitriles can be catalytically performed in the presence of MnO_2 on silica gel. (K.T. Liu et al., Synthesis, 1988, 715):

$$R{-}C{\equiv}N + H_2O \longrightarrow RCONH_2$$

Monoanions and polyanions of nitriles are readily generated as follows:

$$H_3CC{\equiv}N \xrightarrow{t\text{-}BuLi} Li_2CHCN \underset{}{\overset{t\text{-}BuLi}{\rightleftarrows}} Li_2C{=}C{=}N{-}Li$$

Such anions are useful in organic synthesis for alkylations, arylations, acylations and several other reactions. A comprehensive review is available (S. Arseniyadis, K. Kyler and D.S. Watt, Org. React., 1984, 31, 1).

Nitriles form complexes with a number reagents such as BF_3, $AlCl_3$, $SbCl_5$ and $(EtO)_3BF_4$. These complexes are usually repesented as simple donor-acceptor complexes, for example, R-CN→$SbCl_5$. In most cases, however, they probably have more complicated structures, frequently dimeric and sometimes with more than one RCN ligand on the metal atom. Such complexes are useful for synthetic purposes. Recent studies show that Ln^{3+} and Yt^{3+} as complexing ion catalysts (J.H. Forsberg et al., J. Org. Chem., 1987, 52, 1017; J.H. Forsberg et al., J. Het. Chem., 1988, 25, 767):

$$R_1—C\equiv N \quad + \quad R_2—NH_2 \xrightarrow{Ln^{3+}} R_1C(=NH)NH—R_2$$

Reaction of nitrilium salts with tertiarty amides affords N-acyl amidinium salts (J.C. Jochims and R. Abu-El-Halawa, Synthesis, 1990, 488; J.C. Jochims and M.O. Glocker, Chem. Ber., 1990, 123, 1537). The last-mentioned paper contains several entries to other interesting reactions involving nitrilium ions:

$$R_1—C\equiv \overset{\oplus}{N}—R_2 \; SbCl_6^{\ominus} \quad + \quad R_3\,CON(R_4)_2 \longrightarrow R_1C(=O)N(R_2)C(R_3)=\overset{\oplus}{N}(R_4)_2 \; SbCl_6^{\ominus}$$

N-Methylacetonitrilium fluoroborate, conveniently prepared form acetonitrile and trimethyloxonium fluoroborate, is a useful mild electrophile (S.C. Eyley, R.G. Gilaes and H. Heaney, Tetrahedron Letters, 1985, 4649):

$$\text{pyrrole (NH)} \xrightarrow{MeC\equiv \overset{\oplus}{N}-Me \;\; BF_4^{\ominus}} \text{2-pyrrolyl–}C(Me)=N\text{–}Me \xrightarrow{H_2O} \text{2-pyrrolyl–}C(=O)Me$$

Evidence for the involvement of diprotonated nitriles $(RC^{\oplus}{=}N^{\oplus}H_2)$ during aromatic substitution reactions has recently been obtained (M. Yato, T. Ohwada and K Shudo, J. Am. Chem. Soc., 1991, 113, 691).

(i) *Imidoyl chlorides*

This class of compounds has been discussed in a monograph (H. Ulrich "The Chemistry of Imidoyl Halides", Plenum Press, New York 1968. Methods for the trapping, mostly intramolecular, of imidoyl chlorides formed in situ from nitriles and acyl chlorides in the presence of HCl has been surveyed by Yanagida and Komori (S. Yanagida and S. Komori, Synthesis, 1972, 189):

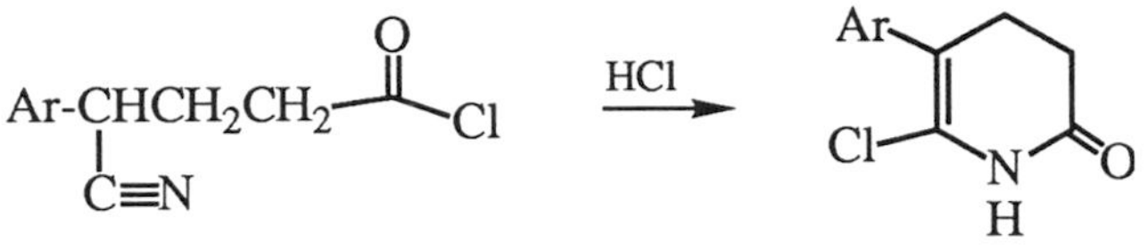

$$\text{Ar-CH(C}{\equiv}\text{N)CH}_2\text{CH}_2\text{COCl} \xrightarrow{\text{HCl}} \text{(cyclic product: Ar, Cl, N-H, =O)}$$

Imidoyl chlorides react with $AgNO_3$ to yield N-nitroamides. The reaction pathway involves initial formation of an imidate, which subsequently rearranges (E. Carvalho, J. Iley and E. Rosa, Chem. Comm., 1988, 1249):

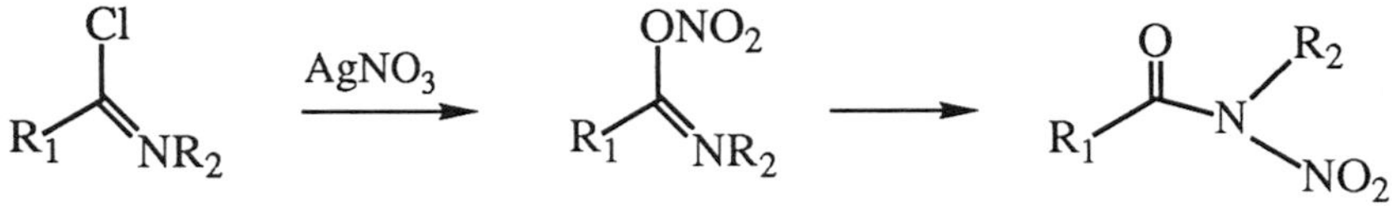

$$R_1C(Cl){=}NR_2 \xrightarrow{AgNO_3} R_1C(ONO_2){=}NR_2 \longrightarrow R_1C(O)N(R_2)NO_2$$

Imidoyl chlorides can be reduced to imines, which can be transformed via acylation and hydrolysis to amides:

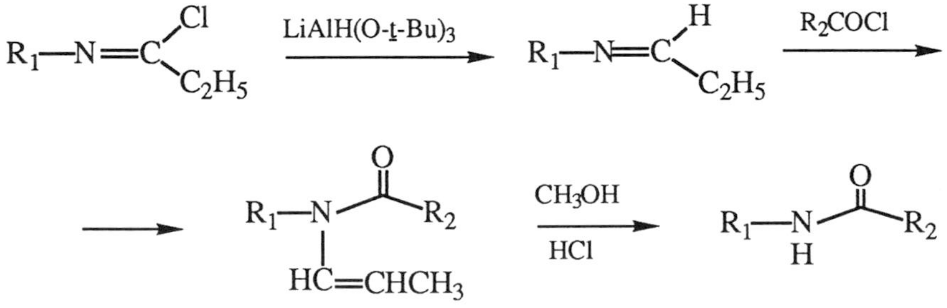

$$R_1{-}N{=}C(Cl)C_2H_5 \xrightarrow{\text{LiAlH(O-t-Bu)}_3} R_1{-}N{=}C(H)C_2H_5 \xrightarrow{R_2COCl}$$

$$\longrightarrow R_1{-}N(HC{=}CHCH_3){-}C(O)R_2 \xrightarrow[\text{HCl}]{\text{CH}_3\text{OH}} R_1{-}NH{-}C(O)R_2$$

Since imidoyl chlorides can be prepared from amides, the reaction sequence represents a general method for acyl exchange in secondary amides (S. Karady et al., Tetrahedron Letters, 1978, 403).

(j) *Alkyl imidates*

A review on imidates is available (D.G. Neilson in "The Chemistry of

Amidines and Imidates", S. Patai ed., Wiley, London, 1975, and an update is due in 1991).

Chiral acyclic imidate esters have been prepared using the two established methods for this class of compounds; i.e. treatment of an ortho ester with an amine, or by alkylation of an amide with Et_3OBF_4. The chiral imidate esters can be deprotonated with various bases to give the corresponding lithio anion derivatives. Alkylations of the lithio derivatives proceed in high synthetic yields and with good to excellent asymmetric induction to give α,α–disubstituted carboxylic acid derivatives (C. Gluchowski et al. J. Org. Chem., 1984, 49, 2650).

Benzyl (and alkyl) trichloroacetimidate is a convenient reagent for selective O-alkylation of hydroxy groups under mild acidic conditions. It is compatible with imide, ester and acetal protecting groups (H.-P. Wessel, T. Iversen and D.R. Bundle, J. Chem. Soc. Perkin I, 1985, 2247).

The thermal rearrangement of imidates to the corresponding amides has been studied by several groups. The migrating group can be alkyl, aryl or acyl (B.C. Challis and A.D. Frenkel, J. Chem. Soc. Perkin II, 1978, 192; D.G. McCarthy and A.F. Hegarty, J. Chem. Soc. Perkin II, 1977, 1085):

$$R_1-C(OR_2)=NR_3 \longrightarrow R_1-C(=O)-NR_2R_3$$

(k) *Amidines*

A fairly recent review discussing the chemistry of amidines has been published (V.G. Granik, Russ. Chem. Rev., 1983, 52, 377). In 1991 a supplement in the Patai series dealing with amidines is expected. Formamide acetals are the most useful reagents for preparation of amidines (G. Simchen in "Advances in Organic Chemistry", H. Böhme and H. Viehe eds, 1979, **9** (1), pp 393-526, Wiley, N.Y.):

$$(R_1)_2N-CH(OR_2)_2 + R_3NH_2 \longrightarrow R_3-N=CHN(R_1)_2$$

This general technique has been used for the preparation of optically active amidines, which are useful as "activators" of a wide range of secondary amines toward metalation and alkylation. At the end of the day the activating group is removed to furnish the optically active α-alkylated amine

(A.I. Meyers et al., Tetrahedron, 1984, 40, 1361; A.I. Meyers et al., Angew. Chem. Int. Ed., 1984, 23, 458; T.K. Highsmith and A.I. Meyers in "Advances in Heterocyclic Natural Product Synthesis", Vol 1, W.M. Pearson ed., J.A.I. Press, London 1991).

For dimethylaminomethylenation, [3-(dimethylamino)-2-azaprop-2-en-1-ylidene]dimethylammonium chloride "Gold´s reagent", (GR) offers possible advantages over other techniques (J.T. Gupton and S.A. Andrews, Org. Synth., 1985, 64, 85, and refs. therein):

$$\text{2,4,6-trichloro-1,3,5-triazine} + 6\,\text{DMF} \xrightarrow[-\,3CO_2]{\text{dioxan, reflux}} 3\ Me_2N{-}CH{=}N{-}CH{=}\overset{\oplus}{N}Me_2\ Cl^{\ominus}$$

$$R{-}NH_2 \xrightarrow{GR} R{-}N{=}CHNMe_2$$

$$RCONH_2 \xrightarrow{GR} RC(=O){-}N{=}CH{-}NMe_2$$

Amidines can be prepared directly from carboxylic acids and amines in the presence of polyphosphoric acid trimethylsilyl ester, (PPSE), (M. Kakimoto et al., Chem. Lett., 1984, 821).

$$R_1COOH + 2\,RNH_2 \xrightarrow{PPSE} [R_1CONHR] \xrightarrow{PPSE} R_1{-}C(=NR){-}NR$$

(l) *Hydroxamic acids,* RCONHOH

A recent comprehensive literature review featuring chemical and biological properties of hydroxamic acids is available (E.B. Pamiago and S. Carvalho, Ciencia e Cultura, 1988, 40, 629). Monographs are also available (H. Kehl ed., "Chemistry and Biology of Hydroxamic Acids", (1982), Karger, N.Y; E. Lipczynska-Kochany, "Some New Aspects of Hydroxamic Acid Chemistry", 1988, Warsaw, Poland).

Hydroxamic acids can occur as many isomers and conformers. Recently, cis-trans isomerism in alkyl hydroxamic acids has been studied by NMR spectroscopy (including ^{15}N studies) (D.A. Brown et al., Mag. Res. Chem., 1988, 26, 970). An interesting theoretical study is also available (N.J. Fitzpatrick and R. Mageswaran, Polyhedron, 1989, 8, 2256).

Traditionally the most common method to synthesize hydroxamic acids is through the acylation of hydroxylamine with esters, anhydrides, or acid chlorides. Formation of various amounts of O-acylated derivatives is sometimes a drawback (H.Grigat and G. Zinner, Arch. Pharm., 1986, 319, 1037, and refs therein). If so, the following method in which secondary amides are first silylated and then oxidised with Mo(v) peroxides is recommended (S.A. Matlin, P.G Sammes and R.M. Upton, J. Chem. Soc. Perkin I, 1979, 2481):

$$R_1C(=O)N(H)R_2 \xrightarrow[2.\ Mo^{V}]{1.\ TMSCl,\ Et_3N} R_1C(=O)N(OH)R_2$$

Thiohydroxamic acids have traditionally been prepared by the thioacylation of hydroxylamines and reaction of hydroximic acid chlorides with H_2S (W. Walter and E. Schaumann, Synthesis, 1971, 111).

A new method involves the use of thiosilanes in the conversion of nitro compounds to thiohydroxamic acids. Presumably the first step in this procedure is thiolation of the anion of the nitroalkane (J.R. Hwu and S.-C. Tsay, Tetrahedron, 1990, 46, 7413):

$$R\text{-}CH_2NO_2 \xrightarrow[2.\ (TMS)_2S;\ H_2O^{\oplus}]{1.\ KH,\ THF} RC(=S)NHOH$$

Further discussion of the preparation of thiohydroxamic acids (and hydroxamic acids) can be found in a monograph (S.R. Sandler and W. Karo, "Organic Funcional Group Preparations", Vol III, 2nd ed., A.P., New York 1989).

(m) *Hydrazidines, amidrazones*

A series of papers on the chemistry of amidrazones has been presented (see R.F. Smith et al., Synth. Commun., 1986, 16, 585 and references therein).

3. THIOCARBOXYLIC ACIDS

A comprehensive review on the chemistry of dithiocarboxylic acid esters has recently been published (S. Kato and M. Ishida, Sulfur Reports, 1988, 4, 155). An earlier review with the title "Synthesis of Dithiocarboxylic acids and Esters" is also available (S.R. Ramadas et al., Synthesis, 1983, 605).

A monograph is also available (S. Scheithauer and R. Mayer, "Thio-and Dithiocarboxylic Acids and Their Derivatives", Heyden & son, Philadelphia 1979. The authors of the last survey have also written a chapter in the Houbel-Weyl series (E5, pp 891-930). A recent book also provides information of interest (A.D. Dunn and W.-D. Rudorf, "Carbon Disulphide in Organic Chemistry", 1989, E. Horwood, Ltd., Chichester).

S-esters of thiocarboxylic acids can be prepared in several ways:

$$\text{N-acylimidazole (RCO–Im)} + R_1SH \longrightarrow RC(=O)SR_1$$

(H. Gais, Angew. Chem. Int. Ed., 1977, 16, 244).

$$RCOOH + (R_1S)_3B \longrightarrow RC(=O)SR_1$$

(A.Pelter et al., J. Chem. Soc. Perkin I, 1977, 1672).

New preparations and applications of thiocarboxylic O-esters have been discussed by B.A. Jones and J.S. Bradshaw.(Chem. Rev., 1984, 84, 17).

Reductive alkylation of esters of dithiocarboxylic acids provides interesting routes to thio acetals (G. Drosten et al., Chem. Ber., 1987, 120, 324):

$$R_1C(=S)SMe \xrightarrow[R_2X]{2e^{\ominus}} RCH(SMe)(SR_2)$$

Monothioformic acid has been synthesized from phenyl formate and sodium hydrogen sulphide and the dynamics of its cis-trans isomerization has been studied (K.I. Lazaar and S.H. Bauer, J. Chem. Phys., 1985, 83, 85).

Thioamides constitute an interesting class of compounds and a list of applications and synthetic methods dealing with these molecules is available (K.E. DeBruin and E.E. Boros, J. Org. Chem., 1990, 55, 6091). A new route to thioamides involving activation of an amine with $(MeO)_2P(S)Cl$, followed by acylation, rearrangement and hydrolysis is also descrribed in this paper:

$$RCOCl + R_1NH\text{-}PS\text{-}(OMe)_2 \longrightarrow RCONR_1PS(OMe)_2 \xrightarrow[CCl_4]{\Delta}$$

$$\longrightarrow RCSNR_1PO(OMe)_2 \longrightarrow RCSNHR_1$$

4. HALOGEN SUBSTITUTED PRODUCTS OF CARBOXYLIC ACIDS

Halogenation of carboxylic acids is undoubtedly the most important method for the preparation of α-halo acids. α-Chloro and α-bromo acids are readily prepared by the classical Hell-Volhard-Zelinsky reaction. α-Iodo acids are available via iodination of ester enolates at low temperature (M.W. Rathke and A. Lindert, Tetrahedron Letters, 1971, 3995):

$$R_1CH_2COOR_2 + C_6H_{11}\text{-}N(Li)\text{-}iPr \xrightarrow[-80^{\circ}C,\ I_2]{THF} R_1CHICOOR_2$$

Harpp has introduced several new procedures for the synthesis of α-halogenoacyl chlorides (D.N. Harpp et al., J. Org. Chem., 1975, 40, 3420):

$$R_1CH_2COCl \xrightarrow[SOCl_2]{NBS} R_1CH(Br)COCl$$

$$R_1CH_2COCl \xrightarrow[SOCl_2]{I_2} R_1CHICOCl$$

An unusual route to α-halo acids has been devised by H.C. Brown and H. Nambu (J. Am. Chem. Soc., 1970, 92, 5790):

$$Et_3B + Cl_2CHCN \xrightarrow{B^{\ominus}} CH_3CH_2\text{—}CH(Cl)\text{—}C{\equiv}N \longrightarrow$$

$$\xrightarrow[H_2O]{HCl} CH_3CH_2\text{—}CH(Cl)\text{—}COOH$$

α-Fluoro acids are usually prepared by substitution reactions. This type of reaction has recently been discussed in a review of modern methods for the introduction of fluorine into organic molecules (J. Mann, Chem. Soc. Rev., 1987, 16, 381). The preparation of 2-fluorostearic acid provides an illustration of the problems involved in this type of synthesis (S. Pogany et al., Synthesis, 1987, 718).

Anions of bulky esters of α-fluoroacetic acid can be stereoselectively reacted with ketones (J.T. Welch and J.S. Plummer, Synth. Commun., 1989, 19, 1081).

Difluoroacetic acid has recently been prepared using the following route:

$$FClC{=}CF_2 \longrightarrow FClCH{-}C(O){-}N(CH_2)_4 \xrightarrow{KF} F_2CH{-}C(O){-}N(CH_2)_4 \xrightarrow[2.\ H^{\oplus}]{1.\ OH^{\ominus}} F_2CHCOOH$$

(M.C. Lutz and W.P. Dailey, Org. Prep. Proced Int., 1987, 19, 468).

Ethyl bromoacetate is a useful reagent in the Reformatsky reaction:

$$BrCH_2COOEt \xrightarrow{Zn} BrZnCH_2COOEt$$

Several reviews on this reaction are available (M.E. Rathke, Org. React., 1975, 22, 571; A. Fürstner, Synthesis, 1989, 571).

The Reformatsky reaction when carried out upon α-bromoesters in the presence of TMSCl gives rise to α-trimethylsilylacetates in good yields (H. Taguchi et al., Bull. Chem. Soc. Japan, 1974, 47, 2529):

$$BrCH_2COOEt \xrightarrow[TMSCl]{Zn} Me_3SiCH_2COOEt$$

A reaction performed under similar conditions, but now with imines present provides access to azetidinones (C. Palomo et al., J. Org. Chem.,1989, 54, 5736):

$$BrCH_2CO_2Et + R_1{-}CH{=}NR_2 \xrightarrow[TMSCl]{Zn} \text{azetidin-2-one (4-}R_1\text{, N-}R_2\text{)}$$

α-Halo carboxylic esters are readily converted into O-silyl ketene acetals by treatment with TMSCl and sodium (W.J. Schulz and J.L. Speier, Synthesis, 1989, 163):

$$R_1R_2C(X)COOR \xrightarrow[Na]{TMSCl} R_1R_2C{=}C(OR)OSiMe_3$$

Trifluoroacetic acid (TFA) as a solvent (including its purification and physical properties) has been discussed in a monograph. (J.B. Milne in "The Chemistry of Nonaqueous Solvents", J.J. Lagowski ed., Academic Press, New York, 1978).

The dissociation of TFA has recently been studied (H. Strehlow and P.H. Hildebrandt, Ber. Bunsenges. Phys. Chem., 1990, 94, 173). TFA and its anhydride (TFAA) are extremely useful and versatile reagents. TFA can be used as medium for nitrations using sodium nitrate (U.A. Spitzer and R. Stewart, J. Org. Chem., 1974, 39, 3936). Trialkylsilanes in TFA selectively reduce the C=O group of arylcarbonyl compounds to CH_2 (C.T. West et al., J. Org. Chem., 1973, 38, 2675).

TFA promotes alkylations of arenes using allylic reactants. In the case of 3-alkoxybutenes the reaction proceeds through the primary, rather than the secondary carbocationic species or its equivalent, to give phenylbut-2-ene (Y. Fujiwara, H. Kuromaru and H. Taniguchi, J. Org. Chem., 1984, 49, 4309):

$$CH_2{=}CH{-}CH(CH_3){-}OR + C_6H_6 \xrightarrow{TFA} C_6H_5CH_2CH{=}CHCH_3$$

TFA reacts with trifluoromethansulphonic acid in the presence of P_2O_5 to give the corresponding mixed anhydride, a particularly reactive agent for trifluoroacetylations at oxygen, nitrogen, carbon, or halogen centers (T.R. Forbus, S.L. Taylor and J.C. Martin, J. Org. Chem., 1987, 52, 4156).

TFAA can be used to "activate" such compounds as imidazoles. The cation formed can subsequently attack suitable aromatics such as 1,3-dimethoxybenzene to form a masked aldehyde (J. Bergman, B. Sjöberg and L. Renström, Tetrahedron, 1980, 36, 2505):

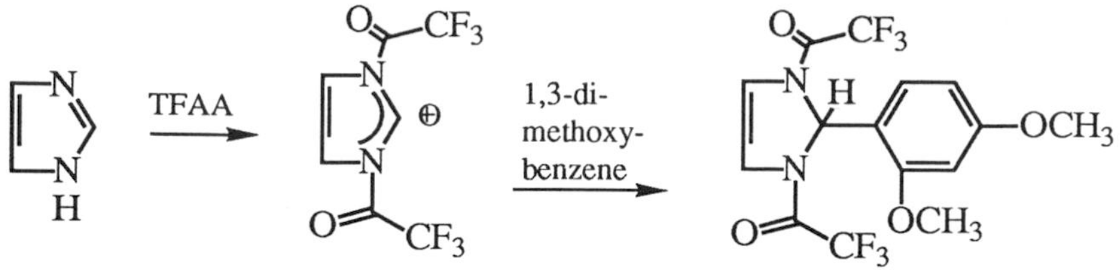

TFAA in conjunction with Bu_4NNO_3 convert amides to the corresponding N-nitroamides (E. Carvalho et al., J. Chem. Res. [S], 1989, 260):

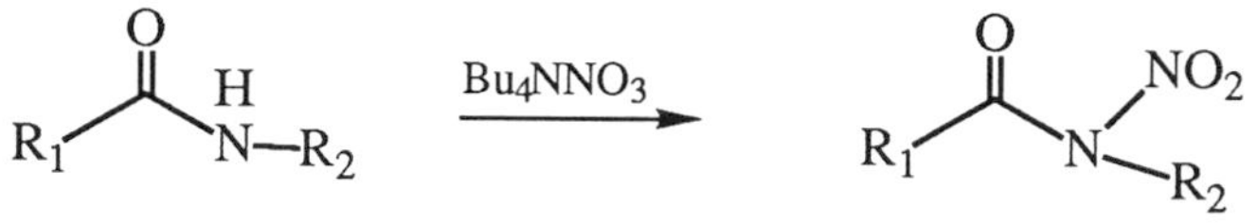

The combination TFAA-H_2SO_4 provides a sulphonylating agent for arenes (T.E. Tyobeka, R.A. Hancock and H. Weigel, Chem. Comm., 1980, 114):

$$H_2SO_4 + 2\,TFAA \longrightarrow (CF_3CO_2)SO_2 + 2\,TFA$$

$$(CF_3CO_2)SO_2 + PhH \longrightarrow PhSO_2Ph + 2\,TFA$$

The peroxide, $(CF_3COO)_2$, can be used for the introduction of CF_3 groups into aromatics. (H. Sawada et al., J. Fluorine Chem., 1990, 46, 909). Reagents of the structure $(C_nF_{2n+1}COO)_2$ can be similarly used (M. Yoshida et al., J. Chem. Soc. Perkin I, 1989, 909).

On heating sodium trifluoroacetate decomposes to difluorocarbene. The intermediate, $CF_3^{\ominus}$, is extremely unstable but it can be trapped by 1,3,5-trinitrobenzene yielding an isolable red Meisenheimer complex (G.P.Stahly, J.Fluorine Chem., 1989, 45, 431):

$$CF_3COO^{\ominus} \longrightarrow CF_3^{\ominus} \longrightarrow CF_2 + F^{\ominus}$$

In the presence of CuI and an aromatic iodide coupling with halide displacement can be effected (R.D. Chambers et al., J. Chem. Soc. Perkin I, 1988, 921):

$$ArI + CF_3CO_2Na \xrightarrow{CuI} ArCF_3$$

5. ORTHO ESTERS

An extensive review on ortho esters and related compounds is available (G. Simchen, in Houben-Weyl, "Methoden der Organischen Chemie", J. Falbe ed., 1985, Vol. E5, pp 3-192, Thieme Verlag, Stuttgart, New York). Ortho esters are extremely useful reagents for a large number of synthetic operations such as preparation of imidates from amines, acylations and generation of heterocyclic rings. In particular, trialkyl orthoacetates and ortho propionates have been employed in regio-and stereocontrolled formation of functionalized alkenes via Claisen-type rearrangements. This reaction was introduced by W.S. Johnson (W.S. Johnson et al., J. Am. Chem. Soc., 1970, 92, 741).

A recent example illustrates the principle (P. Baeckström et al., Synth. Commun., 1990, 20, 423):

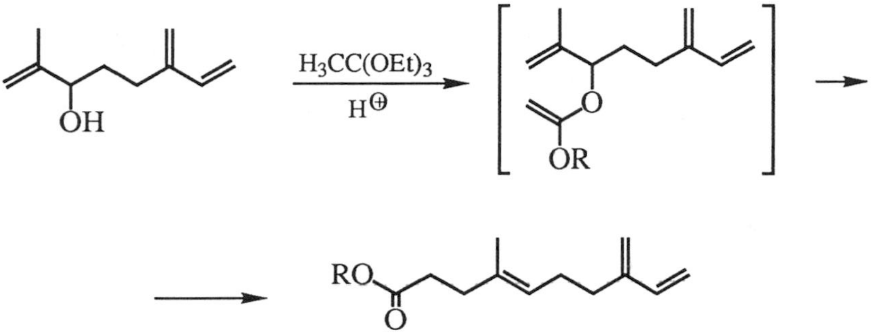

$$\xrightarrow[H^{\oplus}]{H_3CC(OEt)_3}$$

The rearrangement can also be extended to heterocyclic systems (S. Raucher et al., J. Org. Chem., 1979, 44, 1885):

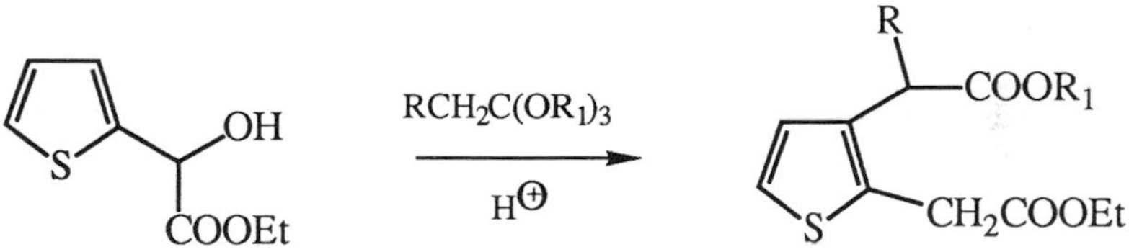

$$\xrightarrow[H^{\oplus}]{RCH_2C(OR_1)_3}$$

The reaction between ortho esters and Grignard reagents to afford acetals is a well-known procedure:

$$RMgX + HC(OR_1)_3 \longrightarrow R\text{-}CH(OR_1)_2.$$

The reactivity of the ortho ester can be enhanced when one of the alkoxy groups can act as a chelator for Mg (R_1 is for example

$OCH_2CH_2OCH_2CH_2OMe$). Dithio-ortho esters can be similarly activated (R.P. Houghton and J.E. Dunlop, Synth. Commun., 1990, 20, 2387).

Aldoximes are converted into nitriles in high yields by treatment with an ortho ester in the presence of a catalytic amount of methanesulphonic acid The primary product can be obtained if the catalyst is omitted (M.M. Rogic et al., J. Org. Chem., 1974, 39, 3424):

$$RCH{=}NOH + R_1{-}C(OEt)_3 \rightleftharpoons RCH{=}NO{-}CR_1(OEt)_2 \xrightarrow{H^{\oplus}}$$

$$\longrightarrow RCN + R_1COOEt + EtOH$$

Ortho esters react readily with diazo componds (M.P. Doyle and M.L. Trudell, J. Org. Chem., 1984, 49, 1196).

$$R_1C(OR)_3 + N_2CH{-}C(=O){-}OR \xrightarrow{\text{Cu-triflate}} R_1C(OR)_2{-}CH(OR){-}C(=O){-}OR$$

It has been known for a long time that ortho esters are useful in so called three-component condensations. A detailed review is available. (W.V. Mezheritskii, E.P. Olekhnovitch and G.N. Dorofeenko, Russ. Chem. Rev., 1973, 42, 896).

An example of such a three-component condensation is given below in which oxindole is converted into its 3-aminoalkylidene derivative:

$$\text{oxindole} + R_1C(OR)_3 + R_2NH_2 \longrightarrow \text{3-}[R_1C(NHR_2){=}]\text{oxindole}$$

(O.S. Wolfbeis, Monatshefte, 1981, 112, 369).

Second Supplements to the 2nd Edition of Rodd's Chemistry of Carbon Compounds, Vol. 1C, edited by M. Sainsbury
© 1992 Elsevier Science Publishers B.V., Amsterdam

Chapter 9b

UNSATURATED MONOBASIC ACIDS

R.B.Miller

1. Introduction

The literature on unsaturated monobasic acids has been dominated in recent years by the numerous investigations on the biologically important fatty acids, such as the C_{20} acid - arachidonic acid; the C_{18} acids - oleic, linoleic and linolenic; the ω-3 fatty acids; and related compounds. In fact so much work has been undertaken on transformations of the eicosanoids to prostaglandins and leukotrienes and their associated biological activity that a number of journals are now devoted to these subjects.

(a) General references

A number of monographs have appeared covering a wide range of topics on unsaturated monobasic acids, mainly associated with their inclusion as fatty acids. General monographs on fatty acids and their biochemistry include: "Polyunsaturated Fatty Acids" ed. W.-H, Kunau and R. T. Holman, American Oil Chemists' Society, Champaign, Ill., 1977; "Fatty Acids" ed. E. H. Pryde, American Oil Chemists' Society, Champaign, Ill., 1979; "Fatty Acids; Synthesis and Applications" (N.E. Bednarcyk and W. L. Erickson, Noyes Data Corp., Park Ridge, N. J., 1973); "Fatty Acids Manufacture: Recent Advances" ed. J.C. Johnson, Noyes Data Corp., Park Ridge, N.J., 1980; "Fatty Acids in Industry: Processes, Properties, Derivatives, Applications" ed. R. W. Johnson and E. Fritz, M. Dekker, New York, 1989; "Biochemistry and Pharmacology of Free Fatty Acids" ed. W. L. Holmes and W. M. Bortz, S. Karger, Basel, 1971; "Plant Lipid Biochemistry: The Biochemistry of Fatty Acids and Acyl Lipids with Particular Reference to Higher Plants and Algae" (C. Hitchcock and B.W. Nichols, Academic Press, New York, 1971); "Biosynthesis of Acetate-Derived Compounds" (N.M. Packter, J. Wiley, New York, 1973); "Fatty Acid Metabolism and Its Regulation" ed. S.

Numa, Elsevier, Amsterdam, 1984; and "Fatty Acids and Glycerides" ed. K. Kuksis, Plenum Press, New York, 1978.

Also a number of monographs dealing with the health aspects of unsaturated fatty acids have appeared, including: "The Unsaturated and Polyunsaturated Fatty Acids in Health and Disease" (J.F. Mead and A.J. Fulco, Thomas, Springfield, Ill., 1976); "Dietary Fat Requirements in Health and Development" ed. J. Beare-Rogers, American Oil Chemists' Society, Champaign, Ill., 1988; "The Omega-3 Phenomenon: the Nutrition Breakthrough of the 80's" (D.O. Rudin and C. Felix, Rawson Assoc., New York, 1987); "Omega-3 Fatty Acids in Health and Disease" ed. R.S. Lees and M. Karel, M. Dekker, New York, 1990; "Nutritional Evaluation of Long-Chain Fatty Acids in Fish Oil" ed. S.M. Barlow and M.E. Stansby, Academic Press, New York, 1982; "Health Effects of Polyunsaturated Fatty Acids in Seafoods" ed. A.P. Simopoulas, R.R. Kifer and R.E. Martin, Academic Press, Orlando, 1986; "Fish and Human Health" (W.E.M. Lands, Academic Press, Orlando, 1986); "Seafoods and Fish Oil in Human Health and Disease" (J.E. Kinsella, M. Dekker, New York, 1987); "Omega-6 Essential Fatty Acids: Pathophysiology and Roles in Clinical Medicine" ed. D.F. Horrobin, Wiley-Liss, New York, 1990; "The Lipids of Human Milk" (R.G. Jensen, CRC Press, Boca Raton, Fla., 1989); and "Icosanoids and Cancer" ed. H. Thaler-Dao, A. Crastes de Paulet and R. Paoletti, Raven Press, New York, 1984.

There were numerous conferences on aspects of fatty acid chemistry whose proceedings have been published, including: "The Essential Fatty Acids: Miles Symposium 1975..." ed. W.W. Hawkins, Miles Laboratories, Rexdale, Ont., 1976; "Golden Jubilee International Congress on Essential Fatty Acids and Prostaglandins" ed. R.T. Holman, Pergamon, Oxford, 1982; "Clinical Uses of Essential Fatty Acids: Proceedings of the First Efamol Symposium Held in London, England in November, 1981" ed. D.F. Horrobin, Eden Press, Montreal, 1982; "Essential Fatty Acids, Prostaglandins and Leukotrienes: The Second International Congress Held in London, March 1985" ed. W.W. Christie et al., Pergamon, Oxford, 1986; "Proceedings of the AOCS Short Course on Polyunsaturated Fatty Acids and Eicosanoids" ed. W.E.M. Lands, American Oil Chemists' Society, Champaign, Ill., 1987; "World Conference on Biotechnology for the Fats and Oils Industry: Proceedings" ed. T.H. Applewhite, American Oil Chemists' Society, Champaign, Ill., 1988; and "Fatty Acid Oxidation: Clinical, Biochemical, and Molecular Aspects: Proceedings of the International Symposium on... Held in Philadelphia in 1988" Liss, New York, 1990.

(b) Natural occurrence and isolation of unsaturated acids

Fungal cultures produced C_8 and C_9 polyacetylenic acids. From the culture fluids of Psilocybe merdaria, Kuehneromyces mutabilis, Russula vesca, and Ramaria flava was obtained 2,4,6-octatriynoic acid, while 4,5,8-nonatriynoic acid was isolated from Serpula lacrymans (M.T.W. Hearn et al., J. Chem.Soc., Perkin Trans., 1973, **1**, 2785).

Pyrolysis of castor oil produced 10-undecenoic acid (G.Das, R.K. Trivedi and A.K. Vasishtha, J.Am. Oil Chem. Soc., 1989, **66**, 938; A.P. Kudchadker, S.A. Kudchadker and D.K. Singh in Encyl. Chem. Process. Des., Vol 6, ed. J. J. McKetta and W. A. Cunningham, M. Dekker, New York, 1978, p. 401).

Mass spectral analysis of the ether extracts of adult and larval thrips (H. japonicus) indicated the presence of (Z)-5- and (E)-3-dodecenoate (K. Haga et al., Appl. Entomol. Zool., 1989, **24**, 242).

Aje, body fat of the insect coccid Llaveia axin, was examined for fatty acids. In addition to a variety of known acids, 3,5,7,9,11-dodecapentaenoic acid and 5,7,9,11,13-tetradecapentaenoic acid were found for the first time (J. Cason, R. Davis and M. H. Sheehan, J. Org. Chem., 1971, **36**, 2621).

From the volatiles of western grape leaf skelentonizer, Harrisina brillans, was isolated *sec*-butyl (Z)-7-tetradecenoate, which showed sex pheromonal activity, and also the isopropyl ester. This was the first report of an isopropyl or *sec*-butyl unsaturated ester from the volatiles of a lepidoteran species (J. Myerson, W.F. Haddon and E.L. Soderstrom, Tetrahedron Lett., 1982, **23**, 2757).

Hexadecatrienoic and hexadecatetraenoic acids were isolated from Japanese sardine oil (Y. Ando, T. Ota and T. Takagi, J. Am. Oil Chem. Soc., 1989, **66**, 1323).

A large number of natural sources have been found for a variety of C_{18} unsaturated acids. Oleic acid, (Z)-9-octadecenoic acid, has been found in large quantity in seed oil extracted from Moringa concanensis (A. Sengupta, C. Sengupta and P. Das, Chem. Ind. (London), 1974, 211), from processed castor oil (G. Lakshminarayana et al., Res. Ind., 1988, **33**, 12) and from lipolysis of olive oil and tallow by the lipase from oat seeds (G. Piazza, Biotechnol. Lett., 1989, **11**, 487). A mixture of oleic acid and linoleic acid, (Z),(Z)-9,12-octadecadienoic acid, was isolated from oleic safflower seed oil (M.J. Diamond and G. Fuller, J. Am. Oil Chem. Soc., 1970, **47**, 362), wild apricot kernel oil (K.K. Argarwal et al., J. Oil Technol. Assoc. India, 1974, **6**, 67) and inbred lines of sunflower (G. Andrich et al., Agrochimica, 1985, **29**, 276). Linoleic acid was the major fatty acid constituent in the seed kernels of Aphananthe aspera Planch (T. Tanaka, S. Ihara and Y. Koyama, J. Am. Oil. Chem. Soc., 1977, **54**, 269). A mixture of oleic, linoleic, and linolenic (octadecatrienoic acid) acids was isolated from Philipine lumbang oil (V.P. Arida and A.U.A. Co, Philipp. J. Sci., 1982, **111**, 125). γ-Linolenic acid, (Z),(Z),(Z)-6,9,12-octadecatrienoic acid, was isolated from a variety of microbiological sources including: Mucor ambiguus (H. Fuykuda and H. Morikawa, Appl. Microbiol. Biotechnol., 1987, **27**, 15), Mucor FB-

354 (Y.C. Shin and H.K. Shin, Han'guk Sikp'um Kwahakhoechi., 1988, **20,** 724), Mortierella strains (O. Susuki in "World Conference on Biotechnology for the Fats and Oil Industry: Proceedings" ed. T.H. Applewhite, American Oil Chemists' Society, Champaign, Ill., 1988, p. 110) and Mucor rouxii (L. Hansson, M. Dostalek and B. Soerenby, Appl. Microbiol. Biotechnol., 1989, **31**, 223). A study of the influence of

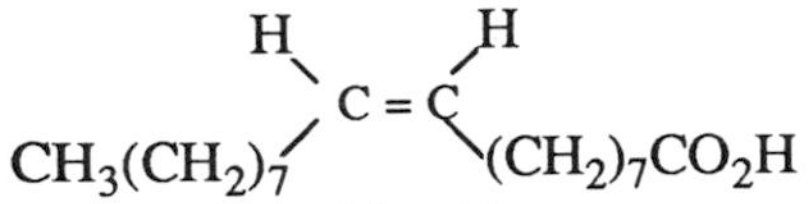

$$CH_3(CH_2)_7CH{=}CH(CH_2)_7CO_2H$$

oleic acid

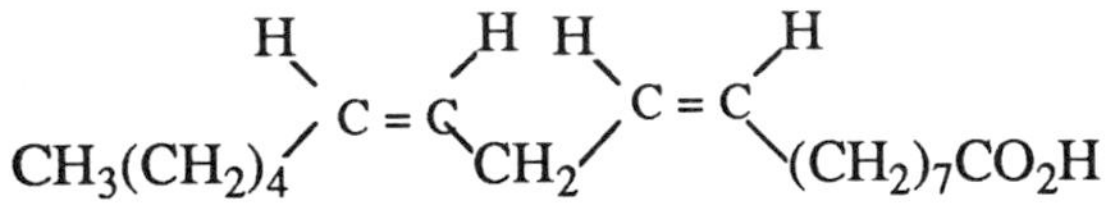

$$CH_3(CH_2)_4CH{=}CHCH_2CH{=}CH(CH_2)_7CO_2H$$

linoleic acid

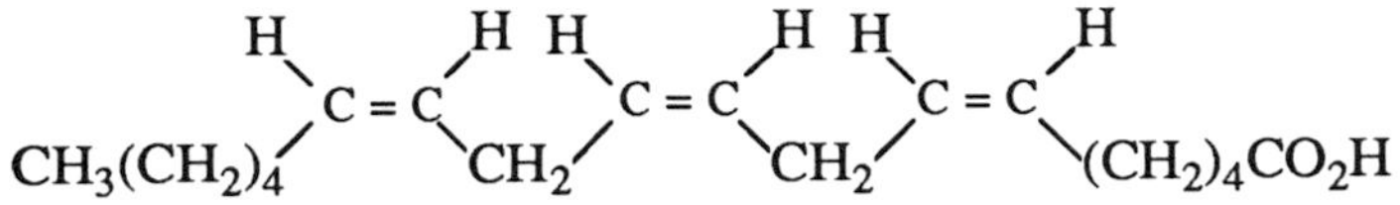

$$CH_3(CH_2)_4CH{=}CHCH_2CH{=}CHCH_2CH{=}CH(CH_2)_4CO_2H$$

γ-linolenic acid

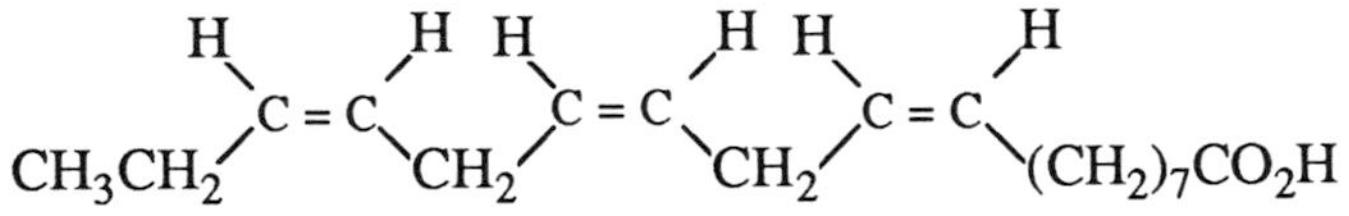

$$CH_3CH_2CH{=}CHCH_2CH{=}CHCH_2CH{=}CH(CH_2)_7CO_2H$$

α-linolenic acid

$$CH_3(CH_2)_4CH{=}CHCH_2CH{=}CHCH_2CH{=}CHCH_2CH{=}CH(CH_2)_3CO_2H$$

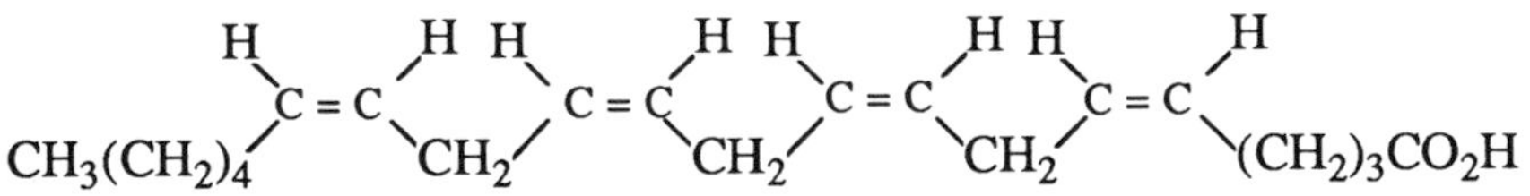

arachidonic acid

different carbon sources, showed that lactose, glycerol or starch gave unsaturated lipids with a high content of γ-linolenic acid in a variety of fungi (J. Sajbidor, M. Certik and S. Dobronova, Biotechnol. Lett, 1988,

10, 347). Other sources of high γ-linolenic acid content were the protozoa Tetrahymena rostrata (Y, Gosselin, G. Lognay and P. Thonart, *ibid.*, 1989, **11**, 423) and microalgae (M. Hirano et al., Appl. Biochem. Biotechnol., 1990, **24-25**, 183). Black currant seed oil was a source for not only γ-linolenic, but also α-linolenic acid, (Z),(Z),(Z)-9,12,15-octadecatrienoic acid, and small amounts of stearidonic acid, (Z),(Z),(Z),(Z)-6,9,12,15-octadecatetraenoic acid (H. Traitler, H.J. Wille and A. Studer, J. Am. Oil Chem. Soc., 1988, **65**, 755). α–Linolenic acid was also isolated from linseed oil (A. Grandgirard et al., J. Am. Oil Chem. Soc., 1987, **64**, 1434). Parinaric acid (9,11,13,15-octadecatetraenoic acid) was isolated from the seed kernels of Parinarium glaberrium (L.A. Sklar, B.S. Hudson and R.D. Simoni, Methods Enzoymol., 1981, **72**, 479). The root bark of Paramacrolobium caeruleum yielded Z- and E-isomers of 7-octadecen-9-ynoic acid and 5-octadecen-7,9-diynoic acid (A. Patil et al., J. Nat. Prod., 1989, **52**, 153). The red marine alga Liagora farinosa lamouroux had (Z),(Z),(Z)-7,9,12-octadecatrien-5-ynoic acid as one of its toxic constituents (J.V. Paul and W. Fenical, Tetrahedron Lett., 1980, **21**, 3327).

Because of the potential economic importance, the natural production of several C_{20} unsaturated acids has been studied by fermentation methods. Peanut oil, when added to the growth medium of arachidonic acid-producing filamentous fungi, enhanced the production of dihomo-γ-linolenic acid, (Z),(Z),(Z)-6,9,12-eicosatrienoic acid (S. Shimizu et al., Agric. Biol. Chem., 1989, **53**, 1437). The mycelial dihomo-γ-linoleic acid content of an arachidonic acid-producing fungus, Mortierella alpina 1S-4, increased with an accompanying marked decrease in the arachidonic acid, (Z),(Z),(Z),(Z)-5,8,11,14-eicosatetraenoic acid, content on cultivation with sesame oil; the resultant mycelia were a rich source of dihomo-γ-linolenic acid (S. Shimizu et al., J. Am. Oil Chem. Soc., 1989, **66**, 237). This same fungus cultivated with potato paste (N. Totani and K. Oba, Appl. Microbiol. Biotechnol., 1988, **28**, 135) or olive and soybean oils (Y. Shinmen et al., *ibid.*, 1989, **31**, 11) produced mainly arachidonic acid. Of various other microorganisms screened for production of arachidonic acid, the fungus strain Mortierella elongata 1S-5 gave the highest productivity (H. Yamada, S. Shimizu and Y. Shinmen, Agric. Biol. Chem., 1987, **51**, 785). The feasibility of growing Porphyridium biomass under various conditions for the production of arachidonic acid was investigated (T.J. Ahern, S. Katoh and E. Sada, Biotechnol. Bioeng., 1983, **25**, 1057; A. Vonshak, Z. Cohen and A. Richmond, Biomass, 1985, **8**, 13). Eicosapentaenoic acid ((Z),(Z),(Z),(Z),(Z)-5,8,11,14,17-isomer) was produced from several strains of Mortierella fungi when grown at low temperature, i.e. not at physiological growth temperature (S. Shimizu et al. Biochem. Biophys. Res. Commun., 1988, **150**, 335; S. Shimizu et al., J. Am. Oil Chem. Soc., 1988, **65**, 1455). The same fungi converted oil containing α-linolenic acid to eicosapentaenoic acid (S. Shimizu et al., *ibid.*, 1989, **66**, 342). Several microalgae were investigated as potential sources for eicosapentaenoic acid (Z. Cohen in "World Conference on Biotechnology for the Fats and Oils Industry: Proceedings" ed. T.H. Applewhite, American Oil Chemists' Society, Champaign, Ill., 1988, p. 285; P. Behrens et al., Novel Microb. Prod. Med. Agric., ed. A.L. Demain, Elsevier, Amsterdam, 1989, p. 253). Also, marine Chlorella minutissima was shown to contain a high amount of eicosapentaenoic acid (A. Seto, H.L. Wang and C.W. Hesseltine, J. Am. Oil Chem. Soc., 1984, **61**, 892).

ω–3–Polyunsaturated fatty acids were found in various fish oils. When cod roe was incubated with porcine pancreatic phospholipase A_2, the liberated fatty acids were mainly ω-3-eicosapentaenoic and docosahexaenoic acids (D.R. Tocher, A. Webster and J.R. Sargent, Biotechnol. Appl. Biochem., 1986, **8,** 83). All (Z) 4,7,10,13,16,19-docosahexaenoic acid was purified from cod liver oil by non-chromatographic means (S.W. Wright, E.Y. Kuo and E.J. Corey, J. Org. Chem., 1987, **52,** 4399). A complete series of even numbered carbon chain polyenic fatty acids with methylene interrupted double bonds having 20-36 carbons were found in phosphatidyl choline from bovine retina (M.I. Aveldano and H. Sprecher, J. Biol.Chem., 1987, **262**, 1180).

Other natural sources that have been shown to contain significant amounts of mixtures of unsaturated fatty acids include Turkish tobacco (T. Chuman and M. Noguchi, Agric. Biol. Chem., 1977, **41**, 1021) and Chinese quince (S. Mihara et al., J. Agric. Food Chem., 1987, **35**, 532). Seven species of herbaceous seed oils (M.S. Ahmad, Fette. Seifen. Anstrichm., 1978, **80**, 353) and 64 genotypes of toria (K.L. Ahuja et al., Plant Foods Hum. Natr., 1989, **39**, 155) have also been investigated.

(c) Methods of Analysis and Purification of Unsaturated Acids

A variety of techniques have been used to analyze unsaturated acids. One of the major ones has been gas-liquid chromatography ("Improved Procedures for the Gas Chromatographic Analysis of Resin and Fatty Acids in Kraft Mill Effluents", National Council of the Paper Industry for Air and Stream Improvement, New York, 1975). A number of stationary phases have been utilized; for example, all the methyl undecynoates and methyl (Z)-undecenoates were studied on both polar and nonpolar columns (M.S.F.L.K. Jie and C.H. Lam, J. Chromatogr., 1974, **97**, 165). Similarly, five trimethylene-interrupted methyl octadecadiynoates were analyzed (M.S.F.L.K. Jie, *ibid*, 1975, **109**, 81), as were all the dimethylene-interrupted methyl octadecadiynoates (C.H. Lam and M.S.F.L.K. Jie, *ibid.*, 1975, **115**, 559) and all the methyl (E),(E)-octadecadienoates (*ibid*, 1976, **121**, 303). Fused-silica capillary gas chromatography has been used to examine the esterified fatty acids from mouse biological samples (D.R. Wing et al., *ibid,* 1986, **368**, 103).

Gas chromatography-mass spectral analysis has also been extensively used for analysis of unsaturated acids. Oleic acid and cis-vaccenic acid have been studied as their pyrrolidine derivatives (K. Satouchi and K. Saito, Biomed. Mass Spectrom., 1979, **6**, 144). The basic profiles of organic acids in urine have been studied (H.M. Liebich and C. Foerst, J. Chromatogr., 1990, **525,** 1). Standard mass spectroscopy using various derivatization techniques has also been used to determine the structure of several organic acids in urine (J. Greter, S. Lindstedt and G. Steen, Adv. Mass Spectrom., 1980, **8B**, 1362).

Numerous NMR techniques and studies have been used for analysis of unsaturated acids. For example, the proton NMR spin-tickling technique has been applied to a series of C-13 labelled acids (J.L. Marshall and R. Seiwell, Org. Magn, Reson., 1976, **8**, 419). C-13 NMR has been used to study acetylenic fatty acids (F.D. Gunstone et al., Chem. Phys. Lipids, 1976, **17**, 1) and conformations of C_{18} unsaturated acids (M. Pons and D. Chapman, Magn. Reson. Chem., 1986, **24**, 612).

Other techniques used to analyze unsaturated acids include Raman spectroscopy (J.E. Davies et al., Chem. Phys. Lipids, 1975, **15**, 48) and thermal analysis (F.I. Khattab, N.A. Al-Ragehy and A.K.S. Ahmad, Thermochim. Acta, 1984, **73**, 47).

The purification of unsaturated acids has been carried out by a variety of physical and chemical methods. A review with 100 references has appeared describing the practical and theoretical aspects of emulsion separation technology and its application to the separation of oleic and stearic acids (N.O.V. Sonntag, Surfactant Sci. Ser., 1988, **28**, 169). Numerous chromatographic techniques have been utilized: isocratic chromatography on a Radial Pak A cartridge (J.P. Roggero and S.V. Coen, J. Liq. Chromatogr., 1981, **4**, 1817), gas-liquid chromatography (D.P. Schwartz and D.G. Gadjeva, J. Am. Oil Chem. Soc., 1988, **65**, 378), and zeolites in a fixed-bed column system (M. Arai, H. Fukuda and H. Morikawa, J. Ferment. Technol., 1987, **65**, 569). Super critical fluid CO_2 techniques have been used to fractionate menhaden oil fatty acid ethyl esters to obtain eicosapentaenoic acid and docosahexaenoic acid (W.B. Nilsson et al., J. Am. Oil Chem. Soc., 1988, **65**, 109).

Crystallization techniques have been used to obtain highly purified unsaturated acids ("Crystallization and Polymorphism of Fats and Fatty Acids" ed. N. Garti and K. Sato, M. Dekker, New York, 1988). Oleic and linoleic acids can be purified by anaerobic low temperature recrystallization from acetonitrile (R.L. Arudi, M.W. Sutherland and B.H.J. Bielski, J. Lipid Res., 1983, **24**, 485). The features of the solvent crystallization of α, β and γ polymorphs of ultra-pure oleic acid were examined in acetonitrile and decane (K. Sato and M. Suzuki, J. Am. Oil Chem. Soc., 1986, **63**, 1356). A five step method for oleic acid purification based on selective crystallization has been developed (E. Mularczyk and J. Drzymala, Sep. Sci. Technol., 1989, **24**, 151). A comparison of fractional crystallization with urea adduct formation for the purification of a series of unsaturated fatty acids (C_{16}, C_{17} and C_{18}) has been conducted (B. Aurousseau and D. Bauchart, J. Am. Oil Chem. Soc., 1980, **57**, 125). Crystallization and urea adduct formation have also been used for isolation of pure oleic, linoleic and linolenic acids and their methyl esters from natural sources (F.D. Gunstone et al., J. Sci. Food Agric., 1976, **27**, 675). A chemical method based on iodolactonization has been developed to purify arachidonic acid, docosahexaenoic acid and eicosapentaenoic acid (E. J. Corey and S.W. Wright, Tetrahedron Lett.,1984, **25**, 2729; *ibid*, J. Org. Chem., 1988, **53**, 5980).

2. Synthesis of Unsaturated Acids

The preparation of unsaturated acids has been carried out by a variety of methods, usually employing standard synthetic techniques. Thus unsaturation can be introduced into a chain by either elimination or reduction; a carboxylic acid group can be added to an unsaturated system, or two smaller units can be coupled to form the desired product. This latter process can be either by alkylation or by some reaction that will form a carbon-carbon double bond in the process. In the sections that follow, these approaches will be demonstrated for specific structural types of unsaturated acids. Also a section is included on the synthesis of labelled unsaturated acids which are important for studying a variety of biological problems.

(a) Ethylenic Acids

One of the most utilized methods of synthesizing ethylenic acids has been formation of a carbon-carbon double bond by Wittig olefination or some modification of this reaction. For example, the shortest total synthesis of methyl arachidonate involved C_3 homologations by Wittig reactions to give the 4 *cis* double bonds starting from hexanal (J. Viala and M. Santelli, J. Org. Chem., 1988, **53**, 6121). In general, the phosphonium ylide can either be part of the alkyl portion:

$$RCH{=}PPh_3 + OHC(CH_2)_mCO_2R' \rightarrow RCH{=}CH(CH_2)_mCO_2R'$$

(A. Starratt, Chem. Phys. Lipids, 1976, **16**, 215; T. Shirakawa et al., Dev. Food Sci., 1988, **18**, 915) or attached to a fragment containing a carboxylic ester (R. Chong, R.R. King and W.B. Whalley, J. Chem. Soc. C, **1971**, 3566; M.G. Hussain and F.D. Gunstone, Bangladesh J. Sci. Ind. Res., 1977, **12**, 215; *ibid*, 1978, **13**, 51) or a carboxylic acid:

$$RCOR' + Ph_3P{=}CH(CH_2)_mCO_2R'' \rightarrow RR'C{=}CH(CH_2)_mCO_2R''$$

(U.T. Bhalerao, J.J. Plattner and H. Rappoport, J. Am. Chem. Soc., 1970, **92**, 3429; D.W. Knight and D. Ojhara, Tetrahedron Lett., 1981, **22**, 5101; *ibid*, J. Chem. Soc., Perkin Trans. 1, **1983**, 955; P.L. Mena, O. Pilet and C. Djerassi, J. Org. Chem., 1984, **49**, 3260; J. Ackroyd, S. Jones and F. Scheinmann, J. Chem. Res., Synop., **1987**, 344; H. Kaga, M. Miura and K. Orito, J. Org. Chem., 1989, **54**, 3477).

A similar Wittig reaction employed $RO_2CCH{=}P(OH)_3$ and (E)-$RO_2CCH{=}CHCH{=}P(OH)_3$ as the ylides (B.Vig, R. Kanwar and S. S. Dahiya, J. Indian Chem. Soc., 1979, **56**, 935). On the other hand, the Horner-Wittig reaction always has the phosphonate salt on the fragment bearing the carboxylate group to give α, β-unsaturated esters or acids:

$$RCOR' + (R''O)_2PO\bar{C}HCO_2R''' \longrightarrow RR'C{=}CHCO_2R'''$$

(B. Vig and A.C. Mahajan, Indian J. Chem., 1972, **10**, 564; O.P. Vig et al., *ibid*, 1975, **13**, 1358; L. Lombardo and R.J.K. Taylor, Synth, Commun., 1978, **8**, 463; *ibid*, Synthesis, **1978**, 131; S. D. Sharma et al., Indian J. Chem., Sect. B., 1979, **18B**, 81; P. Coutrot, M. Snoussi and P. Savignac, Synthesis, **1978**, 133; J. M. Clough and G. Pattenden, Tetrahedron, 1981, **37**, 3911: D. Brittelli, J. Org. Chem., 1981, **46**, 2514; P. Coutrot and A. Ghribi, Synthesis, **1986**, 790).

Another much utilized method has been the malonic ester synthesis or some variation of it. Here the olefinic portion was in the alkylating agent and the carboxyl group was derived from the malonic ester portion:

$$RX + CH_2(CO_2R')_2 \xrightarrow{\text{base}} RCH(CO_2R')_2 \xrightarrow[\text{2) decarboxylation}]{\text{1) hydrolysis}} RCH_2CO_2R'$$

The alkylating agents have been methanesulfonates (F. Spener and H. K. Mangold, Chem. Phys. Lipids, 1973, **11**, 215; R. Van der Linde et al., Recl. Trav. Chim. Pays-Bas, 1975, **94**, 257) or halides (O.P. Vig et al., Indian J. Chem., Sect. B., 1978, **16B**, 114; *ibid*, 1979, **17B**, 558; *ibid*, 1981, **20B**, 970; T. Fukuda et al., Bull, Chem. Soc. Japan, 1981, **54**, 3530; G. Cardinale, J. A. M. Laan and J. P. Ward, Recl. Trav. Chim. Pays-Bas, 1987, **106**, 62). The Knoevenagel condensation has been used to produce either α, β-unsaturated acids (M. Muramatsu et al., J. Label. Compounds, 1972, **8**, 305) or β,γ-unsaturated acids (N. Ragoussis, Tetrahedron Lett, 1987, **28**, 93) depending on the reaction conditions:

$$RR'CHCHO + CH_2(CO_2H)_2 \longrightarrow RR'CHCH{=}CHCO_2H \text{ or } RR'C{=}CHCH_2CO_2H$$

The enolate anion of a variety of substituted acetate esters or their equivalents have been reacted with carbonyl compounds to give α,β-unsaturated esters or acids. The dianion of $HSCH_2\,CO_2Et$ reacted with aldehydes and upon subsequent treatment with ethyl chloroformate and a trivalent phosphorus compound gave E isomers of α,β-unsaturated esters (S. Matsui, Bull. Chem. Soc. Japan, 1984, **57**, 426). The enolates of acylsilanes have been reacted with aldehydes to give α, β-unsaturated acylsilanes which could be converted to the corresponding carboxylic acids (J. A. Miller and G. S. Zweifel, J. Am. Chem. Soc., 1981, **103**, 6217):

$$RCH(SiMe_3)COSiMe_3 \xrightarrow[\text{2) R'CHO}]{\text{1) base}} R'CH{=}CRCOSiMe_3 \longrightarrow R'CH{=}CRCO_2H$$

α–Alkylacrylic acids have been prepared by the reaction of dianions of carboxylic acids with formaldehyde followed by pyrolysis (P.E. Pfeffer, E. Kinsel and L. S. Silbert, J. Org. Chem., 1972, **37**, 1256) or by masking the carboxylate group as an oxazoline (S. Serota et al., *ibid*, 1981, **46**, 4147):

$$RCH_2\text{-(4,4-dimethyl-2-oxazolinyl)} \xrightarrow{(CH_2O)_x} HOCH_2\text{-}RCH\text{-(4,4-dimethyl-2-oxazolinyl)} \xrightarrow[\text{2) hydrolysis}]{\text{1) heat}} CH_2{=}CRCO_2H$$

Alkylation of ester dienolates can take place at either the α-position or γ-position depending upon conditions. Thus lithium dienolates in one study gave exclusively α-alkylation while the copper dienolates gave γ-alkylation (J. A. Katzenellenbogen and A. L. Crumrine, J. Am. Chem. Soc., 1976, **98**, 4925). Use of palladium-phosphine complexes to catalyze the alkylation of lithium ester dienolates with allylic chlorides or bromides gave mainly α-alkylation (G. Mignani et al., Tetrahedron Lett., 1989, **30**, 2383). When esters of 3-methylglutaconates were condensed with aldehydes and then decarboxylated, they gave 5-alkyl-3-methyl-2,4-pentadienoates (C.A. Hendrick et al., J. Org. Chem., 1975, **40**, 1). Treatment of 1,3-dienes with carboxylic acids in the presence of sodium or lithium naphthalenide and TMEDA gave γ,δ-unsaturated acids (T. Fujita et al., Synthesis, **1979**, 310; T. Fujita, J. Chem. Technol. Biotechnol., 1982, **32**, 476):

$$RR'CHCO_2H + R''CH{=}C(R''')CH{=}CH_2 \longrightarrow R''CH_2R'''C{=}CHCH_2CRR'CO_2H$$

Another approach to ethylenic acids has been the opening of lactone rings by organometallic reagents by S_N2' or S_N2 substitution. When an alkenyl group was substituted on the oxygen-bearing carbon of a lactone, reaction with an organocuprate or copper-catalyzed Grignard reagent resulted in a substituted ethylenic acid by S_N2' reaction (B. M. Trost and T. P. Klun, J. Org. Chem., 1980, **45**, 4256; T. Fujisawa et al., Tetrahedron Lett., 1982, **23**, 3583; *ibid*, Chem. Lett., **1981**, 1307; M. Kawashima, T. Sato and T. Fujisawa, Bull. Chem. Soc. Japan, 1988, **61**, 3255):

$$CH_2{=}CH{-}\overset{(CH_2)_n}{\langle\;\;\rangle}C{=}O \xrightarrow{RM} RCH_2CH{=}CH(CH_2)_nCO_2H$$

n = 1, 2, or 3

The alkenyl moiety could also be endocyclic (F. Fujisawa, K. Umezu and M. Kawashima, Chem. Lett., **1984**, 1795). Similarly, allenic acids were formed from β-ethynyl-β-propiolactones on reaction with copper-catalyzed Grignard reagents (T. Sato, M. Kawashima and T. Fujisawa, Tetrahedron Lett., 1981, **22**, 2375). Diketene has been reacted with a variety of organometallic reagents with Ni, Co or Pd catalysis to give 3-substituted-3-butenoic acids (K. Itoh, M. Fukui and Y. Kurachi, J. Chem. Soc., Chem. Commun. **1977**, 500; T. Fujisawa et al., Bull. Chem. Soc. Japan, 1982, **55**, 3555; Y. Abe et al., Chem Pharm. Bull., 1983, **31**, 1108; Y. Abe et al., *ibid*, 4346):

$$\text{(diketene: } CH_2{=}\text{oxetan-2-one)} \xrightarrow[\text{catalyst}]{RM} CH_2{=}CRCH_2CO_2H$$

If the organometallic reagent, cuprate or copper-catalyzed Grignard, contained an ethylenic unit, then β-propiolactones have been opened to give unsaturated acids. The position of the double bond depended upon the structure of the organometallic reagent. Alkenyl cuprates gave 4-alkenoic acids (T. Fujisawa et al., Chem Lett., **1980**, 1123), while homoallylic reagents gave 6-alkenoic acids (T. Sato et al., Tetrahedron Lett., 1980, **21**, 3377; T. Fujisawa, *ibid*, 2553).

Conjugate addition of organometallic reagents to α,β-unsaturated carboxylic acid derivatives have been used to produce ethylenic acids. Use of vinyl organometallics gave 4-alkenoic acids or derivatives (E. Piers et al., Can. J. Chem., 1984, **62**, 1; W. Oppolzer and T. Stevenson, Tetrahedron Lett., 1986, **27**, 1139), while use of allyl silanes gave 5-alkenoic acid derivatives (A. Jellal and M. Santelli, *ibid*, 1980, **21**, 4487). Enol phosphates of β-keto esters reacted with organocuprates to give 3-substituted-2-alkenoates (F. W. Sum and L. Weiler, Tetrahedron, Suppl.,

1981, 303) as did α,β-alkynyl esters (B. S. Pitzele, J. S. Baran and D. H. Steinman, J. Org. Chem., 1975, **40**, 269).

A number of organometallic reagents underwent coupling reactions to give ethylenic acids. 1-Octene reacted with manganese(III) acetate in the presence of a copper(II) catalyst to give mainly 4-decenoic acid together with a small amount of 3-decenoic acid (W. J. De Klein, Recl. Trav. Chim. Pays-Bas, 1975, **94**, 151). Allyl 3-butenoate was rearranged by nickel and rhodium catalysts to mixtures of 3,6- and 2,6-heptadienoic acids; the isomeric ratio was dependent on the solvent, the temperature and the ligands (G.P. Chiusoli, G. Salerno and F. Dallatomasina, J. Chem. Soc., Chem. Commun., **1977**, 793; U. Bersellini et al., Fundam. Res. Homogeneous Catal., 1979, **3**, 893). 1, 3-Dienes reacted under rhodium catalysis with 3-alkenoic acids to give 3,6-dienoic acids which are not readily accessible by other means (G. Salerno, F. Gigliotti and G. P. Chiusoli, J. Organomet. Chem., 1986, **314**, 231); allenes reacted similarly to give dienoic acids with the major product derived from reaction at the terminal carbon atom (G. Salerno et al., *ibid*, 1986, **317**, 373):

$$Me_2C{=}C{=}CH_2 + CH_2{=}CHCH_2CO_2H \xrightarrow{\text{Rh cat.}} \begin{array}{c} Me_2C{=}CHCH_2CH{=}CHCH_2CO_2H \\ + \\ Me_2C{=}CMeCH{=}CHCH_2CO_2H \end{array}$$

Grignard reagents of ω-unsaturated alkyl halides reacted with the bromomagnesium salt of ω-bromo fatty acids to give ω-unsaturated acids (S. B. Mirviss, J. Org. Chem., 1989, **54**, 1948):

$$CH_2{=}CH(CH_2)_nMgX + Br(CH_2)_mCO_2MgBr \rightarrow CH_2{=}CH(CH_2)_{m+n}{-}CO_2H$$

Organometallic reagents and other methods have been used to introduce the carboxylate group onto an unsaturated framework. The most straight forward approach has been the reaction of a Grignard reagent with CO_2 (M.I. Dawson et al., J. Med. Chem., 1977, **20**, 1396; C.-L. Yeh et al., Tetrahedron Lett., **1977**, 4257), although other organometallics have been utilized such as organolithium reagents (G. Cahiez, D. Bernard and J.F. Normant, Synthesis, **1976**, 245), organocopper reagents (A. Alexakis and J. F. Normant, Tetrahedron Lett., 1982, **23**, 5151; M. Furber, R. J. K. Taylor and S. C. Burford, *ibid* , 1985, **26**, 3285), and organoboron reagents (M. Deng, Y. Tang and W. Xu, *ibid*, 1984, **24**, 1797):

$$RM + CO_2 \rightarrow RCO_2H$$

Allenic lithium reagents also reacted with CO_2 to give 2, 3-alkadienoic acids (J.C. Clinet and G. Linstrumelle, Synthesis, **1981**, 875). Electrochemical methods have also been utilized for reactions with CO_2. A series of α-substituted acrylic acids were obtained when terminal alkynes were electrochemically reduced in the presence of nickel catalyst and CO_2 (E. Dunach and J. Perichon, J. Organomet. Chem., 1988, **352**, 239; E. Dunach, S. Derien and J. Perichon, *ibid*, 1989, **364**, C33):

$$RC{\equiv}CH \xrightarrow[\text{electro. red.}]{CO_2,\ \text{Ni cat.}} CH_2{=}CRCO_2H$$

Also the electrochemical reduction of butadiene in the presence of CO_2 in acetonitrile gave 3-pentenoic acid along with other products (W. J. M. Van Tiborg and C. J. Smit, Recl.: J. R. Neth. Chem. Soc., 1981, **100**, 437).

Carbonylation, with CO, could also introduce a carboxylate group. When 1,3-butadiene was reacted with CO in the presence of a palladium catalyst in tert-butanol, tert-butyl 3,8-nonadienoate was obtained (J. Tsuji and H. Yasuda, J. Organometal. Chem., 1977, **131**, 133):

$$CH_2{=}CHCH{=}CH_2 \xrightarrow[\text{t-BuOH}]{\text{CO, Pd cat.}} CH_2{=}CH(CH_2)_3CH{=}CHCH_2CO_2\text{t-Bu}$$

Potassium alkoxides have been directly carbonylated by reaction with CO at high temperature and pressure (V. Rantenstrauch, Helv. Chim, Acta, 1987, **70**, 593):

$$RC(OK)MeCH{=}CH_2 \xrightarrow[\substack{120 - 130^\circ \\ 425 - 440\ \text{bar}}]{CO,\ C_6H_6} RC(CO_2K)MeCH{=}CH_2$$

Another one-carbon elongation process for introducing the carboxylic acid group involved S_N2 displacement of a halide by cyanide followed by hydrolysis (L. Ahlquist et al., Chem. Scr., 1971, **1**, 237):

$$RX \xrightarrow{KCN} RCN \xrightarrow{\text{hydrolysis}} RCO_2H$$

Also the Arndt-Eistert reaction allowed one-carbon elongation of an ethylenic acid (T. Hudlicky and J. P. Sheth, Tetrahedron Lett., **1979**, 2667).

Another approach to ethylenic acids has been the use of [3.3] sigmatropic rearrangements, usually some variant of the Claisen rearrangement. The ester enolate, as the free anion or the silylketene acetal, Claisen process gave a high degree of stereoselectivity (R. E.

Ireland, H. R. Mueller and A.K. Willard, J. Am. Chem. Soc., 1976, **98**, 2868; J.A. Katzenellenbogen and K.J. Christy, J. Org. Chem., 1974, **39**, 3315; J. Boyd, W. Epstein and G. Frater, J. Chem. Soc., Chem. Commun., **1976**, 380; K-K. Chan et al., J. Org. Chem., 1976, **41**, 3497):

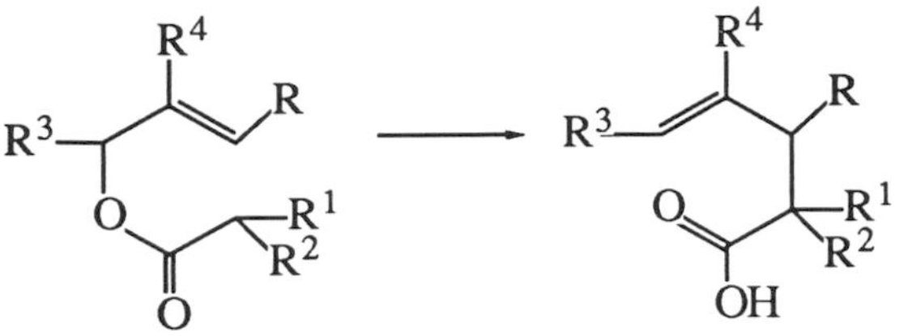

Bromoacetals, prepared from an allylic alcohol, upon treatment with base gave allyl 1-ethoxyvinyl ethers which underwent Claisen rearrangement (P.S. Lidbetter and B.A. Marples, Synth. Commun., 1986, **16**, 1529). The aza-Claisen rearrangement has been employed to enantioselectively prepare 3-substituted 4-pentenoic acids using oxazolidines as chiral auxillary agents (M.J. Kurth and O.H.W. Decker, J. Org. Chem., 1985, **50**, 5769). Allenic esters could be prepared using the ortho-ester Claisen rearrangement of propargylic alcohols (K. Mori, T. Nukada and T. Ebata, Tetrahedron, 1981, **37**, 1343):

$$\mathrm{RCHOHC{\equiv}CH} \xrightarrow[\mathrm{EtCO_2H,\ heat}]{\mathrm{MeC(OEt)_3}} \mathrm{RCH{=}C{=}CHCH_2CO_2Et}$$

A number of elimination procedures have been used to introduce the carbon-carbon double bond into a chain of a carboxylic acid. α,β-Unsaturated acids and esters could be prepared by arylselenylation, oxidation and elimination of the selenoxide (K.B. Sharpless, R.F. Lauer and A.Y. Teranishi, J. Am. Chem. Soc., 1973, **95**, 6137; J. Tsuji et al., Tetrahedron Lett., **1977**, 1917):

$$\mathrm{RCH_2CH_2CO_2R'} \longrightarrow \mathrm{RCH_2CH(SeAr)CO_2R'} \longrightarrow \mathrm{RCH{=}CHCO_2R'}$$

A modification of this approach used saturated aldehydes where the aldehyde group was oxidized *in situ* to a carboxylic acid (F. Outurquin and P. Claude, Synthesis, **1989**, 690). Pyrolysis of α-sulfoxide acids or esters also gave α,β-unsaturated compounds (J.Nokami et al., Tetrahedron Lett., 1980, **21**, 4455; T. Nakai et al., Chem. Lett., **1981**, 1289; H. Ishibashi et al., Synth. Commun., 1989, **19**, 857). Base promoted elimination of halides and alcohol derivatives gave ethylenic acids; chlorides (S. Dolezal, Collect. Czech. Chem. Commun., 1972, **37**,

3117), iodides (D.R. Howton, Org. Prep. Proced. Int., 1974, **6**, 175; A.V.R. Rao, J.S. Yadav and C.S. Rao, Tetrahedron Lett., 1986, **27**, 3297) and tosylates (T. Nakai et al., *ibid*, 1981, **22**, 69) have been employed. Stereospecific elimination of $Ph_2PO_2^-$ from lactones substituted with a diphenylphosphine oxide group gave ethylenic acids (D. Levin and S. Warren, J. Chem. Soc., Perkin Trans. 1, **1988**, 1799):

$$RCH(POPh_2)CH\text{(lactone: }O\text{–}C(=O)\text{–}CH_2\text{)}(CH_2)_m \longrightarrow RCH{=}CH(CH_2)_{m+1}CO_2H$$

β-Lactones underwent Lewis acid catalyzed rearrangement to give 2,3-dialkyl-3-alkenoic acids (T.H. Black and S.L. Maluleka, Synth. Commun., 1989, **19**, 2885):

$$\text{β-lactone (Me, H, R', }CH_2R\text{)} \xrightarrow{MgBr_2} RCH{=}CHR'CHMeCO_2H$$

Base-promoted fragmentation of oxothiolanecarboxylates gave α,β-unsaturated esters (P.G. Baraldi, J. Chem. Soc., Perkin Trans. 1, **1984**, 2501):

$$\text{oxothiolane (}CO_2Me\text{, R', R)} \xrightarrow{\text{base}} RCH{=}CHR'CO_2Me$$

Palladium-catalyzed fragmentation of β-acetoxy acids introduced a carbon-carbon double bond (B.M. Trost and J.M.D. Fortunak, Tetrahedron Lett., 1981, **22**, 3459):

$$HO_2CCRR'CH(OAc)CR''{=}CH \sim CO_2Et \xrightarrow{\text{Pd cat.}} RR'C{=}CHCR''{=}CH \sim CO_2Et$$

Ring opening of siloxycycloalkyl hydroperoxides, obtained from cycloalkenyl silyl ethers, gave ω-unsaturated acids (I. Saito et al., *ibid*, 1983, **24**, 4439):

$$\text{HOO, OSiR}_3\text{-cyclo-}(CH_2)_n \xrightarrow[\text{FeSO}_4]{\text{Cu(OAc)}_2} CH_2{=}CH(CH_2)_{n+1}CO_2H$$

Allylic sulfones were converted to dienoic acids upon treatment with base (P. A. Grieco and D. Boxler, Synth. Commun. 1975, **5**, 315):

$$RR'C{=}CHCH_2SO_2CH_2CO_2Me \xrightarrow[CCl_4]{KOH} RR'C{=}CHCH{=}CHCO_2H$$

The dianion of saturated acids was dehydrogenated with DDQ to give α, β -unsaturated acids (G. Cainelli, E. Cardillo and A.U. Ronchi, J. Chem. Soc., Chem. Commun., **1973**, 94).

Another general route to ethylenic acids has employed the stereospecific reduction of a carbon-carbon triple bond. *Cis* (Z) double bonds have been prepared by partial hydrogenation, usually using a Lindlar catalyst, a poisoned palladium system:

$$RC{\equiv}C(CH_2)_mCO_2R' \xrightarrow[\text{Lindlar cat.}]{H_2} \underset{H}{\overset{R}{>}}C{=}C\underset{H}{\overset{(CH_2)_mCO_2R'}{<}}$$

(G. I. Myagkova et al., Zh. Org. Khim., 1970, **6**, 1568; W. H. Kunau, H. Lehmann and G. Gross, Hoppe-Seyler's Z. Physiol. Chem., 1971, **352**, 542; D. E. Ames and S. H. Binns, J. Chem. Soc., Perkin Trans. 1, **1972**, 255; L. Heslinga, H. J. J. Pabon and D. A. Van Dorp. Recl. Trav. Chim. Pays-Bas, 1973, **92**, 287; R. Klok, N. J. G. Egmond and H. J. J. Pabon, *ibid*, 1974, **93**, 222; G. J. N. Egmond, H. J. J. Pabon and D. A. Van Dorp, *ibid*, 1977, **96**, 172; L. A. Yakusheva et al., Bioorg. Khim., 1982, **8**, 422; H. A. Parish, Jr., R. D. Gillion and W. P. Purcell, Lipids, 1983, **18**, 894). Hydroboration has also been used for reduction of the triple bond, but usually the catalytic procedure gave higher yields (W. J. De Jarlais and E. A. Emken, Lipids, 1986, **21**, 662). The *trans* (E) double bonds have been prepared from chemical reduction of the carbon-carbon triple bond, usually using lithium or sodium in liquid ammonia:

$$RC{\equiv}C(CH_2)_mCO_2R' \xrightarrow[NH_3(l)]{\text{Li or Na}} \underset{H}{\overset{R}{>}}C{=}C\underset{(CH_2)_mCO_2H}{\overset{H}{<}}$$

(F. D. Gunstone and M. LieKenJie, Chem. Phys. Lipids, 1970, **4**, 1).

Reductive cleavage of 2-substituted-1,3-cyclohexanediones has been used to prepare ethylenic acids (N. Polgar et al., J. Chem. Soc. C., **1971**, 870):

$$\text{2-}(CH_2CH{=}CHR)\text{-5-}CH_3\text{-1,3-cyclohexanedione} \xrightarrow[N_2H_4]{NaOH} RCH{=}CH(CH_2)_4CH(CH_3)CH_2CO_2H$$

Many oxidation procedures have been used for the formation of the carboxylic acid group. Primary alcohols have been oxidized to the corresponding acid using Jones reagent (H. Harada and K. Mori, Agric. Biol. Chem., 1989, **53**, 1439) or electrochemically at a nickel hydroxide electrode (J. Kaulen and H. J. Schaefer, Synthesis,**1979**, 513). Aldehydes have been oxidized to the corresponding acid using silver oxide--Tollen's oxidation (T. Kajiwara et al., Agric. Biol. Chem., 1977, **41**, 1481), an ether-aqueous chromic acid two phase system (C. M. Paleos and N. Mimicos, J. Colloid Interface Sci., 1978, **66**, 595), and hydrogen peroxide catalyzed by molybdenum (B. M. Trost and Y. Masuyama, Tetrahedron Lett., 1984, **25**, 173). α, β-Unsaturated aldehydes have been oxidized to the corresponding acids with sodium chlorite-hydrogen peroxide (B. S. Bal., W. E. Childers, Jr. and H. W. Pinnick, Tetrahedron, 1981, **37**, 2091; E. Dalcanale and F. Montanari, J. Org. Chem., 1986, **51**, 567). Hydroboration-oxidation of silyl-substituted enynes afforded tetrasubstituted olefins containing the carboxyl group (G. Zweifel and W. Leong, J. Am. Chem. Soc., 1987, **109**, 6409).

A number of oxidative cleavage procedures have been utilized for degrading systems to ethylenic acids. Fatty acids containing remote unsaturation have been converted to the next lower homologue by α-hydroxylation and then oxidative cleavage using chromium trioxide-sodium periodate (T.A. Hase and K. McCoy, Synth. Commun. 1979, **9**, 63):

$$RCH_2CO_2H \xrightarrow[O_2]{LDA} RCH(OH)CO_2H \xrightarrow[ACOH]{CrO_3\text{ - }NaIO_4} RCO_2H$$

Allylcarboxylic acids have been prepared by oxidative bisdecarboxylation of substituted α-hydroxymalonic acids; either aqueous sodium periodate or cerium ammonium nitrate have been used as the oxidant (M.F. Salomon, S.N. Pardo and R.G. Salomon, J. Am. Chem. Soc., 1980, **102**, 2473; *ibid,* 1984, **106**, 3797). Acylcarbalkoxytriphenylphosphoranes have been oxidately cleaved to carboxylic acids by alkaline sodium perchlorate (M.P. Cooke, Jr., J. Org. Chem., 1983, **48**, 744):

$$RCOC(=PPh_3)CO_2Et \xrightarrow{\text{alk. NaOCl}} RCO_2H$$

Oxidative electrolysis of a cyclic γ-ketal carboxylic acid gave an olefinic acid (P.G.M. Wuts and M.C. Cheng, J. Org. Chem., 1986, **51**, 2844):

CO_2H R O O

$$\longrightarrow CH_2 = CR(CH_2)_3CO_2H$$

Bissulfenylation of allylic thiolcarbamates followed by ethanolysis gave α,β-unsaturated esters (T. Nakai, T. Mimura and T. Kurokawa, Tetrahedron Lett., **1978**, 2895):

$$RR'C = CHCH_2SCONMe_2 \xrightarrow[Me_2S_2]{LDA} \xrightarrow{\text{ethanolysis}} RR'C = CHCO_2Et$$

Electrolysis of 10-undecenoic acid in a mixture with monomethyl succinic acid under netrual conditions followed by hydrolysis gave 12-tridecenoic acid (Naser-ud-Din, Pak. J. Sci. Ind. Res., 1976, **19**, 132).

(b) Acetylenic Acids

Alkynyl coupling reactions have been used extensively for the preparation of acetylenic acids. Cadiot-Chodkiewicz coupling of 1-haloacetylenes and terminal alkynes gave conjugated diynoic acids; the carboxylic acid group could either be on the haloacetylene (T.V. Kulik and A.Y. Il'chenko, Zh. Org.Khim. 1989, **25**, 728) or on the terminal alkyne (G.I. Myagkova et al., *ibid.*, 1970, **6**, 1568; D.E. Ames and S.H. Binns, J. Chem. Soc., Perkin Trans. 1, **1972**, 255; A. Singh and J.M. Schnur, Synth. Commun., 1986, **16**, 847):

$$RC{\equiv}CX + HC{\equiv}C(CH_2)_nCO_2H$$
$$RC{\equiv}CH + XC{\equiv}C(CH_2)_nCO_2H$$
$$\longrightarrow RC{\equiv}C\text{-}C{\equiv}C(CH_2)_nCO_2H$$

Oxidative coupling of two terminal alkynes, one containing a carboxylate group, also gave conjugated diynoic acids (S.G. Morris, J. Am. Oil Chem. Soc., 1971, **48**, 376). Metal salts of terminal acetylenes could react by S_N2 displacement of alkyl halides to give acetylenic acids; the carboxylate group could be either on the acetylene portion or the alkyl halide portion (W.H. Kunau, H. Lehmann and R. Gross, Hoppe-Seyler's Z. Physiol. Chem, 1971, **352**, 542; W. Bos and H.J.J. Pabon, Recl.

Trav. Chim. Pays-Bas, 1980, **99**, 141; N. Gilman and B. Holland, Synth. Commun., 1974, **4**, 199; R.I. Fryer, N.W. Gilman and B.C. Holland, J. Org. Chem., 1975, **40**, 348; K. Eiter, et al., Justus Liebigs Ann. Chem. **1978**, 658; W.J. DeJarlais and E.A. Emken, Synth. Commun., 1980, **10**, 653; L.A. Yakusheva et al., Bioorg. Khim., 1982, **8**, 422; H.A. Parish, Jr., R.D. Gilliom and W.P. Purcell, Lipids, 1983, **18**, 894; Y.Y. Belosludtsev et al., Bioorg. Khim., 1988, **12**, 1425):

$$\left.\begin{array}{l} RCH_2C{\equiv}CM + X(CH_2)_nCO_2M \\ RCH_2X + MC{\equiv}C(CH_2)_nCO_2M \end{array}\right\rangle \longrightarrow RCH_2C{\equiv}C(CH_2)_nCO_2H$$

β-Propiolactones reacted with alkynyl aluminum reagents to give 4-alkynoic acids (M. Shinoda et al., Tetrahedron Lett., 1986, **27**, 87):

$$\text{(R'-substituted }\beta\text{-propiolactone)} \xrightarrow[\text{PhMe, -35°}]{RC{\equiv}CAlMe_2} RC{\equiv}CCHR'CH_2CO_2H$$

α,β-Unsaturated acyl cyanides reacted with alkynyl silanes in the presence of titanium tetrachloride to give 3-alkynylacyl cyanides which could be hydrolyzed to the 3-alkynoic acids (A. Jellal, J.P. Zahra and M. Santelli, *ibid*., 1983, **24**, 1395).

Metal salts of terminal acetylenes reacted with CO_2 to give 2-alkynoic acids (E.J. Corey and P.L. Fuchs, *ibid,* **1972**, 3769; W.N. Smith and E.D. Kuehn, J. Org. Chem. 1973, **38**, 3588; G. Friour et al., Bull. Soc. Chim, Fr., **1979**, 515):

$$RC{\equiv}CM \xrightarrow{CO_2} RC{\equiv}CCO_2H$$

The carboxylate group of non-conjugated acetylenic acids could be introduced by cyanide displacement of a halide containing the carbon-carbon triple bond followed by hydrolysis (F.D. Gunstone and M. Lieken Jie, Chem. Phys. Lipids, 1970, **4**, 1; G. Holan and D.F. O'Keefe, Tetrahedron Lett., **1973**, 673). The malonic ester synthesis could also be used to prepare non-conjugated acetylenic acids (R. Klok, G.J.N. Egmond and H.J.J. Pabon, Recl. Trav. Chim. Pays-Bus, 1974, **93**, 222). Alkylation of t-butyl lithioacetate with iodides containing methylene-interrupted acetylenic bonds gave polyacetylenic acids (W. Bos and H.J.J. Pabon, *ibid*, 1980, **99**, 141).

A modified Wittig reaction using an α-iodophosphonium ylide could be used to prepare 2-alkynoic esters which could be hydrolyzed to the corresponding acids (J. Chenault and J.F.E. Dupin, Synthesis, **1987**, 498):

$$RCHO + Ph_3P = C(I)CO_2Et \xrightarrow[\text{2) hydrolysis}]{\text{1) } K_2CO_3} RC{\equiv}CCO_2H$$

Vicinial dibromides have been dehydrobrominated with base to give acetylenic acids (L.J. Zakhrkin and I.M. Churilova, Izv. Akad. Nauk SSSR, Ser. Khim., **1984**, 2635; P. Vinczer, Org. Prep. Proced. Int., 1989, **21**, 232). Treatment of 4,4-dihalopyrazolones, derived from β-keto esters, with dilute aqueous base gave 2-alkynoic acids (A. Silveira, Jr. et al., J. Am. Oil Chem. Soc. 1971, **48**, 661):

$$RCOCH_2CO_2R' \longrightarrow \longrightarrow \text{(4,4-dihalopyrazolone: R–C=N–NH–C(=O)–CX}_2\text{)} \xrightarrow{\text{base}} RC{\equiv}CCO_2H$$

Careful control of the reaction conditions allowed oxidation of primary alkynols to alkynoic acids using chromium reagents (B.C. Holland and N.W. Gilman, Synth. Commun., 1974, **4**, 203; G. Just, et al., *ibid*, 1979, **9**, 613). The haloform reaction has been used to oxidize α-acetylenic methyl-substituted ketones and secondary alcohols to 2-alkynoic acids (L.I. Vereshchagin et al., Izv. Vyssh. Ucheb. Zaved., Khim. Khim. Teknol., 1970, **13**, 214).

(c) Enynoic Acids

Carboxylic acids containing both carbon-carbon double and triple bonds, enynoic acids, have been prepared by the same methods used for synthesis of ethylenic and acetylenic acids. Organometallic reagents have been coupled with alkyl halides, where the carboxylate group could be on either component. The organometallic reagent was usually a metalated terminal acetylene (L. Heslinga, H.J.J. Pabon and D.A. Van Dorp, Recl. Trav. Chim. Pays-Bas, 1973, **92**, 287; R. Sood, M. Nagasawa and C.J. Sih, Tetrahedron Lett., **1974**, 423; T. Otsuki, R.F. Brooker and M.O. Funk, Lipids, 1986, **21**, 178; W.J. DeJarlais and E.A. Emken, *ibid,* 662):

$$RC{\equiv}CM + BrCH_2CH{=}CH(CH_2)_nCO_2H \longrightarrow RC{\equiv}CCH_2CH{=}CH(CH_2)_nCO_2H$$

$$RCH{=}CH(CH_2)_nBr + MC{\equiv}C(CH_2)_mCO_2H \longrightarrow RCH{=}CH(CH_2)_nC{\equiv}C(CH_2)_mCO_2H$$

Alkylation of the lithium dianion of 2-butynoic acid with alkenyl halides followed by esterification gave enynoic esters (C.C. Shen and C. Ainsworth, Tetrahedron Lett., **1979**, 83):

$RR'C{=}CHCH_2X + LiCH_2C{\equiv}CCO_2Li \xrightarrow{} \xrightarrow{R''X} RR'C{=}CHCH_2CH_2C{\equiv}CCO_2R''$

Coupling of metal acetylides with vinyl halides gave conjugated enynoic acids (G. Struve and S. Seltzer, J. Org. Chem; 1982, **47**, 2109; V. Ratovelomanana and G. Linstrumelle, Tetrahedron Lett., 1984, **25**, 6001):

$$RC{\equiv}CM + XCH{=}CH(CH)_nCO_2Me \longrightarrow RC{\equiv}CCH{=}CH(CH_2)_nCO_2Me$$

$$RCH{=}CHX + MC{\equiv}C(CH_2)_nCO_2Me \longrightarrow RCH{=}CHC{\equiv}C(CH_2)_nCO_2Me$$

Vinyl tin reagents coupled with allenic halides to give methylene-separated enynoic acids after hydrolysis (E.J. Corey et al., *ibid*, 2419):

$$RCH{=}CHSnBu_3 + CH_2{=}C{=}CBr(CH_2)_nC(OR')_3 \dashrightarrow RCH{=}CHCH_2C{\equiv}C(CH_2)_nCO_2H$$

Organometallic reagents possessing both ethylenic and acetylenic bonds reacted with CO_2 to give enynoic acids (H. Priebe and H. Hopf, Angew. Chem., 1982, **94**, 299).

Wittig olefination could also be used to prepare enynoic acids; the triple bond and the carboxylate group could either be on the aldehyde or phosphonium ylide component (E.R.H. Jones et al., J. Chem. Soc. C., **1971**, 1156; G.V.M. Sharma, D. Rajagopal and E.S. Rao, Synth. Commun., 1989, **19**, 3181).

Elimination reactions to introduce the ethylenic bond have also been used to prepare enynoic acids (C.F. Garbers et al., J. Chem. Soc. C., **1971**, 1878; A.K. Saksena et al., Tetrahedron Lett., 1985, **26**, 6423).

Oxidation of enyne aldehydes with chromium or silver reagents gave the corresponding enynoic acids (A. Kobayashi, Y. Shibata and K. Yamashita, Agric. Biol. Chem., 1975, **39**, 911; R. Vlakhov et al., Synth. Commun., 1986, **16**, 509).

(d) Labelled Unsaturated Acids

A number of labelled unsaturated monobasic acids have been prepared in order to study mechanistic and biological problems. These labels vary from the hydrogen isotopes, deuterium and tritium, to carbon isotopes, C-13 and C-14, to halogens, to radiolabels of iodine and tellurium.

Deuterium labelled unsaturated acids have been prepared for a wide variety of substrates. (E)- and (Z)-2-Deuterio-3-alkenoic acids were prepared by photoisomerization of the corresponding 2-alkenoic acids using CH_3OD, D_2O and EtOAc(18:1:1) as solvent (S. Safe and C. Penney, J. Label. Compounds, 1971, **7**, 341). (E)-2-[2-$^2H_1$]-Decenoic and (E)-2-[3-$^2H_1$]-decenoic acids have been prepared and converted to the corresponding decenoyl thiol esters (K. Saito et al., Eur. J. Biochem.,

1981, **116**, 581). A facile synthesis of (E)-[11,11,12,12,12-$^2H_5$]-9-dodecenoic acid has been carried out (C. Loefstedt and M. Bengtsson, J. Chem. Ecol., 1988, **14**, 903). A number of deuterium labelled C_{18} unsaturated acids have been prepared: (Z)-[11-$^2H_2$]-9-octadecenoic acid (oleic acid) (W.P. Tucker, S.B. Tove and C.R. Kepler, J. Label. Compounds, 1971, **7**, 11; *ibid*, 137; S.B. Farren, E. Sommerman and P.R. Cullis, Chem. Phys. Lipids, 1984, **34**, 279; A.P. Tulloch, *ibid*, 1979, **25**, 225), methyl (Z)-[18-$^2H_3$], [17-$^2H_2$], [16-$^2H_2$], [14-$^2H_2$] and [12-$^2H_2$]-9-octadecenoates (methyl oleates) (A.P. Tulloch, *ibid*, 1979, **23**, 69), (Z)-[8-$^2H_2$], [7-$^2H_2$], [6-$^2H_2$], [5-$^2H_2$], [4-$^2H_2$] and [3-$^2H_2$]-9-octadecenoates (oleates) (A.P. Tulloch, *ibid*, 225), methyl (E)- and (Z)-[13,14-$^2H_2$], [14,15-$^2H_2$] and [14,14,15,15-$^2H_4$]-10-octadecenoates and (E)- and (Z)-[14,14,15,15,17,18-$^2H_6$]-11-octadecenoic acids (W.J. DeJarlais, E.A. Emken and W.R. Miller, J. Labelled Compd. Radiopharm., 1982, **19**, 1135), (Z),(Z)-[9,10,12,13-$^2H_4$]-9,12-octadecadienoic acid (linoleic acid) (L. Crombie and D.O. Morgan, J. Chem. Soc., Chem. Commun., **1987**, 503), the four geometric isomers of [9,10-$^2H_2$]-12,15-octadecadienoic acid (H. Rakoff and E.A. Emken, J. Labelled Compd. Radiopharm., 1982, **19**, 19), methyl (Z),(Z),(Z)-[15,16-$^2H_2$] and [6,6,7,7-$^2H_4$]-9,12,15-octadecatrienoates (H. Rakoff, Lipids, 1988, **23**, 280), and (Z),(Z),(Z),(Z)-[9,10,11,12,13,14,15,16-$^2H_8$]-9,11,13,15-octadecatetraenoic acid (parinaric acid) (M.M. Goerger and B.S. Hudson, J. Org. Chem., 1988, **53**, 3148). Also a number of deuterium labelled arachidonic acids ((Z),(Z),(Z),(Z)-5,8,11,14-eicosatetraenoic acid) have been prepared. The labelling sites have been [20,20,20-$^2H_3$]- (G. Prakash et al., J. Labelled Compd. Radiopharm., 1989, **27**, 539), [19,19,20,20,20-$^2H_5$]- (B.E. Marron et al., J. Org. Chem., 1989, **54**, 5522), and [5,6,8,9,11,12,14,15-$^2H_8$]- (P.M. Woollard, C.N. Hensby and P.T. Lascelles, J. Chromatogr., 1978, **166**, 411; U.H. Do et al., Lipids, 1979, **14**, 819; D.F. Tabe, M.A. Phillips and W.C. Hubbard, Prostaglandins, 1981, **22**, 349; *ibid*, Methods Enzymol., 1982, **86**, 366; M. Dawson et al., J. Labelled Compd. Radiopharm, 1987, **24**, 291).

Unsaturated acids labelled with the radioactive hydrogen isotope tritium have been prepared by heterogeneous isotopic exchange with 100% gaseous tritium in solution (V.P. Schevchenko and N.F. Myasoedou, J. Labelled Compd. Radiopharm., 1982, **19**, 95). The tritium labelled arachadonic acid, (Z),(Z),(Z),(Z)-[5,6-$^3H_2$]-5,8,11,14-eicosatetraenoic acid, has been prepared by several groups (H.-H. Tai et al., Biochemistry, 1980, **19**, 1989; H.H. Tai and C.J. Sih, Methods Enzymol., 1982, **86**, 99; G.I. Myagkova et al., Bioorg. Khim., 1987, **13**, 415).

Carbon isotopes have been incorporated into unsaturated fatty acids by a variety of methods. C-13 compounds have been prepared for studies using mass spectral or NMR techniques for analysis. Monoenoic fatty acids labelled at the carboxyl group have been made by a three step process involving decarboxylation of the unlabelled fatty acid to the next lower iodide, displacement of iodide by $^{13}CN^-$, and hydrolysis (A. Tunlid et al., J. Microbiol. Methods, 1987, **7**, 77):

$$RCO_2H \longrightarrow RI \xrightarrow{Na^{13}CN} R\text{-}^{13}CN \xrightarrow{\text{hydrolysis}} R\text{-}^{13}CO_2H$$

Another approach to 1-^{13}C labelled acids has been developed in which radicals generated by photolysis of esters derived from N-hydroxy-2-thiopyridone reacted with electrophilic C-13 labelled isocyanides to give adducts which could be hydrolyzed to amides and in turn to the desired labelled acids; this process has been applied to oleic, linoleic and arachidonic acids (D.H.R. Barton, N. Ozbalik and B. Vacher, Tetrahedron, 1988, **44**, 3501):

$$R\text{—}CO_2\text{—}N(\text{2-thioxopyridyl}) \xrightarrow[R'N^{13}C]{h\nu} R'N{=}^{13}C(R)\text{—}S\text{—}(\text{2-pyridyl}) \longrightarrow R^{13}CONHR' \longrightarrow R^{13}CO_2H$$

C-13 labelled cyanide has been used in the synthesis of methyl [1-^{13}C]-9-octadecynoate and (Z)-[1-^{13}C]-9-octadecenoic acid (oleic acid) (T.W. Whaley et al., J. Labelled Compd. Radiopharm., 1981, **18**, 1593); also (Z),(Z)-[1-^{13}C]-9,12-octadecadienoic acid (linoleic acid) has been prepared using this labelled reagent (J.R. Campbell and C.H. Clapp, Bioorg. Chem., 1989, **17**, 281), as well as [1-^{13}C]-(Z),(Z),(Z),(Z)-5,8,10,14-eicosatetraenoic acid (arachidonic acid) (U.H. Do et al., Lipids, 1979, **14**, 819). Double labelled [1,2-$^{13}C_2$]-2,4-dimethyl-4-pentenoic acid has been prepared using a Claisen rearrangement protocol (M.H. Block and D.E. Cane, J. Org. Chem., 1988, **53**, 4923).

Unsaturated monobasic acids labelled with radioactive C-14 have been prepared, usually from C-14 labelled carbon dioxide or cyanide. A chromatographic method has been developed for isolating the methyl esters of uniformly labelled fatty acids prepared by methanolysis of the total lipids of Chlorella pyrenoidosa which had been grown in the presence of $^{14}CO_2$ (S.S. Radwan, J. Chromatogr., 1982, **234**, 463). An approach to carboxyl-labelled carboxylic acids has been developed utilizing the exchange reaction between the carboxyl group of sodium or potassium salts of the acids containing an α-hydrogen and $^{14}CO_2$ (A. Szabolcs, J. Szammer and L. Noszko, Tetrahedron, 1974, **30**, 3647). Some specific C-14 labelled acids or esters which have been prepared include: methyl [1-^{14}C] and [3-^{14}C]-2-methylene-hexadecanoates (L.E. Weaner and D.C.

Hoerr, J. Labelled Compd. Radiopharm. 1986, **23**, 355);[1-^{14}C]-5,9,13-trimethyl-4,8,12-tetradecatrienoic acid (farnesylacetic acid) (K. Nishioka, I. Nakatsuka and H. Kanamaru, Radioisotopes, 1988, **37**, 133); [1-^{14}C]-(E)-11-octadecenoic acid (by a biosynthetic preparation from [1-^{14}C]-linoleic acid) (W. W. Christie, M.L. Hunter and C.G. Harfoot, J. Label. Compounds, 1973, **9**, 483); geometric and positional isomers of [1-^{14}C]-octadecenoic acids (A.J. Valicenti, F.J. Pusch and R.T. Holman, Lipids, 1985, **20**, 234); [1-^{14}C]-8,11-octadecadienoic acid, [1-^{14}C] and [3-^{14}C]-10,13-eicosadienoic acid, [1-^{14}C]-11,14-eicosadienoic acid, [1-^{14}C]-7,10,13-eicosatrienoic acid and [1-^{14}C]-8,11,14-eicosatrienoic acid (J. Budny and H. Sprecher, Biochim. Biophys. Acta, 1971, **239**, 190); [1-^{14}C]-5,8,14-eicosatrienoic acid, [1-^{14}C]-5,11,14-eicosatrienoic acid, [1-^{14}C]-5,8,14-eicosatriynoic acid and [1-^{14}C]-5,11,14-eicosatriynoic acid (R.W. Evans and H. Sprecher, Chem. Phys Lipids, 1985, **38**, 327); [1-^{14}C]-(Z),(Z),(Z),(Z)-5,8,11,14-eicosatetraenoic acid (arachidonic acid) (Y.Y. Liu and M. Minich, J. Labelled Compd. Radiopharm., 1988, **25**, 635); methyl [1-^{14}C]-7,10,13,16-docosatetraenoate (J.R. Neergaard, J.G. Coniglio and H.E. Smith, *ibid*, 1982, **19**, 1063); [1-^{14}C]-7,10,13,16,19-docosapentaenoic acid (H. Sprecher and S.K. Sankarappa, Methods Enzymol., 1982, **86**, 357); and [1-^{14}C]-2-tetracosenoic acid (D.W. Johnson and A. Poulos, Biomed. Environ. Mass Spectrom., 1989 (Vol. Date 1988), **18**, 603).

Some unsaturated acids have been intramolecularly labelled with isotopes of both hydrogen and carbon. For example, [11,11-$^{2}H_2$, 1-^{14}C]-6,9,12-octadecatrienoic acid has been synthesized (M. Hamberg, Biochim. Biophys. Acta, 1984, **793**, 129), as has (10S)- and (10R)-[10-^{3}H, 1-^{14}C]-5,8,11,14,17-eicosapentaenoic acid (S. Hammarstroem, J. Biol. Chem., 1983, **258**, 1427).

Oxygen-18 labelling of eicosanoids, including arachidonic acid, have been carried out using exchange reactions (R.C. Murphy and K.L. Clay, Methods Enzymol., 1982, **86**, 547; J.Y. Westcott, K.L. Clay and R.C. Murphy, Biol. Mass Spectrom., 1985, **12**, 714).

A variety of halogen substituted unsaturated acids have been prepared. The fluorine compounds were usually synthesized for use in biological studies. A facile synthesis of various types of fluorine-containing α,β-unsaturated acids was developed using 2-trifluoromethylacrylic acid as a key reagent (T. Fuchikami, Y. Shibata and Y. Suzuki, Tetrahedron Lett., 1986, **27**, 3173). Di- and trifluorovinyllithium reagents have also been used as reagents for preparation of unsaturated fluorine-substituted acids (J.P. Gillet, R. Sauvetre and J.F. Normant, Synthesis, **1986**, 355). The ultrasound-assisted carboxylation of RCF=CHI gave β-fluoro-α,β-unsaturated acids (H. Yamanaka et al., Nippon Kagaku Kaishi, **1986**, 1321):

$$RCF{=}CHI \xrightarrow[\text{ultrasound}]{Zn,CO_2} RCF{=}CHCO_2H$$

Some specific fluorinated compounds that have been prepared include: 2,2-difluoro-4-decenoic acid (H. Greuter, R.W. Lang and A.J. Romann, Tetrahedron Lett., 1988, **29**, 3291) and (E)- and (Z)-16-fluoro-9-hexenoic acid (G.D. Prestwich, K.A. Plavcan and M.E. Melcer, J. Agric. Food Chem., 1981, **29**, 1018). Also a number of substituted arachidonic acids have been prepared with the fluorine substituent(s) in the following positions: 5-fluoro-(T. Taguchi et al., Chem. Parm. Bull., 1987, **35**, 1666); 2,2-difluoro-(T. Morikawa et al., *ibid*, 1989, **37**, 813); 7,7-, 10,10- and 13,13-difluoro- (P.Y. Kwok et al., J. Am. Chem. Soc., 1987, **109**, 3684); and 20,20,20-trifluoro- (Y. Tanaka et al., Arch. Biochem. Biophys., 1988, **263**, 178). Other halogen-containing unsaturated acids have also been prepared. A convenient synthesis of α-chloro-α,β-unsaturated acids used a Horner-Wittig approach (P. Savignac, M. Snoussi and P. Coutrot, Synth. Commun., 1978, **8**, 19):

$$(EtO)_2P(O)CH(Cl)CO_2H \xrightarrow[\text{2) RCOR'}]{\text{1) BuLi}} RR'C{=}C(Cl)CO_2H$$

Esters of 7-halo-5-heptynoic acid have been prepared where the halogen can be bromine or iodine (F. Theil et al., J. Prakt. Chem., 1985, **327**, 917; R.K. Haynes et al., Aust. J. Chem., 1987, **40**, 273). ω-Iodoacetylenic fatty acids have been prepared by treating the acetylenic acid with iodine in sodium hydroxide (A. Ueno and T. Maeda, Yakugaku Zasshi, 1970, **90**, 1578).

Unsaturated acids containing a radioactive iodine label have been used for a variety of biological studies. A comparison of three methods for the preparation of ω-^{123}I-labelled fatty acids by iodione for bromine exchange has been carried out (P. Laufer et al., J. Labelled Compd. Radiopharm., 1981, **18**, 1205). 14-[^{125}I]Iodo-9-tetradecynoic acid was prepared by iodine exchange of the non-radioactive compound with Na^{125}I (C.A. Otto et al., *ibid*, 1347). Iodine-123 and -125 containing unsaturated acids have been used for myocardial imaging. Specific compounds that have been prepared include: 16-[^{123}I]iodo-9-hexadecenoic acid (G.D. Robinson, Jr. and A.W. Lee, J. Nucl. Med., 1975, **16**, 17; G.D. Robinson, Jr. and F.W. Zielinski, J. Labelled Compd. Radiopharm., 1977, **13**, 220; F. Richie et al., Radiochem. Radioanal. Lett., 1982, **53**, 225), 18-[^{125}I]iodo-17-octadecenoic acid (F.F. Knapp et al., J. Med. Chem., 1984, **27**, 94); (E)-19-[^{123}I and ^{125}I]iodo-18-nonadecenoic acid and (E)-19-[^{123}I and ^{125}I]iodo-3-methyl-18-nonadecenoic acid (M.M. Goodman, A.P. Callahan and F.F. Knapp, Jr., *ibid*, 1985, **28**, 807); and (E)-19-[^{123}I and ^{125}I]iodo-3,3-dimethyl-18-nonadecenoic acid (M.M. Goodman et al., Nucl. Med. Biol., 1989, **16**, 813). The tellurium [^{123m}Te] labelled unsaturated fatty acids have also been used for myocardial imaging. The types of acids that have been prepared include a series of 16-tellura[^{123m}Te]substituted-9-hexadecenoic acids (G.P. Basmadjian et al., J. Labelled Compd. Radiopharm., 1979, **16**, 160; S.L. Mills et al., *ibid*, 1981, **18**, 721; G.P. Basmadjian, S.L. Mills and R.D. Ice, Nucl. Med. Commun., 1982, **3**, 150) and a series of fatty acids where the tellurium has replaced a methylene group at positions 6, 9, 11 or 17 (F.F. Knapp, Jr. et al., J. Nucl. Med., 1981, **22**, 988). Combinations of radioactive halogen labels on tellurium substituted fatty acids have also been studied as myocardial imaging agents. 18-[^{82}Br]Bromo-5-tellura-17-octadecenoic acid has been prepared for this purpose (P.C. Srivastava et al., J. Med. Chem., 1985, **28**, 408), as was a series of I-125 vinyliodo-alkenyl acids with tellurium substitution at various positions along the chain [F.F. Knapp, Jr. et al., *ibid*, 1983, **26**, 1293; *ibid*, 1984, **27**, 57; P.C. Srivastava, F.F. Knapp, Jr. and G.W. Kabalka, Phosphorus Sulfur, 1988 (Vol. Date 1987), **38**, 49).

3. Biological Studies of Unsaturated Monobasic Fatty Acids

Because of their importance in a number of biological systems and the significant biological activity of a number of their metabolites, such as prostaglandins and leukotrienes, unsaturated monobasic fatty acids have been the subject of extensive biological studies. In the sections that follow some studies in the areas of biosynthesis, biological activity, biological oxidation and biological reduction will be described.

(a) Biosynthesis

While the general biosynthetic route to fatty acids involving acetate, acyl coenzymes, etc. has been well established, more recent studies have focused on processes such as chain elongation, chain shortening, desaturation, etc. in specific systems. A review described the interaction of three compartments in a leaf cell necessary for the formation of oleic acid from acetate by the fatty acid synthetase enzymes (P.K. Stumpf et al., Dev. Plant Biol., 1982, **8**, 3). A disrupted spinach chloroplast preparation readily synthesized linolenic acids from acetate under anaerobic conditions but oleic acid was not a precursor; the probable immediate precursor of linolenic acid was (Z),(Z),(Z)-7,10,13-hexadecatrienoic acid (B.S. Jacobson, C.G. Kannangara, and P.K. Stumpf, Biochem. Biophys. Res. Commun., 1973, **51**, 487). An in vitro study of chloroplasts from young olive tree leaves showed high incorporation of labelled acetate in monounsaturated fatty acids, especially oleic acid, but a low synthetic rate for α-linolenic acid. This latter in vitro observation, in comparison with that of the in vivo conditions, suggested the existence of a cooperation between chloroplasts and other parts of the cell to synthesize α-linolenic acid (L.M. Daza and J.P. Donaire, Physiol. Plant., 1982, **54**, 207). Using chloroplasts which actively synthesized palmitic and oleic acids from labelled acetate, it was shown that the inner envelope membrane was more labelled than the outer one, and the proportion of oleic acid was higher in the outer membrane (J.P. Dubacq et al., Dev. Plant Biol., 1984, **9**, 311). A study on the effect of changes in the environment on lipid metabolism in the brown alga Fucus serratus has shown that light stimulated the incorporation of labelled acetate into oleic and, especially, linoleic acid as did lowering the incubation temperature from 15° to 4°C (K.L. Smith and J.L. Harwood, J. Exp. Bot., 1984, **35**, 1359).

A number of unsaturated fatty acids were formed in biological systems by an elongation process. Such a system was identified in 21-day-old rat liver and brain mitrochondria and shown to elongate a multitude of saturated and unsaturated acyl coenzymes ranging in chain length from C_{12} to C_{22}. Acetyl-CoA, but not malonyl-CoA, was the immediate precursor of the C_2 addition unit (S.C. Boone and S.J. Wakil, Biochemistry, 1970, **9**, 1470). A study indicated that cytochrome b_5 was involved in the elongation of hepatic microsomal fatty acids (S.R. Keyes and D.L. Cinti, J. Biol. Chem., 1980, **255**, 11357). Evidence has indicated that human platelets have the capacity to elongate γ-linolenic acid to dihomo-γ-linolenic acid (M.M.G. de Bravo, M.E. de Tomas and O. Mercuri, Biochem. Int., 1985, **10**, 889). The elongation of endogenous arachidonoyl-CoA to C_{22} and C_{24} tetraenoic acids was examined in newborn swine cerebral microsomes, and the presence of two parallel pathways was suggested (S. Yoshida and M. Takeshita, J. Neurochem., 1986, **46**, 1353). Swine cerebral microsomes have also been used to study the elongation of eicosenoyl-CoA (T. Saitoh, S. Yoshida and M. Takeshita, Biochem. Int., 1988, **16**, 871) and the inhibitory effect of this acyl coenzyme on the elongation of long-chain saturated fatty acids (*ibid*, Biochim. Biophys. Acta, 1988, **960**, 410). Subcellular fractions from developing seeds of mustard (Sinapis alba), honesty (Lunaria annua), and nasturtium (Tropaeolum majus) synthesized very long-chain Z monounsaturated fatty acids by elongation from oleoyl-CoA and malonyl-CoA (D.J. Murphy and K.D. Mukherjee, FEBS Lett., 1988, **230**, 101).

Many systems used a combination of elongation and desaturation for the biosynthetic pathway to unsaturated fatty acids. In rat epidymal adipose tissue, synthesis of palmitoleic acid and oleic acid involved desaturase activity, whereas eicosaenoic acid and eicosadienoic acid were the products of elongation of endogenous unsaturated fatty acids (H. Kanoh and D.B. Lindsay, Biochem. J., 1971, **125**, 38p). Rat liver microsomes have been used to measure the rates of chain elongation and desaturation of acids in the linoleate, oleate and palmitoleate biosynthetic pathways. These studies were designed to determine whether there is a relation between rates of conversion and the type of unsaturated fatty acids found in rat liver lipids (J.T. Bernert, Jr. and H. Sprecher, Biochim. Biophys. Acta, 1975, **398**, 354). Cultured rat kidney cells possessed active acyl Δ6-desaturase and elongase which facilely converted linoleic acids to eicosatetraenoic acids (J.C. Chern and J.E. Kinsella, *ibid,* 1983, **750**, 465). Hexacosatetraenoic acid metabolism was studied in vivo in neonatal rat brain and the products included C_{28-36} tetraenoic and C_{26-28} pentaenoic very-long-chain fatty acids which were formed by elongation and desaturation of the substrate (S.B. Robinson, D.W. Johnson and A. Poulos, Biochem. J., 1990, **267**, 561). In the guinea pig, the proportion of labelled arachidonic acid formed from labelled linoleic acid was progressively higher in the maternal liver (<2%), the placenta, and the fetal liver and reached about 25% in the fetal brain. In cultured neuroblastoma cells, the conversion of linoleic acid to arachidonic acid could be altered by other fatty acids in a manner supporting a concerted action of the modulating fatty acid on the desaturation and chain elongation enzymes (H.W. Cook and M.W. Spence, Biochim. Biophys. Acta, 1987, **918**, 217). The biosynthesis of unsaturated fatty acids in marine sponges has been studied. Labelled short-chain fatty acid 10-methylhexadecanoic acid as well as its 10R and 10S antipodes were incorporated into the marine sponge Aplysina fistularis and transformed in situ into 22-methyl-5,9-octacosadienoic acid by elongation and desaturation (D. Raederstorff et al., J. Org. Chem., 1987, **52**, 2337). Studies with the marine sponge Jaspis stellifera showed that the very-long-chain branched fatty acids 25-methyl-5,9-hexacosadienoic acid and 24-methyl-5,9-hexacosadienoic acid, as well as the straight-chain 5,9-hexacosadienoic acid, originated from the short-chain precursors 13-methyltetradecanoic acid, 12-methyltetradecanoic acid and palmitic acid, respectively. These results confirmed that methyl branching did not occur after chain elongation; the unusual desaturation at positions 5 and 9 probably took place after chain elongation (N. Carballeira et al., *ibid*, 1986, **51**, 2751). Labelling and incorporation experiments have completely elucidated the biosynthesis of the most characteristic sponge fatty acids, (Z),(Z)-5,9-hexacosadienoic acid and (Z),(Z),(Z)-5,9,19-hexacosatrienoic acid (S. Hahn et al., J. Am. Chem. Soc., 1988, **110**, 8117).

The desaturation process has been studied in a variety of systems. The cytochrome P 450-mediated desaturation of valproic acid to its hepatotoxic metabolite, 2-n-propyl-4-pentenoic acid, was examined in liver microsomes from rats, mice, rabbits and humans. Based on labelling studies, it was proposed that 4-ene and 4-hydroxy valproic acid were products of a common cytochrome P 450-dependent metabolic pathway in which a carbon-centered radical at C-4 served as the key intermediate (A.E. Rettie et al., J. Biol. Chem., 1988, **263**, 13733). Desaturation of palmitic acid was investigated in a rat liver enzyme system; both (E)-2-hexadecenoic acid and (Z)-9-hexadecenoic acid (palmitoleic acid) were formed (M. Nakano and Y. Fujino, Agric. Biol. Chem., 1975, **39**, 707). Rat liver microsomes have also been used to study the desaturation of linoleic acid to γ-linolenic acid (A. Catala and R.R. Brenner, An. Asoc. Quim. Argent., 1975, **60**, 149; T-C. Lee et al., Biochim. Biophys. Acta, 1977, **489**, 25). The desaturation of a number of positional isomers of (E)-octadecenoic acids were investigated using liver microsomal enzymes from essential fatty acid-deficient rats. Each positional isomer desaturated at a unique rate but the site of desaturation was in the 9-position, indicating action of Δ9-desaturase (M.M. Mahfouz, A.J. Valicenti and R.T. Holman, *ibid*, 1980, **618**, 1; R.T. Holman and M.M. Mahfouz, Prog. Lipid Res., 1981, **20**, 151). The Δ5-desaturation of 8,11,14-eicosatrienoic acid to arachidonic acid was studied in rat liver microsomes, and it was shown that Δ5-desaturation of fatty acids in vitro required the participation of a peripheral component of cytosolic origin (A.I. Leikin and R.R. Brenner, Lipids, 1989, **24**, 101). Rates of desaturation of fatty acids incorporated into rat brain microsomal membranes were compared with those of fatty acids added exogenously; the rate of desaturation was greater for the membrane-bound substrate than for added fatty acids (E.E. Aeberhard, M. Gan-Elepano and J.F. Mead, *ibid*, 1981, **16**, 705).

Plant systems have also been used to study desaturation. From immature safflower seeds, the stearoyl-acyl carrier protein desaturase used in the biosynthesis of oleic acid was purified (T.A. McKeon and P.K. Stumpf, J. Biol. Chem., 1982, **257**, 12141). Sunflower seeds have been used for studying the desaturation of oleic acid (C.P. Rochester and D.G. Bishop, Dev. Plant Biol., 1982, **8**, 57; J.G. Silver et al., J. Exp. Bot., 1984, **35**, 1507). Oleic acid desaturation has also been studied in young winter wheat (Triticum aestivum) root tissue (C. Willemot and J. Labrecque, Plant Physiol., 1982, **70**, 1526) and in young leaves of Pisum sativum (D.J. Murphy, K.D. Mukherjee and I.E. Woodrow, Eur. J. Biochem., 1984, **139**, 373). A homogenate of marigold seeds was used to study the biosynthesis of calendic acid, (8E),(10E),(12Z)-8,10,12-octadecatrienoic acid, from labelled linoleic and oleic acids (L. Crombie and S.J. Holloway, J. Chem. Soc. Perkin Trans. 1, **1985**, 2425).

A number of agents have been shown to have significant effects on the desaturation process. The inhibitory effect of positional isomers of (E)-octadecenoic acid on the desaturation of palmitic acid to palmitoleic acid (Δ9-desaturase), linoleic acid to γ-linolenic acid (Δ6-desaturase) and 8,11,14-eicosatrienoic acid to arachidonic acid (Δ5-desaturase) in rat liver microsomes has been studied (M.M. Mahfouz, S. Johnson and R.T. Holman, Lipids, 1980, **15**, 100); the effect of insulin on these same three desaturases in rat liver has also been investigated (I.N.T. de Gomez, M.J.T. de Alaniz and R.R. Brenner, Acta Physiol. Pharmacol. Latinoam., 1985, **35**, 327). Insulin effects on Δ5-desaturase in humans has also been studied (S. El Boustani et al., Metab. Clin. Exp., 1989, **38**, 315). The ability of seven phenoxyacetic acid derivatives to induce microsomal stearoyl-CoA Δ9-desaturase activity in rat liver was compared with that of clofibric acid (Y. Kawashima, N. Hanioka and H. Kozuka, J. Pharmacobio-Dyn., 1984, **7**, 286). 2-Octynoyl-CoA has been shown to be a mechanism-based inhibitor of pig kidney medium-chain acyl-CoA dehydrogenase (P.J. Powell and C. Thorpe, Biochemistry, 1988, **27**, 8022).

Beside desaturation, enzymatic allylic rearrangement has been observed. In the conversion of (E)-2-decenoic acid to (Z)-3-decenoic acid, labelling studies have shown that the pro-(4R) hydrogen atom is lost (J.M. Schwab and J.B. Klassen, J. Chem. Soc., Chem. Commun., **1984**, 296; *ibid*, J. Am. Chem. Soc., 1984, **106**, 7217).

Some systems produced specific unsaturated fatty acids by degradation of longer-chain fatty acids. The mitochondrial fraction of rat liver metabolized 6,9,12,15,18-tetracosapentaenoic acid, 4,7,10,13,16,19-docosahexaenoic acid, 4,7,10,13,16-docosapentaenoic acid and 4,7,10,13-docosatetraenoic acid to 5,8,11,14-eicosatetraenoic acid, 5,8,11,14,17-eicosapentaenoic acid, 5,8,11,14-eicosatetraenoic acid, and 5,8,11-eicosatrienoic acid, respectively (W.H. Kunau and B. Couzens, Hoppe-Seyler's Z. Physiol. Chem., 1971, **352**, 1297). Labelling studies have shown that in the sex pheromone gland of the moth Trichoplusia ni that (Z)-11-hexadecenoate was chain-shortened to (Z)-9-tetradecenoate, which in turn was chain-shortened to (Z)-7-dodecenoate (L.B. Bjostad and W.L. Roelofs, Science, 1983, **220**, 1387).

(b) Biological Activity

Unsaturated monobasic acids showed a wide range of biological activities, some due to the acid itself while others were due to metabolic products such as prostaglandins, leukotrienes or other oxidation products formed *in situ*. Unsaturated fatty acids inhibited the progesterone biosynthesis from cholesterol in human term placenta mitochondrial fraction by their action on the cholesterol side-chain mixed-function oxidase system but not on the NADPH-regeneration system (B. Tialowska, J. Klimek and L. Zelewski, Acta Biochim. Pol., 1980, **27**, 257). Unsaturated fatty acyl coenzyme-A, but not free fatty acids, inhibited cholesterol synthesis in vitro in a rat liver postmitochondrial supernatant system, the more unsaturation in the acyl coenzyme-A, the higher the level of inhibition (F.H. Faas, W.J. Carter and J.O. Wynn, Biochim. Biophys. Acta, 1977, **487**, 277). Terminal acetylenic fatty acids (C_5 through C_{12}) showed weak and broad antifungal activity, with activity increasing as the chain length increased (M. Nakatani et al., Bull. Chem. Soc. Jpn., 1986, **59**, 3535). Among various unsaturated fatty acids examined, (E)-2-tridecenoic acid was the most potent inhibitor of gastric secretion in rats; the chain length and degree of unsaturation were discussed in relation to the antisecretory activity of these compounds (T.

Mimura et al., J. Pharmacobio-Dyn., 1983, **6**, 527). Minimal inhibitory concentrations of a series of fatty acids against a cariogenic bacterium, Streptococcus mutans, were determined by a tube dilution technique. Among unsaturated fatty acids, (Z)-10-heptadecenoic acid, (Z)-6-octadecenoic acid, (Z)-11-octadecenoic acid, and (Z),(Z)-9,12-octadecadienoic acid had potent antibacterial activity (M. Hattori et al., Chem. Pharm. Bull., 1987, **35**, 3507). Studies have shown that fatty acids were an important source of energy for corpus cardiacum-stimulated trehalose synthesis in the fat body of the American cockroach, Periplaneta americana (G.E. McDougall and J.E. Steele, Insect Biochem., 1988, **18**, 591).

The biological activity of a number of members of the C_{18} and C_{20} families of unsaturated monobasic acids has been studied. The amount of oleic acid transported to the lymph in the rat small intestine correlated with an increase in chylomicron production, whereas very-low-density lipoproteins production became saturated (P. Tso, M.B. Lindstroem and B.Borgstroem, Biochim, Biophys. Acta, 1987, **922**, 304). While oleic acid caused no hemodynamic changes in dogs, linoleic acid had cardiotoxic effects, but the effect was less than that of a leukotoxin metabolite of linoleic acid (A. Fukushima et al., Cardiovasc. Res., 1988, **22**, 213). Exogenous arachidonic acid stimulated glucose-induced insulin release by isolated, perfused hamster pancreatic islets; this effect was apparently derived from newly synthesized prostaglandins (T. Hirano et al., Endocrinol. Jpn., 1984, **31**, 549). The teratogenic action of both phospholipase A_2-inhibitory protein (PLIP) and glucocorticoids was reversed by arachidonic acid, presumably as the precursor of prostaglandins and thromboxanes, suggesting that PLIP mediated the effects of glucocorticoids by inhibiting phospholipase A_2 (C. Gupta et al., Proc. Natl. Acad. Sci. U.S.A., 1984, **81**, 1140). The release of arachidonic acid and its subsequent metabolism was an apparent early requirement for the initiation of cell cycle traversal by epidermal growth factor in BALB/c 3T3 fibroblasts (R.D. Nolan, R.M. Danilowicz and T.E. Eling, Mol. Pharmacol., 1988, **33**, 650). Arachidonate-induced aggregation of human platelets suspended in buffer might depend on product(s) of lipoxygenase rather than of cyclooxygenase, and was hence insensitive to inhibition by acetysalicylate compared with arachidonate-induced aggregation of human platelets suspended in plasma (W.J. McDonald-Gibson et al., Prostaglandins, Leukotrienes Med., 1984, **15**, 1). Platelet aggregation and serotonin release, induced by high concentrations of arachidonic acid and other unsaturated fatty acids (oleic, linoleic and linolenic acids), were examined in washed human platelet suspensions in the absence of albumin; the mechanism of these actions on platelets was independent of prostaglandin endoperoxides, thromboxane A_2, and, perhaps, phosphatidic acid and 1,2-diacylglycerol (Y. Hashimoto et al., Biochim. Biophys. Acta, 1985, **841**, 283). Eicosapentaenoic acid inhibited ADP- and collagen-induced platelet aggregation in humans and rats in vivo; the effect was greater in humans (I. Morita et al., Int. Congr. Ser.-- Excerpta Med., 1983, **623**, 144). An emulsion of 1,2,3,-trieicosapentaenoylglycerol was suggested to be suitable for those patients who need both intravenous feeding and prevention of thrombosis, such as postoperative patients, by studies in rabbits (M. Urakaze et al., Thromb. Res., 1986, **44**, 673). A similar study showed that an injectable emulsion of 1,2,3-tridocosahexaenoylglycerol may be used for patients having immediate risk of thrombosis or for those who need docosahexaenoic acid but cannot take it orally (T. Hamazaki et al., Lipids, 1987, **22**, 1031). Eicosatetraynoic acid, an in vitro inhibitor of eicosanoid metabolism, suppressed DNA synthesis in human PC-3 cells derived from a metastatic prostatic adenocarcinoma (K.M. Anderson et al., Prostate, 1988, **12**, 3). Several 7,7- and 10,10-dimethyl analogues of arachidonic acid were shown to inhibit the formation of the anaphylactic slow reacting substance in rat peritoneal cells by antagonizing phospholipase A_2 rather than Δ^5-lipoxygenase activity (N. Cohen et al., Prostaglandins, 1984, **27**, 553).

A number of dietary studies have been conducted on the biological effects of unsaturated monobasic acids. The results of a study of fatty acid metabolism and the contribution of dietary fatty acids to milk cholesteryl ester (CE) and phospholipid (PL) in normal lactating women showed that the influence on fatty acid composition of milk triglyceride, PL and CE occurred 8-10 hours after dietary fat was consumed, and that the same fatty acid pool was apparently used for synthesis of these milk constituents (E.A. Emken et al., J. Lipid Res., 1989, **30**, 395). Arachidonate-stimulated aortic prostacyclin production was markedly enhanced in aortic segments from rats raised on diets containing olive oil, relative to diets containing safflower, soy oil and lard; this unique stimulation suggested an effect of the extraordinarily high mono:polyunsaturated fatty acid ratio, or alternatively, of a still-to-be identified substance(s) in olive oil (J.W. Blankenship et al., Prostaglandins, Leukotrienes Essent. Fatty Acids, 1989, **36**, 31). Red blood cell membranes, from blood collected from humans fed diets with either 3% or 15% polyunsaturated fatty acids for 80 days, were used to estimate human red blood cell liability to lipid peroxidation in vitro (C.G. Fraga et al., Lipids, 1990, **25**, 111). Comparative effects of a semi-synthetic diet containing supplements of corn oil, no fat, linoleate or (E),(E)-linoleate on rat blood coagulation parameters were studied, and the only diets that appeared to produce abnormal hematological and hemostatic properties were those containing (E),(E)-linoleate (G. Raccuglia and O.S. Privett, *ibid*, 1970, **5**, 85). Cats were fed purified diets that were either deficient in essential fatty acids or that provided linoleate with or without arachidonate. Linoleate prevented testis degeneration in male cats, which was observed in animals fed the essential fatty acid diet. In contrast, female cats that were fed diets lacking arachidonate were unable to bear live kittens, whether linoleate was provided in the diet or not. Thus, linoleate appeared to meet the requirements for spermatogenesis in males, but dietary arachidonate was essential for adequate reproduction in female cats (M.L. MacDonald et al., J. Nutr., 1984, **114**, 719). In normal, hypertensive, and hyperlipemic humans, diets supplemented with linoleic acid or α-linolenic acid resulted in an increase of the corresponding fatty acids in serum lipids but only a slight augmentation of the prostaglandin precursors arachidonic acid and eicosapentaenoic acid. This latter observation was a characteristic finding in humans, but different from preferred laboratory animals, for instance, rats (P. Singer et al., Prostaglandins, Leukotrienes Med., 1986, **24**, 173). The effects of a dietary α-linolenic acid deficiency on reproduction and postnatal growth in rats were studied during three successive gestations and four successive generations (P. Guesnet, G. Pascal and G. Durand, Reprod. Nutr. Dev., 1986, **26**, 969). Supplementation of patients with α-linolenic acid deficiency with ethyl α-linolenate followed by a purified fish oil began to normalize symptoms within ten days (K.S. Bjerve et al., Am. J. Clin. Nutr., 1989, **49**, 290). A diet including mold oil from lipid-accumulating fungus (Mortierella ramanniana anglispora) containing γ-linolenic acid showed an inhibitory effect on thrombus formation in the rat microvessels induced by the light-fluorescent dye method (T. Nakahara et al., Thromb. Res., 1990, **57**, 371). Studies have suggested that in certain nutritional states where the liberation of eicosapentaenoic acid from human endothelial cells was accompanied by that of endogenous arachidonic acid, substantial amounts of prostaglandin I_3 could contribute to the prostacyclin-like activity of the vessel wall (J.C. Bordet, M. Guichardant and M. Legarde, Biochem. Biophys. Res. Commun., 1986, **135**, 403). The cytotoxic effect of arachidonic acid on human breast cancer MCF-7 cells was studied and the role of dietary polyunsaturated fatty acids in prevention of breast cancer discussed (A. Najid, J.L. Beneytout and M. Tixier, Cancer Lett., 1989, **46**, 137). An essential fatty acid deficient diet was administered to rats to determine the feasibility of using a model of endogenous arachidonic acid deficiency to study the role of prostaglandins in the kidney, but the results indicated that the usefulness of this animal model may be limited (A.R. Sinaiko, R.T. Holman and T.P. Green, Prostaglandins, Leukotrienes Med., 1985, **19**, 87). Larvae of the mosquito Culex pipiens were reared in synthetic media containing suboptimal levels of arachidonic acid such that emerging adults had reduced flight abilities, and studies showed that supplementation with prostaglandins failed to bring flight performances up to the levels attained with optimal concentrations of arachidonic acid in the diet, indicating that prostaglandins could not substitute for all arachidonic acid functions (R.H. Dadd and J.E. Kleinjan, J. Insect Physiol., 1988, **34**, 779). The possible use of diets rich in eicosapentaenoic acid to induce leukotriene formation, thereby decreasing inflammatory reactions, was discussed (T. Terano, J.A. Salmon and S. Moncada, Prostaglandins, 1984, **27**, 217; *ibid*,

Biochem. Pharmacol., 1984, **33**, 3071). Apparently, dietary docosahexaenoic acid is retroconverted to eicosapentaenoic acid in man, which was quickly transformed, like dietary eicosapentaenoic acid itself, to prostaglandin I_3 (S. Fisher et al., Prostaglandins, 1987, **34**, 367).

(c) Biological Oxidations

Many of the most active metabolites of unsaturated monobasic acids are products of enzymatic oxidation. The two most studied pathways are the cyclooxygenases, which lead to products such as prostaglandins and thromboxins, and the lipoxygenases, which lead to open chain hydroxylated products such as leukotrienes and monohydroperoxy compounds. Most of these studies have been carried out on arachidonic acid metabolites, but other C_{20} acids along with the C_{18} family of oleic, linoleic and linolenic acids have also been studied.

A review with 21 references concerning the conversion of dietary eicosapentaenoic acid to prostaglandins and leukotrienes in man by different cells has been published (P.C. Weber et al., Prog. Lipid Res., 1986, **25**, 273). In human blood platelet microsomal fractions, arachidonic acid was metabolized by the cyclooxygenase pathway to thromboxane B_2, prostaglandin E_2 and 12-hydroxyheptadecatrienoic acid and by the lipoxygenase pathway to (8Z), (10E), (14E)-12-hydroxy-8,10,14-eicosatrienoic acid (A.G.E. Wilson et al., Prostaglandins, 1979, **18**, 409). In platelet-depleted human monocytes, the major product of arachidonic acid metabolism in response to inflammatory particles was thromboxane A_2, a cyclooxygenase product, while in response to the calcium ionophore A23187 increased levels of both leukotriene B_4 and C_4, lipoxygenase products, appeared (N.A. Pawlowski et al., J. Allergy Clin. Immunol., 1984, **74**, 324). This latter stimulant's effect on arachidonic acid metabolism was also studied with suspensions of human blood leukocytes and blood platelets where nine metabolites were characterized (B. Frutean de Laclos, P. Braquet and P. Borgeat, Prostaglandins, Leukotrienes Med., 1984, **13**, 47) and with cultured human skin fibroblasts where prostaglandin E_2 was the major product of the cyclooxygenase pathway and 15-hydroxy-5,8,11,13-eicosatetraenoic acid was the main lipoxygenase product (B. Mayer et al., Biochim, Biophys. Acta, 1984, **795**, 151). Isolated pancreatic islets from rats have been studied under several sets of conditions and shown to synthesize a profile of arachidonate lipoxygenase and cyclooxygenase products (J. Turk et al., *ibid,* 1984, **794**, 110; *ibid*, 125). The gonadotroph-enriched pituitary cells from rats, as opposed to the somatotroph- and lactotroph-enriched cell fractions, produced oxygenated arachidonic acid metabolites in which cyclooxygenase products were more prevalent than lipoxygenase products (J.Y. Vanderhoek et al., Prostaglandins, Leukotrienes Med., 1984, **15**, 375). Studies on several reproductive systems showed a range of cyclooxygenase and lipoxygenase products from arachidonic acid metabolism: rabbit fetal membranes (W.J. Elliot et al., Prostaglandins, 1984, **27**, 27), rat uterine homogenates (A. Morgan, R.G. McDonald-Gibson and T.F. Slater, Biochem. Soc. Trans., 1984, **12**, 837), and human amniotic membrane obtained after vaginal delivery (K. Kinoshita, K. Satoh and S. Sakamoto, Biol. Res. Pregnancy Perinatol., 1984, **5**, 61). The retina of various species have been investigated for arachidonic acid metabolism and both cyclooxygenase and lipoxygenase pathways were found to be operative: rat retina (D.L. Birkle and N.G. Bazan, Biochim. Biophys. Acta, 1984, **795**, 564), bovine retina (*ibid*, Prostaglandins, 1984, **27**, 203), and rabbit retina and iris (Y. Preud'homme, D. Demolle and J.M. Boeynaems, Invest. Ophthamol. Visual Sci, 1985, **26**, 1336). Isolated rat intestine was used to demonstrate that laxation by phenolphthalein involved the formation of lipoxygenase and cyclooxygenase products from exogenous arachidonic acid (F. Capasso, I.E. Tavares and A. Bennett, Eur. J. Pharmacol., 1984, **106**, 419). Studies showed that intact cultured calf aortic endothelial cells metabolized arachidonic acid by three different routes: incorporation into triglycerides and phospholipids in a ratio of about 2:1; formation of lipoxygenase products, e.g. 12-hydroxy-5,8,10,14-eicosatetraenoic acid; and formation of cyclooxygenase products, e.g. 6-ketoprostaglandin $F_{1\alpha}$ and prostaglandin $F_{2\alpha}$ (H. Kuehn et al., Prostaglandins, Leukotrienes Med., 1985, **17**, 291). Rat, rabbit and bovine aortae were examined for lipoxygenase and cyclooxygenase products from arachidonic acid and linoleic acid (C.D. Funk and W.S. Powell, J. Biol. Chem., 1985, **260**, 7481). It was shown that the lipoxygenase and cyclooxygenase products from arachidonic acid metabolism in pulmonary airway epithelial cells was species dependent in human, dog, and sheep airways (M.J. Holtzman et al., Biochim. Biophys. Acta, 1988, **963**, 401).

Numerous studies have been carried out on the inhibitory effect of various types of drugs on lipoxygenase and cyclooxygenase metabolic pathways for arachidonic acid. While several nonsteroidal anti-inflammatory drugs inhibited cyclooxygenase but had little or no effect on the lipoxygenase activity, antimalarial drugs and various antioxidants inhibited both enzymes (S.A. Saeed, S. Khan and M.A.S. Khan, Pak. J. Pharmacol., 1984, **1**, 49). Aspirin and indomethacin have been investigated as inhibitors in a variety of species and cell types (J. Chang et al., Inflammation, 1984, **8**, 143; J.R. Walker and J. Harvey, Adv. Inflammation Res., 1984, **6**, 227; M.L. Dahl and P. Uotila, Prostaglandins, Leukotrienes Med., 1984, **16**, 95; S. Hewertson, R.G. McDonald-Gibson and T.F. Slater, Biochem. Soc. Trans., 1984, **12**, 835; K. Punnonen, P. Uotila and E. Mantyla, Res. Commun. Chem. Pathol. Pharmacol., 1984, **44**, 367; R.F. Myers et al. Inflammation, 1985, **9**, 91). Also the effects of various flavonoids (Y. Kimura, H. Okuda and S. Arichi, Planta. Med., **1985**, 132) and stilbenes (*ibid*,

Biochim. Biophys. Acta, 1985, **834**, 275) on rat peritoneal polymorphonuclear leukocyte lipoxygenase and cyclooxygenase activities were studied. Dexamethasone inhibited the formation of products from both cycloxygenase and lipoxygenase pathways in cultured human ocular trabecular cells (R.N. Weinreb et al., Int. Congr. Ser. - Excerpta Med., 1984, **636**, 213). Various dehydroarachidonic acids were studied as inhibitors and the 5,6- and 11,12-isomers were shown to cause irreversible inhibition of leukotriene and prostaglandin biosynthesis, respectively (E.J. Corey and J.E. Munroe, J. Am. Chem. Soc., 1982, **104**, 1752). In rat microphages, eicosapentaenoic acid inhibited the formation of cyclooxygenase products to a greater extent than lipoxygenase products (H. Saito et al., Int. Congr. Ser. - Excepta Med., 1983, **623**, 162).

The specific formation of prostaglandins from unsaturated monobasic acids via the cyclooxygenase pathway has been the subject of numerous investigations. One showed that prostaglandin precursors of all (Z) 8,11,14-trienoates and 5,8,11,14-tetraenoates were restricted to acids with 19, 20, or 21 carbon atoms. All (Z) 8,11,14-eicosapolyenoates with extra double bonds at positions 2,3, or 4 were also precursors while an additional double bond toward the end of the chain, especially at position 18, led to nearly complete loss of activity (R.K. Beerthuis et al., Recl. Trav. Chim. Pays-Bas, 1971, **90**, 943). The production of prostaglandins from arachidonic acid by guinea pig lung has been reviewed (T.E. Eling and M.W. Anderson, Agents Actions, 1976, **6**, 543). Ram seminal vesicle microsomes produced prostaglandin E_2 and D_2 and 15-ketoprostaglandin E_2 from arachidonic acid (P.R. Ravikumar et al., J. Pharm. Sci., 1979, **68**, 1302). Canine gastrointestinal tract microsomes were studied for the formation of 6-ketoprostaglandin $F_{1\alpha}$ and prostaglandin E_2 from arachidonic acid (L.E. LeDuc and P. Needleman, Adv. Prostaglandin Thromboxane Res., 1980, **8**, 1515). These two prostaglandins were also found as metabolites of arachidonic acid in minced rabbit pericardium (A. Honda et al., Biochim, Biophys. Acta, 1984, **794**, 403). ADP stimulated the synthesis of prostaglandin I_2, as reflected by the release of 6-ketoprostaglandin $F_{1\alpha}$, in bovine aorta endothelial cells (A. Van Coevorder and J.M. Boeynaems, Prostaglandins, 1984, **27**, 615). In cultured smooth muscle cells of rat aorta, four diuretic agents significantly enhanced the formation of prostacyclin from exogenous arachidonic acid (B. Dorian et al., Biochem. Parmaclo., 1984, **33**, 2265). On melittin stimulation, cultured malignant human epidermal keratinocytes produced a number of prostaglandin products (R.H. Rice and L. Levine, Biochem. Biophys. Res. Commun., 1984, **124**, 303). Human umbilical arteries converted arachidonic acid into a variety of prostaglandins and 11- and 15-hydroxyeicosatetraenoic acids by the cyclooxygenase pathway (B.N.Y. Setty, M.J. Stuart and R.W. Walenga, Biochim. Biophys. Acta, 1985, **833**, 484). It was shown that simply incubating ^{14}C-labelled arachidonic acid in homogenates of human reproductive tissues and measuring the amount of radioactivity transformed into various prostaglandins gave only qualitative conclusions (V. Dimov, N.J. Christensen and K. Green, *ibid*, 1983, **754**, 38). A study of the variation of prostaglandin products from arachidonic acid metabolism as a function of gestation in rabbit placenta microsomes was conducted (M.H. Bloch et al., Prostaglandins, 1985, **29**, 203). Thrombin, histamine and ionophore A 23187 stimulated human endothelial cells to synthesize prostaglandins (S.L. Hong et al., Thromb. Res., 1985, **38**, 1). Formation of prostacyclin and thromboxane A_2 was increased in chlosterol-fed rabbit atherosclerotic aortic segments relative to nonatherosclerotic segments upon incubation with arachidonic acid (J.L. Mehta et al., Proc. Natl. Acad. Sci. U.S.A., 1988, **85**, 4511). Prostaglandins of the E and F series were formed from arachidonic acid in male and female houseflies; during mating, arachidonic acid was transferred from males to females where it could be metabolized to prostglandin $F_{2}\alpha$ (E.J. Wakayama, J.W. Dillwith and G.J. Blomquist, Insect Biochem., 1986, **16**, 895). A review, with 11 references, on the formation of prostaglandins and thromboxanes from 5,8,11,14,17-eicosapentaenoic acid has appeared (D.F. Reingold and P. Needleman, Trends Pharmacol. Sci., 1980, **1**, 359).

A variety of compounds have been shown to inhibit the formation of prostaglandins from arachidonic acid by disruption of the cyclooxygenase pathway, including indomethacin (A.G.E. Wilson et al., Prostaglandins, 1979, **18**, 409); mepacrine (A. Raz, Thromb. Haemostasis, 1983, **50**, 784); nitroglycerin and other organic nitrates (C. Wirthumer-Hoche, K. Silberbauer and H. Sinzinger, Prostaglandins, Leukotrienes Med., 1984, **15**, 317); isoxicam (A Bennett et al., Int. Congr. Symp. Ser. -- R. Soc. Med., 1984, **75**, 3; T.A. Pugsley et al., Drug Dev. Res., 1985, **5**, 171); aspirin (C. Bonne et al., Ann. Allergy, 1985, **54**, 158); verapamil (M.L. Dahl and P. Uotila, Prostaglandins, Leukotrienes Med., 1985, **17**, 191); and interferon (D. Boraschi et al., J. Immunol., 1984, **132**, 1987). A number of unsaturated monobasic acids have been shown to inhibit arachidonic acid's metabolism to prostaglandins, including 12-hydroperoxy-5,8,10,14-eicosatetraenoic acid (Y. Hashimoto et al., Biochem. Biophys. Res. Commun., 1985, **130**, 781); linoleic acid and docosahexaenoic acid (K.C. Srivastava, Prostaglandins, Leukotrienes Med., 1985, **17**, 319); and a variety of naturally occurring conjugated octadecatrienoic acids (D.H. Nugteren and E. Christ-Hazelhof, Prostaglandins, 1987, **33**, 403).

The other major biological oxidation pathway for arachidonic acid is the lipoxygenase route which initially gives a hydroperoxy compound which is reduced to a hydroxy derivative; this process is site specific. Also leukotrienes and lipoxins are produced by these enzymes. One of the most studied systems is the 5-lipoxygenase pathway. Thus the lipoxygenase of potato afforded 5(S)-hydroperoxyeicosatetraenoic acid upon enzymatic oxidation of arachidonic acid; this product was a key biological precursor of 5-hydroxyeicosatetraenoic acid and the initial product in the leukotriene pathway (E.J. Corey et al., J. Am. Chem. Soc., 1980, **102**, 1435; *ibid*, 1983, **105**, 4093). In guinea pig lungs, it was shown that arachidonic acid was a metabolic precursor of slow reacting substance of anaphylaxis by way of a 5-lipoxygenase pathway (S. Watanabe-Kohno and C.W. Parker, J. Immunol., 1980, **125**, 946). In MC-9 mast cells, it was shown that the 5-lipoxygenase metabolism of arachidonic acid required hydroperoxide for activation, and cellular levels of hydroperoxides might be an important factor regulating leukotriene synthesis (R.W. Bryant et al., Prostaglandins, 1986, **32**, 615). The arachidonate 5-lipoxygenase from porcine leukocytes was purified to near homogeneity by immunoaffinity chromatography with a monoclonal anti-5-lipoxygenase antibody (N. Ueda et al., J. Biol. Chem., 1986, **261**, 7982). Using rat basophilic leukemia cell homogenates, 5,8,11-eicosatrienoic acid was converted to leukotriene A_3 by the 5-lipoxygenase-leukotriene pathway. Further studies showed that fatty acids which have double bonds at positions 5 and 8 were readily converted to the 5-hydroperoxide, but an additional double bond at position 11 was necessary for leukotriene biosynthesis (B.A. Jakschik, A.R. Morrison and H. Sprecher, J. Biol. Chem., 1983, **258**, 12797; Y. Wei et al., Prostaglandins, 1985, **29**, 537). The effect of 5-lipoxygenase activity stimulated by the ionophore A 23187 on a variety of cells has been studied in human neutrophils (F.F. Sun and J.C. McGuire, Biochim. Biophys. Acta, 1984, **794**, 56); human blood leukocytes (M. Nadeau et al., Can. J. Biochem. Cell. Biol., 1984, **62**, 1321); cloned murine mast cells (M.W. Musch et al., Prostaglandins, 1985, **29**, 405); human promyelocytic leukemia (HL-60) cells before and after N,N-dimethylformamide-induced differentiation (A.P. Agins et al., Biochem. Biophys. Res. Commun., 1985, **126**, 143); human dispersed lung cells (J. Harvey et al., Br. J. Pharmacol., 1985, **86**, 417); human polymorphonuclear leukocytes (J.A. Salmon, L.C. Tilling and S. Moncada, Prostaglandins, 1985, **29**, 377); and rat isolated intestine (F. Capasso et al., *ibid*, **30**, 119). A number of compounds have been studied as inhibitors of the 5-lipoxygenase pathway for arachidonic acid metabolism: various coumarins, including esculetin, daphnetin and fraxetin (Y. Kimura et al., Biochim. Biophys. Acta, 1985, **834**, 224); the nontoxic, anti-inflammatory seleno-organic compound, ebselen (H. Safayhi, G. Tiegs and A. Wendel, Biochem. Pharmacol., 1985, **34**, 2691); and BW 755C (H. Shoam and E. Razin, Biochim. Biophys. Acta, 1985, **837**, 1). Also a number of arachidonic acid analogues have been shown to be inhibitors, including cyclopropyl derivatives such as (7E), (9E), (11Z), (14Z)-*trans*-5,6-methanoeicosatetraenoic acid (Y. Arai et al., Chem. Pharm. Bull., 1982, **30**, 379; *ibid*, J. Med. Chem., 1983, **26**, 72; *ibid*, Adv. Prostaglandin, Thromboxane, Leukotriene Res., 1983, **11**,

169) and various 7,7-dimethyl derivatives of C_{20} dienoic, trienoic and tetraenoic acids (J. Ackroyd et al., Tetrahedron Lett., 1983, **24**, 5139; C.D. Perchonock et al., *ibid*, 2457; J. R. Pfister and D.V. KrishnaMurthy, J. Med. Chem., 1983, **26**, 1099).

The 12-lipoxygenase pathway for arachidonic acid metabolism has also been studied. This route has been used to provide a convenient source of 12(R)-hydroxy-5,8,10,14-eicosatetraenoic acid from human platelets (J.C. McGuire et al., Prep. Biochem., 1978, **8**, 147; W.C. Pickett and R.C. Murphy, Prostaglandins Res. Stud. Ser., 1981, **1**, 73; F.F. Sun, Methods Enzymol., 1981, **72**, 435) and the corresponding 12-hydroperoxy compound from porcine leukocytes (S. Kitamura et al., Eur. J. Biochem., 1988, **176**, 725). The arachidonate 12-lipoxygenase was purified to near homogeneity from the cytosol of porcine leukocytes by immunoaffinity chromatography using a monoclonal antibody against the enzyme (C. Yokoyama et al., J. Biol. Chem., 1986, **261**, 16714). Other systems that have been studied for the formation of 12-hydroxyeicosatetraenoic acid from arachidonic acid by the 12-lipoxygenase pathway included rat uterus (A. Morgan et al., Assoc. Int. Cancer Res. Symp. [Volume Date 1983], 1984, **3**, 255; H. Thaler-Dao et al., Prostaglandins, Leukotrienes Med., 1985, **18**, 59), rat aorta smooth muscle cells in conjunction with platelet-derived growth factor (J. Nakao et al., Life Sci., 1985, **37**, 1435), two models of infarcted canine myocardium (E.R. McCluskey et al., Prostaglandins, 1985, **29**, 387), and bovine cornea microsomes (R.C. Murphy et al., J. Biol. Chem., 1988, **263**, 17197).

A 15-lipoxygenase pathway for the metabolism of arachidonic acid has been found in some systems. Thus rabbit reticulocyte lipoxygenase produced 15-hydroperoxy-5,8,11,13-eicosatetraenoic acid as the major product; 15-oxygenated derivatives were also the major products from eicosatrienoic acids (R.W. Bryant et al., J. Biol. Chem., 1982, **257**, 6050). In human neutrophils incubated with supratherapeutic concentrations of nonsteroidal anti-inflammatory agents, arachidonic acid metabolism to (5Z),(8Z),(11Z),(13E)-15(S)-hydroxyl-5,8,11,13-eicosatetraenoic acid was proportional to lactate dehydrogenase release (J. McGuire et al., *ibid*, 1985, **260**, 8316). Soybean lipoxygenase has been shown to metabolize arachidonic acid by a 15-lipoxygenase pathway (S. Laakso, Lipids, 1982, **17**, 667; A.R. Morrison, W. Brown and N. Tank, Prostaglandins, 1984, **27**, 753; E.J. Corey and R. Nagata, Tetrahedron Lett., 1987, **28**, 5391). 14,15-Dehydroarachidonic acid (E.J. Corey and H. Park, J. Am. Chem. Soc., 1982, **104**, 1750) and (10Z),(13Z)-12-methylidene-10,13-nonadecadienoic acid (E.J. Corey and M. D'Alarcao, Tetrahedren Lett. 1986, **27**, 3589) have been shown to be irreversible inhibitors of soybean lipoxygenase.

Other biological oxidation systems of arachidonic acid have also been studied. In rat liver microsomes, oxidation by a NADPH-dependent oxygenase gave a number of oxygen containing products (J. Capdevila et al., Proc. Nat. Acad. Sci. U.S.A., 1982, **79**, 767). Among other oxidation products from rabbit liver microsomes were four epoxyeicosatrienoic acids (E.H. Oliw, F.P. Guengerich and J.A. Oates, J. Biol. Chem., 1982, **257**, 3771). In cells derived from the medullary portion of the thick ascending limb of the loop of Henle of the rabbit kidney, a cytochrome P 450-dependent monooxygenase system metabolized arachidonic acid to 5,6-epoxyeicosatrienoic acid and 11,12-dihydroxyeicosatrienoic acid (M. Schwartzmann et al., Nature (London), 1985, **314**, 620). BW 755C was shown to be an inhibitor of the hepatic monooxygenase system (J. Turk et al., Biochim. Biophys. Acta, 1985, **835**, 1). Two dihydroxy metabolites, 11,12-dihydroxy-5,8,14-eicosatrienoic acid and 14,15-dihydroxy-5,8,11-eicosatrienoic acid, were obtained from arachidonic acid by a low-speed rabbit renal cortical supernatant (E.H. Oliw et al., J. Biol. Chem., 1981, **256**, 9924). Two hydroxy epoxide intermediates in the formation of 8,11,12- and 10,11,12-trihydroxyeicosatrienoic acids from arachidonic acid by the high-speed supernatant of rat lung have been identified as 8-hydroxy-11,12-epoxy-5,9,14-eicosatrienoic acid and 10-hydroxy-11,12-epoxy-5,8,14-eicosatrienoic acid, respectively (C.R. Pace-Asciak, E. Granstroem and B. Samuelsson, *ibid*, 1983, **258**, 6835). Another study has shown the formation of a wide range of trihydroxy metabolites from arachidonic acid, including 11,12,19- and 11,12,20-trihydroxy-5,8,14-eicosatrienoic acid and 14,15,19- and 14,15,20-trihydroxy-5,8,11-eicosatrienoic acid from rabbit renal cortex and NADPH (E.H. Oliw and J.A. Oates, Prostaglandins, 1981, **22**, 863). Lipoxins, trihydroxy eicosanoids containing a conjugated tetraene system, were products of soybean lipoxygenase-catalyzed conversion of arachidonic acid (D.E. Sok et al., Han'guk Saenghwa Hakhoechi, 1988, **21**, 147; H. Kuehn et al., Adv. Exp. Med. Biol., 1988, **229**, 39).

The biological oxidation of a number of 18 carbon unsaturated monobasic acids has also been studied. The movement of α-linolenic acid through the mitochondrial outer membrane to oxidation sites was studied in rat liver and compared to that of linoleic and oleic acids. The overall β-oxidation in total mitochrondria was in the order α-linolenic acid > linoleic acid > oleic acid and was explained by a different spatial arrangement due to the number and position of the double bonds (P. Clouet, I. Niot and J. Bezard, Biochem. J., 1989, **263**, 867). Cell-free extracts of a species of Torulopsis converted oleic acid to 17-hydroxyoleic acid (E. Heinz, A.P. Tulloch and J.F.T. Spencer, Biochim. Biophys. Acta, 1970, **202**, 49). Various microorganisms, including three fungal strains, four yeast strains and six bacterial strains, could be used to transform oleic acid to (9Z)-12-hydroxy-9-octadecenoic acid, ricinoleic acid (K. Soda, Proc.-World Conf. Biotechnol. Fats Oils Ind. (Meeting Date 1987), ed. T.H. Applewhite, American Oil Chemists' Society; Champaign, Ill., 1988, pp. 178-9).

Incubation of linoleic acid with a crude homogenate of tomato gave a 69% yield of 9- and 13-hydroperoxy-10,12-octadecadienoic acids in a 96:4 ratio (J.A. Matthew, H.W.S. Chan and T. Galliard, Lipids, 1977, **12**, 324). These same two products were formed from linoleic acid by prostaglandin endoperoxide synthetase in particulate fractions and homogenates of fetal calf aorta (C.D. Funk and W.S. Powell, Biochim. Biophys. Acta, 1983, **754**, 57). The corresponding 9- and 13-hydroxyoctadecadienoic acids were formed by human umbilical vein endothelial cells (T.L. Kaduce et al., J. Biol. Chem., 1989, **264**, 6823). Rat liver peroxisomes converted linoleic acid to 3-hydroxy-10,12-octadecadienoic acid and 2,3-dehydromethyllinoleic acid by a β-oxidation system (U. Diczfalusy et al., Biochim. Biophys. Acta, 1990, **1043**, 182). The major metabolites from linoleic and linolenic acids when incubated with rabbit hepatic microsomes were the dihydroxy compounds 9,10-dihydroxy-12-octadecenoic acid and 12,13-dihydroxy-9-octadecenoic acid and 9,10-dihydroxy-12,15-octadecadienoic acid, 12,13-dihydroxy-9,15-octadecadienoic acid, and 15,16-dihydroxy-9,12-octadecadienoic acid, respectively (E.H. Oliw, Biochem. Biophys. Res. Commun., 1983, **111**, 644). Hexanal and hexenals [(Z)-3-hexenal and (E)-2-hexenal] were produced by lipoxygenase and hydroxyperoxide lyase in plant tissues from linoleic and linolenic acids, respectively (A. Hatanaka, J. Sekiya and T. Kajiwara, J. Agric. Food Chem., 1983, **31**, 176; J. Sekiya et al., Agric. Biol. Chem., 1976, **40**, 185). The 13-hydroperoxy linolenic acid was the major hydroperoxide metabolite produced from linolenic acid by soybean lipoxygenase (E. Kaplan and K. Ansari, J. Chromatogr., 1985, **350**, 435), while 9-hydroperoxy-γ-linolenic acid was produced by soybean lipoxygenase from γ-linolenic acid (O. Hiruta et al., J. Am. Oil Chem. Soc., 1988, **65**, 1911). The oxidative cleavage of linolenic acid by a protein fraction from mushrooms (Psalliota bispora) gave octadienols from an intermediate, (8E),(12Z),(15Z)-10-hydroperoxy-8,12,15-octadecatrienoic acid (M. Wurzenberger and W. Grosch, Lipids, 1986, **21**, 261). γ-Linolenic acid with soybean lipoxygenase at several pHs gave, upon reduction, 6,13-dihydroxyoctadecatrienoic acid isomers (M.R. Kim and D.E. Sok, Biochem. Biophys. Res. Commun., 1989, **159**, 1154). Studies with linolenic acid showed that double bonds at positions 3, 6 and 9 were required for conversion of polyunsaturated fatty acids into analogous products containing a cyclopentenone ring (B.A. Vick and D.C. Zimmerman, Plant Physiol., 1979, **63**, 490). Allene epoxides have been implicated as intermediates in the formation of ketols and cyclopentenones from linolenic acid (L. Crombie and D.O. Morgan, J. Chem. Soc., Chem. Commun., **1988**, 588; A.R. Brash et al., Proc. Natl. Acad. Sci. U.S.A., 1988, **85**, 3382).

Fatty acids can also be oxidized at the terminal or penultimate positons, ω or ω-1 oxidation. A polypoid mutant of Candida tropicalis converted 9-octadecenoic acid to 9-octadecen-1,18-dioic acid (R. Jiao et al., Kexue

Tongbao (Foreign Language Edit.), 1985, **30**, 1703). Terminal acetylene analogues of fatty acids, such as 11-dodecynoic acid, have been shown to specifically inactivate hepatic cytochrome P 450 enzymes that catalyze ω- and ω-1-hydroylation of lauric acid (P.R. Ortiz de Montellano and N.O. Reich, J. Biol. Chem., 1984, **259**, 4136; S. Shak et al., *ibid*, 1985, **260**, 13023; C.A. CaJacob and P.R. Ortiz de Montellano, Biochemistry, 1986, **25**, 4705; A.S. Muerhoff et al., J. Biol. Chem., 1989, **264**, 749).

Besides the biological oxidations described above, a number of model systems have been investigated in laboratories to correlate with the biological studies. The autoxidation system has attracted particular attention ("Autoxidation of Unsaturated Lipids" ed. H.W.-S. Chen, Academic Press, Orlando, 1987). The air oxidation of arachidonic acid gave a mixture of hydroperoxy arachidonic acid products; one of these was identified as 5-hydroperoxy-6,8,11,14-eicosatetraenoic acid, a proposed intermediate in the biosynthesis of the slow-reacting substances of anaphylaxis (N.A. Porter et al., Biochem. Biophys. Res. Commun., 1979, **89**, 1058). The autoxidation of linoleic and arachidonic acids as a function of cosubstrates was examined and a kinetic expression was derived that was useful in describing polyunsaturated fatty acid oxidation product mixtures (N.A. Porter et al., J. Am. Chem. Soc., 1981, **103**, 6447). The autoxidation of linoleate and arachidonate model membrane systems was studied (H. Weenen and N.A. Porter, *ibid*, 1982, **104**, 5216; K.E. Peers and D.T. Coxon, Chem. Phys. Lipids, 1983, **32**, 49). The product distribution of the autoxidation of methyl linoleate was investigated (F. Haslbeck, W. Grosch and J. Firl, Biochim. Biophys. Acta, 1983, **750**, 185). The autoxidation of linoleic acid gave small amounts of four diastereomeric dihydroxyoctadecadienoic acids (M. Hamberg, *ibid*, **752**, 353). Singlet oxygen oxidation of arachidonic acid gave eight hydroperoxides as initial products (N.A. Porter, J. Logan and V. Kontoyiannidou, J. Org. Chem., 1979, **44**, 3177; J. Terao and S. Matsushita, Agric. Biol. Chem., 1981, **45**, 587). Oxidation of linoleic acid catalyzed by hemoproteins and transition metal ions gave a large number of positional isomers of hydroperoxyoctadecadienoic acid (H.W.S. Chan and V.K. Newby, Biochim. Biophys. Acta, 1980, **617**, 353; J. Terao and S. Matsushita, Agric. Biol. Chem., 1981, **45**, 595). Oxidation of arachidonic acid by hydrogen peroxide and Cu(II) ion gave six hydroperoxyeicosatetraenoic acids (J.M. Boeynaems, J.A. Oates and W.C. Hubbard, Prostaglandins, 1980, **19**, 87; J. M. Boeynaems et al., Anal. Biochem., 1980, **104**, 259).

(d) Biological Reductions

Biological reduction of carbon-carbon double bonds, the effective biohydrogenation, in unsaturated monobasic acids has been studied in a variety of systems. The biochemical properties and substrate specificity of short- and long-chain rat liver microsomal (E)-2-enoyl coenzyme A reductase have been determined (M .N. Nagi et al., Arch. Biochem. Biophys., 1983, **226**, 50). The stereochemical mechanism of the reduction of (Z)-2-octenoyl coenzyme A catalyzed by (Z)-2-enoyl coenzyme A reductase from Escherichia coli was studied and the following features were found: (1) the pro-(4R) H atom of NADPH was incorporated into the C-3 position of octenoyl coenzyme A; (2) the H atom from the medium, water, was introduced into the C-2 position; and (3) the reduction occurred by an anti-addition of hydrogen via a 2-(Si), 3-(Re) attack on the cis-double bond (M. Mizugaki et al., Chem. Pharm. Bull., 1982, **30**, 2155). From a cell-free extract of the rumen anaerobe, Butyrivibrio fibrisolvens, was isolated a fraction containing an electron donor for the reduction of (9Z),(11E)-octadecadienoic acid to (11E)-octadecenoic acid (S. Yamazaki and S.B. Tove, J. Biol. Chem., 1979, **254**, 3812). The enzymatic degradation of (2E), (4Z)-decadienoyl coenzyme A was shown not to involve direct degradation by the β-oxidation cycle, but instead to be reduced by the NADPH-dependent 2,4-dienoyl coenzyme A reductase to 3-decenoyl coenzyme A, which was further degraded via the β-oxidation pathway (D. Cuebas and H. Schulz, *ibid*, 1982, **257**, 14140). The biohydrogenation of α-linolenic acid and γ-linolenic acid by Eubacterium lentum involved migration of the 12-double bond to the 11 position providing 9,11,15- and 6,9,11-octadecatrienoic acids, respectively, followed by reduction of the 9-double bond to give 11,15- and 6,11-octadecadienoic acids, respectively (G. Janssen et al., Biomed. Mass Spectrom., 1985, **12**, 134).

4. Individual Unsaturated Monobasic Acids

This section lists the titles of articles which describe the synthesis of specific unsaturated monobasic acids and is arranged by the number of carbon atoms in the specific acid.

(a) Unsaturated acids with 5-9 carbon atoms

"4-Pentynoic and 5-Hexynoic Acid from Jones Oxidation of 5-Hexynol" (G. Just et al., Synth. Commun., 1979, **9**, 613).

"Transition Metal-Catalyzed Rearrangement of Allyl But-3-enoate to Hepta-2,6-dienoic or Hepta-3,6-dienoic Acids" (G.P. Chiusoli, G. Salerno and F. Dallatomasina, J. Chem. Soc., Chem. Commun., **1977**, 793).

"Synthesis of Ethyl (Z)-4,7-Octadienoate" (O.P. Vig et al., Indian J. Chem., Sect. B, 1981, **20B**, 970).

"The Diels-Alder Reaction with 6-Methylene-7-octenoic Acid, a Functionalized Butadiene" (G. Cardinale, J.A.M. Laan and J.P. Ward, Recl. Trav. Chim. Pays-Bas, 1987, **106**, 62).

"Synthesis of trans-Non-4-en-2-ynoic Acid" (R. Vlakhov et al., Synth. Commun., 1986, **16**, 509).

(b) Unsaturated acids with 10-14 carbon atoms

"Copper(II)-Catalyzed Formation of Decenoic Acids from Manganese(III) Acetate and 1-Octene in Acetic Anhydride/Acetic Acid Mixtures" (W.J. De Klein, Rec. Trav. Chim. Pays-Bas, 1975, **94**, 151).

"Novel Flavor Component for Diary Products: (E)-6-Decenoic Acid" (T. Shirakawa et al., Dev. Food Sci., 1988, **18**, 915).

"A Simple Synthesis of Pellitorine (N-Isobutyl-E,E-2,4-decadieneamide) from the Butadiene Telomer" (J. Tsuji et al., Tetrahedron Lett., **1977**, 1917).

"Chemistry of Fungi. LXI. Synthesis of (±)-Sclerotiorin, (±)-4,6-Dimethyl-trans-2, trans-4-octadienoic Acid , and an Analog of Rotiorin" (R. Chang, R.R. King and W.B. Whalley, J. Chem. Soc. C, **1971**, 3566).

"Synthesis of Compounds Containing the Isoprene Unit. New Stereospecific Synthesis of the Geranyl and Farnesyl Skeleton" [(2E)-Geranic Acid] (G. Cardillo; M. Contento and S. Sandri, Tetrahedron Lett., **1974**, 2215).

"Asymmetric Synthesis of (+) and (-)-Citronellic Acids from Methylheptenone" (E. Stephan, G. Pourcelot and P. Cresson, Chem. Ind. (London), **1988**, 562).

"Synthesis of (±)-Methyl 2,5-Dimethyl-3-vinylhex-4-enoate (Methyl Santolinate)" (J. Boyd, W. Epstein and G. Frater, J. Chem. Soc. Chem. Commun., **1976**, 380).

"Pheromone Synthesis. Part 114. Simple Synthesis of a Mixture of (E)- and (Z)-5-Undecenoic Acid, Sex Pheromone of the Varied Carpet Beetle" (H. Harada and K. Mori, Agric. Biol. Chem., 1989, **53**, 1439).

"Stereospecific Synthesis of Homogeranic and Homoneric Acids" (P. Gosselin, C. Maignan and F. Rouessac, Synthesis, **1984**, 876).

"Simple Synthesis of 11-Dodecynoic Acid from Cyclododecanone" (L.I. Zakhrkin and I.M. Churilova, Izv. Akad. Nauk SSSR., Ser. Khim., **1984**, 2635).

"Synthesis of 5,8,11-Dodecatriynoic Acid and Its Use in the Synthesis of Arachidonic Acid and Related Acids" (R.I. Fryer, N.W. Gilman and B.C. Holland, J. Org. Chem., 1975, **40**, 348).

"Synthesis of trans-3, cis-5-Tetradecadienoic Acid (Megatomoic Acid), the Sex Attractant of the Black Carpet Beetle, and Its Geometric Isomers" (J.O. Rodin, M.A. Leaffer and R.M. Silverstein, *ibid*, 1970, **35**, 3152).

"Stereospecifc Synthesis of (Z,Z)-3,5-Tetradecadienoic Acid, a Component of Attagenus Elongatulus (Casey) Pheromone" (W.J. DeJarlais and E.A. Emken, J. Chem. Ecol., 1987, **13**, 179).

"Pheromone Synthesis. 4. A Synthesis of (±)-Methyl n-Tetradeca-trans-2,4,5-trienoate, an Allenic Ester Produced by the Male Dried Bean Beetle Acanthoscelides Obtectus (Say)" (P.J. Kocienski, G. Cernigliaro and G. Feldstein, J. Org. Chem., 1977, **42**, 353).

"Pheromone Synthesis. XLII. Synthesis of Optically Active Forms of Methyl (E)-2,4,5-Tetradecatrienoate, the Pheromone of the Male Dried Bean Beetle" (K. Mori, T. Nukada and T. Ebata, Tetrahedron, 1981, **37**, 1343).

"Synthesis of Some Acetylenic Acids" [All of the monoalkynoic acids, excluding both terminal and conjugated acetylenic acids, of chain length C_{10} to C_{14}] (N.W. Gilman and B.C. Holland, Chem. Phys. Lipids, 1974, **13**, 239).

(c) Unsaturated acids with 15-19 carbon atoms

"Stereoselective Synthesis of Alkyl (2E,4E)- and (2Z,4E)-3,7,11-Trimethyl-2,4-dodecadienoates. Insect Growth Regulators with Juvenile Hormone Activity" (C.A. Henrick et al., J. Org. Chem., 1975, **40**, 1).

"Approaches to the Synthesis of the Insect Juvenile Hormone Analog Ethyl 3,7,11-Trimethyl-2,4-dodecadienoate and Its Photochemistry" (C.A. Hendrick et al., J. Org. Chem., 1975, **40**, 8).

"Synthesis of Compounds Containing the Isoprene Unit. New Stereospecific Synthesis of the Geranyl and Farnesyl Skeleton" [(2E)-Farnesoic Acid] (G. Cardillo, M. Contento and S. Sandri, Tetrahedron Lett., **1974**, 2215).

"Polyunsaturated Acids. Part I. The Synthesis of Hexadeca-trans-2,trans-4,trans-6-trienoic (16:3) and Hexadeca-trans-2,trans-4,trans-6,trans-8-tetraenoic (16:4) Acid" (M.G. Hussain and F.D. Gunstone, Bangladesh J. Sci. Ind. Res., 1977, **12**, 215).

"A Practical Synthesis of (2E,6E,8E)-N-(2-Methylpropyl)-2,6,8-hexadecatrien-10-ynamide" (G.V.M. Sharma, D. Rajagopal and E.S. Rao, Synth. Commun., 1989, **19**, 3181).

"Fatty Acids. 33. Synthesis of All the Octadecynoic Acids and All the trans-Octadecenoic Acids" (J.A. Barve and F.D. Gunstone, Chem. Phys. Lipids, 1971, **7**, 311).

"Preparation of trans-2-Octadecenoic Acid" (D.R. Howton, Org. Prep. Proced. Int., 1974, **6**, 175).

"Two Geometrical Isomers of Linoleic Acid: Improved Total Syntheses" [(9Z),(12E)- and (9E),(12Z)-Octadecadienoic Acids] (T. Otsuki, R.F. Brooker and M.O. Funk, Lipids, 1986, **21**, 178).

"Semihydrogenation of 1-Chloroacetylenes and the Synthesis of Octadec-cis-10-en-5-ynoic Acid" (M.S.F. Lie, K. Jie and G. Tsang, Chem. Phys. Lipids, 1974, **13**, 21).

"Natural Acetylenes. XXXII. Synthesis of Crepenynic Acid (9-Octadecen-12-ynoic Acid)" (E.R.H. Jones et al., J. Chem. Soc. C, **1971**, 1156).

"1,4-Diene and 1,4-Enyne Synthesis via Dichloronorcarenol Cleavage. Synthesis of Crepenynic Acid" (T.L. Macdonald, Tetrahedron Lett., **1978**, 4201).

"Synthesis of Octadeca-trans-2, cis-9, cis-12-trienoic Acid and Its Evaluation as a Honey Bee Attractant" (A.N. Sarratt and R. Boch, Can. J. Biochem., 1971, **49**, 251).

"Synthesis of Methyl (E,Z,Z)-2 (or 4 or 6),9,12-Octadecatrienoate and Methyl (E,Z,Z)-2 (or 6 or 8),11,14-Eicosatrienoate" (R. Klok et al., Recl. Trav. Chim. Pays-Bas, 1980, **99**, 132).

"Polyunsaturated Acids. Part II. Preparation of Octadeca-trans-2,trans-4,trans-6-trienoic Acid (18:3) and Octadeca-trans-2,trans-4,trans-6,trans-8-tetraenoic (18:4) Acid" (F.D. Gunstone and M.G. Hussain, Bangladesh J. Sci. Ind. Res., 1978, **13**, 51).

"A New, General, and Stereoselective Synthesis of Long Chain Tetraenoic Acids Exemplified by β-Parinaric Acid" [(9E),(11E),(13E),(15E)-Octadecatetraenoic Acid] (T. Hayashi and T. Oishi, Chem. Lett., **1985**, 413).

(d) Unsaturated acids with 20 or more carbon atoms

"Synthesis of 2-Substituted cis-8,cis-11,cis-14-Eicosatrienoic Acids, Precursors for 2-Substituted Prostaglandins" (R. Van der Linde et al., Recl. Trav. Chim. Pays-Bas, 1975, **94**, 257).

"Synthesis of Substituted cis-8,cis-11,cis-14-Eicosatrienoic Acids, Precursors of Correspondingly Substituted Prostaglandins" (L. Heslinga et al., *ibid*, 262).

"Synthesis of 4,5,8-Eicosatrienoic Acids" (J.W. Patterson, Jr. et al., J. Org. Chem., 1983, **48**, 2572).

"Synthesis of Methyl (E,Z,Z)-2 (or 4 or 6),9,12-Octadecatrienoate and Methyl (E,Z,Z)-2(or 6 or 8),11,14-Eicosatrienoate" (R. Klok et al., Recl. Trav. Chim. Pays-Bas, 1980, **99**, 132).

"Synthesis of 5,8,11-Dodecatriynoic Acid and Its Use in the Synthesis of Arachidonic Acid and Related Acids" (R.I. Fryer, N.W. Gilman and B.C. Holland, J. Org. Chem., 1975, **40**, 348).

"Efficient Stereoselective Synthesis of Methyl Arachidonate via C_3 Homologation" (J. Viala and M. Santelli, *ibid*, 1988, **53**, 6121).

"Short, Stereocontrolled Syntheses of Irreversible Eicosanoid Biosynthesis Inhibitors. 5,6-, 8,9-, and 11,12-Dehydroarachidonic Acid" (E.J. Corey and J. Kang, Tetrahedron Lett., 1982, **23**, 1651).

"Synthesis of 8Z,11Z,14Z-Eicosatrien-5-ynoic (5,6-Dehydroarachidonic) Acid" (G.I. Myagkova et al., Bioorg. Khim., 1985, **11**, 1693).

"A New Synthesis of 5,8,11,14-Eicosatetraynoic Acid" (Yu. Yu. Belosludtsev et al., *ibid*, 1986, **12**, 1425).

"Synthesis of 5,8,11,14,17-Eicosapentaenoic Acid" (*ibid*, 1988, **14**, 100).

"Synthesis of Prostaglandin Synthetase Substrate Analogs. 2. (8Z,11Z,14Z)-15-Methyl-8,11,14-eicosatrienoic Acid" (M.I. Dawson et al., J. Med. Chem., 1977, **20**, 1396).

"Synthesis of (8Z,14Z)-13,13-Dimethyleicosa-8,14-dien-11-ynoic Acid as an Inhibitor of Prostaglandin Cyclooxygenase" (C-L. Yeh et al., Tetrahedron Lett., **1977**, 4257).

"Surface Pressure Dependence of Molecular Tilt in Langmuir-Blodgett Films of 22-Tricosenoic Acid" (I.R. Peterson et al., Thin Solid Films, 1988, **161**, 325).

"Phospholipid Studies of Marine Organisms. 7. Stereospecific Synthesis of (5Z,9Z)-, (5Z,9E)-, (5E,9Z)-, and (5E,9E)-5,9-Hexacosadienoic Acid" (P.L. Mena, O. Pilet and C. Djerassi, J. Org. Chem., 1984, **49**, 3260).

Second Supplements to the 2nd Edition of Rodd's Chemistry of Carbon Compounds, Vol. 1C, edited by M. Sainsbury
© 1992 Elsevier Science Publishers B.V., Amsterdam

Chapter 10

CARBON MONOXIDE AND ITS DERIVATIVES, CARBONIC ACID AND ITS DERIVATIVES

I.G.C. COUTTS

1. Reactions of carbon monoxide without transition-metal catalysis

The majority of these reactions involve organometallic reagents. They have been generally surveyed (C. Narayana and M. Periasamy, Synthesis, 1985, 253) and a review has appeared on the carbonylation of organoboranes (H.C. Brown, Organic Synthesis via Boranes, Wiley-Interscience, 1975)

The most studied reactions are the carbonylation of lithium carbanions and nitrogen anions yielding unstable and highly reactive acyl- and carbamoyl lithiums of limited synthetic potential (V. Rautenstrauch and M. Joyeux, Angew. Chem. Int. Ed., 1979, **18**, 83) However, treatment of butyl lithiums at -110° with CO in the presence of ketones or esters gives reasonable yields of α-hydroxyketones or α-diketones respectively (D. Seyferth, R.M. Weinstein and W-L. Wang, J. Org. Chem., 1983, **48**, 1144) Silylmethyl lithiums (**1**) react with CO to give (**2**) which undergo a 1,2-silicon shift to form the *E*-enolates (**3**) (S. Murai *et al.*, J. Amer. Chem. Soc., 1984, **106**, 2440)

$$\underset{(\mathbf{1})}{\text{R-CH(SiMe}_3\text{)Li}} \rightarrow \underset{(\mathbf{2})}{\text{R-CH(SiMe}_3\text{)COLi}} \rightarrow \underset{(\mathbf{3})}{\text{RCH=C(SiMe}_3\text{)OLi}}$$

Carbamoyl lithiums, formed by carbonylation of the lithium salts of secondary amines, give no product with silyl chlorides, but react with trialkyltin chlorides to give trialkylcarbamoylstannanes (**4**) (P. Jutzi and F.W. Schroder, Angew. Chem. Int. Ed., 1971, **10**, 339.; C.M. Lindsay and

D.A. Widdowson, J. Chem. Soc., Perkin Trans., I, 1988, 569)

$$R'_2NCOLi + R_3SnCl \rightarrow R_2'NCOSnR_3 \quad \textbf{(4)}$$

Glyoxylamides and hydroxymalonamides, isolated from reaction of lithium dialkylamides with CO, may arise by carbon-carbon formation between the dialkylamide and carbamoyllithium intermediates (D.G. Perez and N.S. Nudelman, J. Org. Chem., 1988, **53**, 405)

On treatment with elemental sulphur at -78°, carbamoyl lithiums form *S*-alkyl thiocarbamates (T. Mizuno *et al.*, Tetrahedron Lett., 1991, **32**, 6867) In the presence of sulphur, and a tertiary base (eg DBU), alcohols are carbonylated at high pressure to *S*-alkylthiocarbonates, $RO(CO)S^-HN^+R'_3$ (T. Mizuno *et al.*, Tetrahedron Lett., 1988, **29**, 4767), which are converted smoothly by $CuCl_2$ to carbonates $(RO)_2CO$ (T. Mizuno *et al.*, Synthesis, 1989, 636)

Reaction of cation (**5**) with CO and quenching with methanol yields ester (**6**), *E* : *Z* ratio 93 : 7 (D. Farcasiu, R. Rich and K.D. Rose, J. Org. Chem., 1989, **54**, 4582)

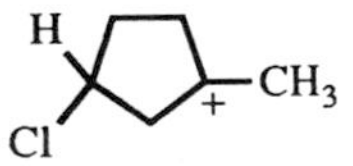

H, Cl, CH_3, +
(**5**)

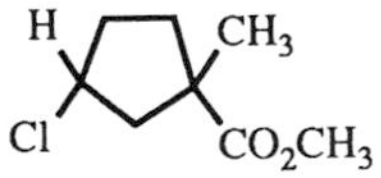

H, Cl, CH_3, CO_2CH_3
(**6**)

The carbonylation of methanol and dimethyl ether, catalysed by superacids (eg $HF\text{-}BF_3$) gives high yields of acetic acid and methyl acetate (A. Bagno, J. Bukala and G.A. Olah, J. Amer. Chem. Soc., 1990, **112**, 4284)

Alkyl or aryl halides react with CO under free-radical conditions to afford aldehydes (I. Ryu *et al.*, J. Amer. Chem. Soc., 1990, **112**, 452, Tetrahedron Lett., 1990, **31**, 6887).

$$RX + CO/Bu_3SnH/AIBN \rightarrow [R\cdot \rightleftharpoons R\dot{C}O] \rightarrow RCHO$$

2. Reactions of carbon monoxide with transition-metal catalysis

As reactions of synthetic importance in industry and the laboratory, carbonylations catalysed by transition metals have been the subject of several reviews. General oversights are provided by R.A. Sheldon, (Chemicals from Synthetic Gas, Reidel, Dordrecht, 1983) and I. Tkatchenko, (Comprehensive Organometallic Chemistry, G. Wilkinson Ed., Pergamon Press, 1982, Vol 8, pp 101-223), while surveys have appeared on the catalysed incorporation of carbon monoxide into olefins (R.H. Crabtree, The Organic Chemistry of the Transition Metals, Wiley, 1987, p 185) and into nickel-allyl complexes (G.P. Chiusoli, Acc. Chem. Research, 1973, 422). There follows a selection of reports illustrating recent advances, or containing useful lists of cited references.

Heck reaction Since the initial announcement (A. Schoenberg and R.F. Heck, J. Org. Chem., 1974, **39**, 3327) that high yields of amides and esters can be obtained from treatment of aromatic bromides or iodides with a catalytic amount of a Pd(O) or Pd(II) species and a primary or secondary amine or alcohol in the presence of carbon monoxide and base, a large number of application have been published. These have been reviewed (R.F. Heck, Palladium Reagents in Organic Synthesis, Academic Press, New York 1985) and have been comprehensively cited (R.J. Perry and S.R. Turner, J. Org. Chem., 1991, **56**, 6573)

The reaction of aryl iodides with CO to give corresponding carboxylic acids is catalysed by inexpensive nickel cyanide, provided a phase-transfer catalyst is present; for the carbonylation of benzyl chlorides under these conditions, the addition of lanthanide chlorides is beneficial. (I. Amer and H. Alper, J. Org. Chem., 1988, **53**, 5147; J. Amer. Chem. Soc., 1989, **111**, 927) Aryl chlorides can be used instead of bromides or iodides if the reaction is catalysed by Pd(O) liganded to bis(diisopropylphosphino) propane (Y. Ben-David, M. Portnoy and D. Milstein, J. Amer. Chem. Soc., 1989, **111**, 8742), or by palladium supported on activated carbon with added $K_2Cr_2O_7$ (V. Dufaud, J. Thivolle-Cazat and J-M. Basset, J. Chem. Soc., Chem. Commun., 1990, 426).

Organic tellurides, available by the addition of PhTeH to arylacetylenes or propargylic alcohols, react with CO (1 bar) in methanol with Pd(II) to give methyl carboxylates in high yield. (K. Ohe *et al.*, J. Org. Chem. 1987, **52**, 4859)

(*Z*)-RCH=CHTePh + CO → (*Z*)-RCH=CHCO_2Me

Chiral 1-arylethyl esters can be carbonylated directly to 2-arylpropanoic acids with some retention of chirality, particularly at low substrate conversion; eg (*S*)-naproxen was obtained with 82% ee at 20% yield. (J.M. Baird *et al.*, J. Org. Chem., 1991, **56**, 1928)

The reaction of ethyl chloroformate, palladium diacetate, triethylamine and carbon monoxide (1 bar, 20°) gives a good yield of diethyl malonate; ^{13}C labelling shows that the methylene group of the malonate comes from the acetate of the palladium catalyst. (D.H.R. Barton *et al.*, Tetrahedron Lett., 1990, **31**, 325). Vinylmercurials are efficiently carbonylated to divinyl ketones at atmospheric pressure using a $[Rh(CO)_2Cl]_2$/LiCl catalyst (R.C. Larock and S.S. Hershberger, J. Org. Chem., 1980, **45**, 3840). Salts of aromatic N-chlorosulphonamidates react with CO, with palladium catalysis, to yield arylsulphonyl isocyanates (G. Besenyei, S. Nemeth and L.I. Simandi, Angew. Chem. Int. Ed., 1990, **29**, 1147)

Arenediazonium tetrafluoroborates, treated with CO, sodium carboxylates and palladium acetate, give mixed acid anhydrides (K. Kikukawa *et al.*, J. Org. Chem., 1981, **46**, 4413); carbonylation of a mixture of a diazonium salt and a tetraalkyltin reagent affords alkylaryl ketones (K. Kikukawa *et al.*, J. Chem. Soc., Perkin Trans. I, 1987, 1511)

ArN_2X + CO + RCO_2Na → $ArCO_2COR$
ArN_2X + CO + R_4Sn → ArCOR

Aryl triflates couple with organostannanes in the presence of CO to give good yields of aryl ketones if the reaction is catalysed by dichloro [1,1'-bis(diphenylphosphino)ferrocene]palladium (II) (A.M. Echavarren and J.K. Stille, J. Amer. Chem. Soc., 1988, **110**, 1557); this coupling reaction is the key step in a synthesis of the macrocyclic diterpene jatrophone (A.G. Gyorkos, J.K. Stille and L.S. Hegedus, J. Amer. Chem. Soc., 1990, **112**, 8465)

Dichloromethane reacts with palladium complexes containing bulky ligands to form intermediates (**7**) which release keten on carbonylation (M. Huser, M-T. Youinou and J.A. Osborn, Angew. Chem. Int. Ed., 1989, **28**, 1386)

$[Pd(PCy_3)_2(dba)] + CH_2Cl_2 \rightarrow ClPd(PCy_3)_2CH_2Cl$ (**7**)

(**7**) + CO $\rightarrow$ $CH_2{=}C{=}O$ (Cy = cyclohexyl, dba = dibenzylideneacetone)

Although the carbonylation of polychloroalkanes usually requires harsh conditions, carbon tetrachloride can react with CO (100 bar, 120°) to give Cl_3CCOCl in a catalytic cycle using ternary metallic salts (E. Monflier *et al.*, J. Chem. Soc., Chem. Commun., 1992, 439)

Methyl formate is selectively converted to acetaldehyde by CO (80 bar, 180°) with ruthenium catalysis (D. Vanhoye *et al.*, Angew. Chem. Int. Ed., 1988, **27**, 683) Isomerisation of methyl formate to acetic acid is best carried out in N-methylpyrrolidone under carbon monoxide pressure in the presence of palladium or cobalt compounds and lithium iodide (G. Jenner, Tetrahedron Lett., 1990, **31**, 3887)

The intermediate (**8**), formed between CO and $Et_2NH_2NiI_2$ can be isolated and used to generate new carbon-carbon bonds. (H. Hoberg and F.J. Fananas, Angew. Chem. Int. Ed. 1985, **24**, 325)

$(Et_2NH)_2Ni(I)CONEt_2 + CO + Et_2NH \rightarrow Et_2NCOCONEt_2$
(**8**)

(**8**) + PhC≡CPh $\rightarrow$ Ph(Et₂NCO)C=C(Ph)Ni(I)(NHEt₂)

The reaction of higher order mixed allylic cyanocuprates with CO at low temperatures, followed by addition of an enone, leads to products of 1,4-acylation which contain an allyl-CO moiety (B.H. Lipshutz and T.R. Elworthy. Tetrahedron Lett., 1990, **31**, 477)

Cinnamyl acetates undergo a cyclocarbonylation catalysed by $PdCl_2(PPh_3)_2$, to 1-acetoxynaphthalenes. From 3-heteroallyl acetates there can be obtained benzofurans, benzothiophenes, indoles and carbazoles, with best results being achieved at 170°/50-70 bar (M. Iwasaki *et al.*, Chem. Lett., 1988, 1159; J. Org. Chem., 1991, **56**, 1922) Hydroformylation of N-allylacetamide over ruthenium catalysts is chelation controlled, giving **(9)** and the double carbonylation product **(10)** with high selectivity (I. Ojima and Z. Zhang, J. Org. Chem., 1988, **53**, 4425)

$$CH_3CONHCH_2CH{=}CH_2 \longrightarrow CH_3CONHCH_2CH(CHO)CH_3 + \text{(10)}$$

(9) **(10)**

Carbonylative coupling of a 2-haloaniline with terminal alkynes in the presence of Pd(O) species gives good yields of 4-quinolones (S. Torii, H. Okumoto and L.H. Xu, Tetrahedron Lett., 1991, **32**, 237)

$$+ \quad RC{\equiv}CH \quad + \quad CO \longrightarrow$$

The direct carbonylation of nitroaromatics to isocyanates or carbamates, avoiding the use of phosgene, is catalysed by a ruthenium carbonyl.

$$ArNO_2 + 3CO + CH_3OH \rightarrow ArNHC(O)OCH_3 + 2CO_2$$

Activation of the nitro group may occur by a one-electron reduction by the Ru(0) complex. (A.J. Kunin, M.D. Noirot and W.L. Gladfelter, J. Amer. Chem. Soc., 1989, **111**, 2739)

A mixed metal cluster-derived catalyst, $[(FePdPt(CO)_4(Ph_2P_2CH_2PPh_2)]_2$ allows the efficient carbonylation of 2-nitrophenol to benzoxazol-2-one, a structural element of several herbicides (P. Braunstein, J. Kervennal and J-L. Richert, Angew. Chem. Int. Ed., 1985, **24**, 768)

An interesting synthesis of isoindolinones **(12)** from aryl halides uses sequential incorporation of molecular nitrogen and carbon monoxide at atmospheric pressure (Y. Uozumi *et al.*, J. Amer. Chem. Soc., 1989, **111**, 3725)

$$TiCl_4 + THF + Mg + N_2 \longrightarrow [THFMg_2Cl_2TiN] \xrightarrow{CO_2} 3THF.Mg_2Cl_2OTiNCO$$

(11)

(11) + [2-(RCH₂CO)C₆H₄–Pd(PPh₃)₄] ⟶ **(12)**

Metal-catalysed carbonylation of cyclopropenes gives vinylketen-metal complexes of considerable synthetic value. The formation of α-pyrones **(13)** from a cyclopropene ester is illustrative:-

$$\text{cyclopropene ester} + CO \xrightarrow{[ClRh(CO)_2]_2} \textbf{(13)}$$

(S.H. Cho and L.S. Liebeskind, J. Org. Chem., 1987, **52**, 2634; this paper contains a comprehensive citation of related results)

Oxidative alkoxycarbonylation of dipropargylamines, catalysed by palladised charcoal, gives pyrrolidines **(14)** which isomerise to pyrrole-3,4-diacetic acids **(15)** (G.P. Chiusoli, M. Costa and S. Reverberi, Synthesis, 1989, 262)

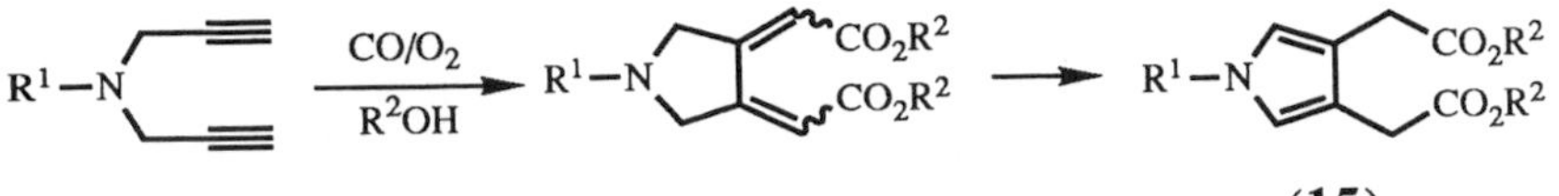

(14) (15)

Pyrroles are also formed by a reaction of CO with zirconocene complexes of imines **(16)** (S.L. Buchwald, M.W. Wannamaker and B.T. Watson, J. Amer. Chem. Soc., 1989, **111**, 776)

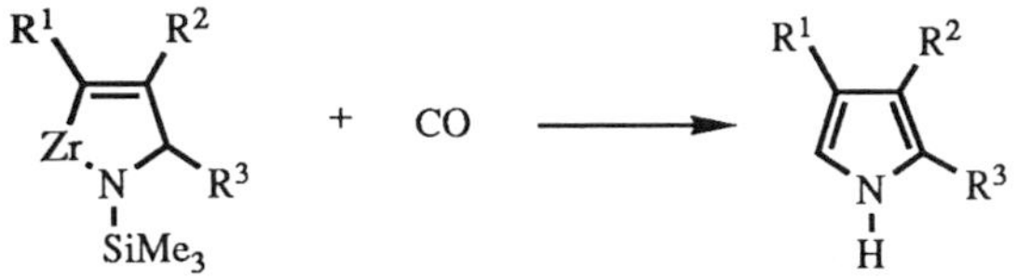

(16)

The carbonylation of zirconabicycles has also been used to form bicyclic enones (E-I Negishi *et al.*, J. Amer. Chem. Soc., 1989, **111**, 3336)

Rhodium (1) complexes catalyse the stereospecific and enantiospecific ring expansion-carbonylation of aziridines to β-lactams, the process occurring with retention of configuration (S. Calet, F. Urso and H. Alper, J. Amer. Chem. Soc., 1989, **111**, 931)

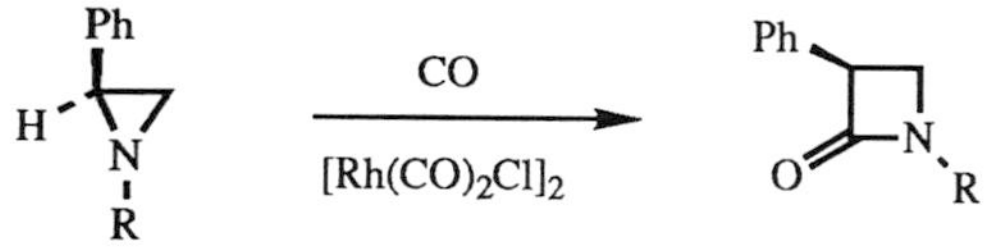

Hexamethyldisilane is oxidised by CO in the presence of a Ni-kieselguhr catalyst to siloxanes (K.P.C. Vollhardt and Z-Y. Yang, Angew. Chem. Int. Ed., 1984, **23**, 460)

$$Me_3Si\text{-}SiMe_3 + CO \rightarrow Me_3SiOSiR_3 + R_3SiOSiR_2OSiR_3.....$$

3. Metal carbonyls

The reactions of carbon monoxide incorporated into metal carbonyls has been reviewed (A.L. Lapidus and M.M. Savel'ev, Russ. Chem. Rev. (Engl. Transl.), 1988, **57**, 17; D.F. Shriver and M.J. Sailor, Acc. Chem. Res., 1988, **21**, 374)

Ruthenium. Tetra-μ^3-carbonyldodecacarbonylhexarhodium, $Rh_6(CO)_{16}$ is used to catalyse selective reductions. Thus 1,5-*cyclo*octadiene is converted to *cyclo*octene by transfer hydrogenation with isopropanol and the rhodium cluster (J. Kaspar, R. Spogliavich and M. Giaziani, J. Organomet. Chem. 1985, **281**, 299) Reaction of diphenylacetylene with CO and aromatic nitro compounds catalysed by $Rh_6(CO)_{16}$, gives high yields of *N*-aryl-2,3-diphenylmaleimides; the CO performs the dual function of reducing and carbonylating agent (A.F.M. Iqbal, Angew. Chem. Int. Ed., 1972, **11**, 634). With the same catalyst it is possible to hydrogenate selectively the pyridine ring of quinolines and isoquinolines under water-gas shift conditions, isoquinolines are further carbonylated to give *N*-formyl-1,2,3,4-tetrahydroisoquinolines as final products (S.-I. Murahashi, Y. Imada and Y. Hirai, Tetrahedron Lett., 1987, **28**, 77)

Chromium. Tetralins are oxidised to α-tetralones by *tert*-butylhydroperoxide and catalytic amounts of $Cr(CO)_6$ (A.J. Pearson and G.P. Han, J. Org. Chem., 1985, **50**, 2791). For other examples of allylic oxidations with this system see A.J. Pearson *et al.*, Tetrahedron Lett., 1984, **25**, 1235; J. Chem. Soc., Perkin Trans. I, 1985, 267 . Complexes between $Cr(CO)_6$ and carbenes undergo cycloadditions with alkynes, giving phenols by CO incorporation (M.F. Semmelhach *et al.*, J. Amer. Chem. Soc., 1987, **109**, 4397 and references therein)

$(CO)_5Cr{=}C(OMe)Ph$ + $R^1{-}C{\equiv}C{-}R^2$ ⟶ (naphthol: OH, R^1, R^2, X)

X = H, OMe.

Cobalt. These metal carbonyls have been reviewed by J.F. Petignani (The Chemistry of the Metal-Carbon Bond, F.R. Hartley Ed., Wiley, 1989, Vol

5, pp 63-106). Among recent reactions mediated by these versatile catalysts are the following.

1,4-Dienes are carbonylated to pentenones (**17**)

(**17**)

This reaction has been used in a synthesis of cuparenone (D. Eilbracht, E. Balss and M. Acker, Chem. Ber., 1985, **118**, 825). Benzylic thiols are converted by CO to phenylacetic acids (S.C. Shin, S. Antelii and H. Alper, J. Org. Chem., 1985, **50**, 141), while benzyl acetates are carbonylated in the presence of Me_3SiH to β-phenethyl alcohols (N. Chatani *et al.*, J. Org. Chem., 1990, **24**, 5924). Similarly, an acetoxy group at the anomeric centre of a glycoside may be selectively replaced by a siloxymethyl group on reaction with CO and a hydrosilane; the conversion proceeds with inversion (N. Chatani *et al.*, J. Org. Chem., 1988, **53**, 3389). Vinyl epoxides are regioselectively carbonylated to β-hydroxy acids if a phase-transfer catalyst and iodomethane are also present in the system. Thus epoxide (**18**) gives (**19**) and (**20**) in a 95:5 ratio, and under similar conditions (**21**) is transformed to (**22**) by a triple carbonylation (H. Alper and S. Calet, Tetrahedron Lett., 1988, **29**, 1763; H. Alper, A. Eisenstat and N. Satyanarayana, J. Amer. Chem. Soc., 1990, **112**, 7060 and references cited therein).

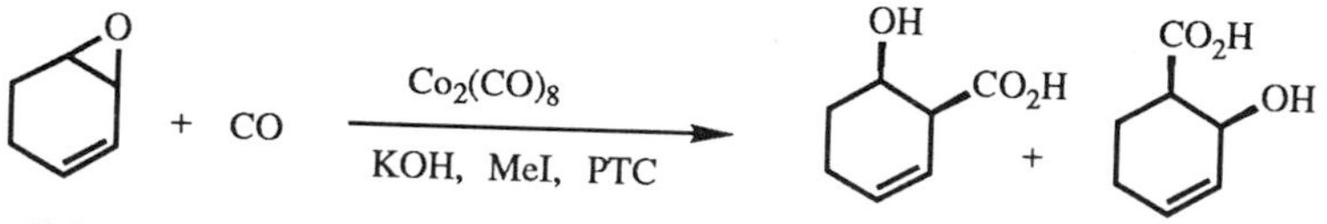

+ CO; $Co_2(CO)_8$; KOH, MeI, PTC

(**18**) (**19**) (**20**)

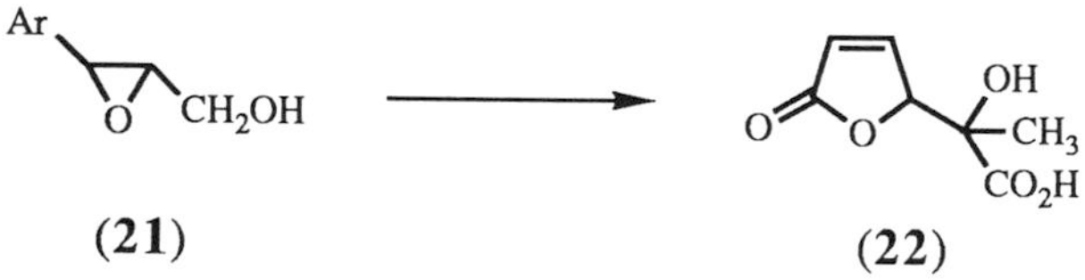

(**21**) (**22**)

The stereoselectivity of $Co_2(CO)_8$ - catalysed carbonylations is very sensitive to the nature of base and solvent present in the reaction, as exemplified by the conversion of both(*E*) and (*Z*)-β-bromostyrene to *E*-cinnamic acid (M. Miura *et al.*, J. Chem. Soc., Perkin Trans. I, 1989, 73)

Iron. Comprehensive reviews have appeared on iron pentacarbonyl (H. Alper, Organic Synthesis *via* metal carbonyls, I. Wender, P. Dino Eds., Wiley, 1977, Vol 2, pp 545-593), the formation and reactions of cyclohexadiene-$Fe(CO)_5$ complexes (A.J. Pearson, Acc. Chem. Research, 1980, **13**, 463) and the reactions of tetracarbonylferrates (J.P. Collman, Acc. Chem. Research, 1976, **9**, 342; J.J. Brumer, Chem. Rev., 1990, **90**, 1041)

An iron carbonyl, $[(C_5H_5)Fe(CO)(PPh_3)]$ has been extensively used as a chiral auxiliary for asymmetric synthesis (S.G. Davies, Aldrichimica Acta, 1990, **23**, 31). The recent increasing interest in the use of metal carbonyl complexes to achieve enantioselective preparations has been surveyed (W.R. Roush and J.C. Park, J. Org. Chem., 1990, **55**, 1143)

4. Isocyanides, isonitriles

Preparation

The development of useful synthetic routes to isocyanides has led to their emerging as increasingly important functional groups.

1. In the most general route to isonitriles, the dehydration of N-monosubstituted formamides, phosgene with triethylamine or pyridine is an efficient combination for large scale high yield reaction; effective, but less hazardous dehydration is achieved if phosgene is replaced by diphosgene (G. Skorna and I. Ugi, Angew. Chem. Int. Ed., 1977, **16**, 259) or triphosgene (H. Eckert and B. Forster, Chem. Int. Ed., 1987, **26**, 894). Formamides are also converted to isonitriles by the Vilsmeier reagent $SOCl_2$ -DMF (H.M. Walborsky and G.E. Nizik, J. Org. Chem., 1972, **37**, 187). Air sensitive epoxyisonitriles **(23)** have been prepared from epoxyformamides (J.E. Baldwin and I.A. O'Neil, Tetrahedron Lett., 1990, **31**, 2047)

OCHNH O $\xrightarrow[\text{Pr}^i_2\text{NEt, -78°}]{(CF_3SO_2)_2O}$ CN O + CN

(23)

2. If the classic Hofmann carbylamine reaction, involving a primary amine, chloroform and strong base, is carried out under phase transfer condition good yields of isonitriles are obtained (W. W. Gokel, R. P. Widera and W. P. Weber., Org. Synth. 1976, **55**, 96)

3. Alkyl halides, especially iodides, are transformed in high yield to isocyanides by reaction with $Me_4NAg(CN)_2$ (L.B. Engemyr *et al.*, Acta. Chem. Scand., 1974, **A28**, 255) Silver cyanide converts *erytho* and *threo*-2-methylthio-3-halobutanes to isonitriles with complete retention of configuration (J.C. Carretero and J.L. Garcia-Ruano, Tetrahedron Lett., 1985, **25**, 3381), and reacts with acyl iodides or imidoyl bromides to give N-acyl or N-imidoylisonitriles (G. Höfle and B. Lange, Angew. Chem. Int. Ed., 1977, **16**, 262, 727)

4. Epoxides and oxetanes undergo ring opening with trimethylsilylcyanide and zinc iodide, yielding trimethylsilyl ethers of β- or γ-hydroxyisonitriles respectively (P.G. Gassman and L.M. Haberman, Tetrahedron Lett., 1985, **26**, 4971)

5. A variety of aryl and heteroaryl isocyanides can be obtained by thermal decomposition of iminoisoxazolones **(24)** (C. Wentrup V Stutz and H. J. Wollweber., Angew. Chem. Int. Ed., 1978, **17**, 688)

Ph N O O + RN=O ⟶ Ph NR N O O $\xrightarrow{\Delta}$ RN≡C

(24)

6. Oxidation of 5-alkyl(aryl)aminotetrazoles with lead tetraacetate or sodium

hypobromite give moderate to good yields of isonitriles (G. Höfle and B. Lange, Angew. Chem. Int. Ed., 1976, **15**, 113)

7. Reductive routes to isonitriles include the reaction of isocyanates with trichlorosilane-triethylamine (J.E. Baldwin *et al.*, J. Chem. Soc., Chem. Commun., 1982, 942) and electroreductive detosylation of mono- and dialkyl derivatives of p-toluenesulphonyl methylisocyanide (U. Hess, H Brosig and W. P. Fehlhammer, Tetrahedron Lett., 1991, **32**, 5229)

Reactions

Addition reactions. Isonitriles can be characterised as rhodium complexes. (A.W. Hanson, A.J. McAlees and A. Taylor, J. Chem. Soc., Perkin Trans 1, 1985,441) Selenium catalyses the addition of elemental sulphur to isonitriles, giving high yields of aryl and isothiocyanates (S-I. Fujiwara *et al.*, Tetrahedron Lett., 1991, **32**, 3503) Carbene complexes react with alkenyl isocyanides to yield ketenimine complexes, which on addition of further isocyanide form stable azafulvene complexes (**25**) (R. Aumann *et al.*, Angew. Chem. Int. Ed., 1987, **26**, 563)

$(OC)_5W{=}C(OEt)Ph$ + 2RCH=CHN≡C ⟶ R, N, R, Ph, N, OEt, $(OC)_5W$

(25)

Trifluoromethylisocyanide reacts with hydrogen halides or trifluoroacetic acid to give N-trifluoromethylmethanimines and formamides (P. Lentz I Brudgam and H Hartl, Angew. Chem. Int. Ed., 1987, **26**, 921)

$$CF_3NHCHO \xleftarrow{CF_3CO_2H} CF_3N{\equiv}C \xrightarrow{HX} CF_3NH{=}CX$$

Sequential treatment of benzotriazolyl isonitriles with primary amines and Grignard reagents constitutes an efficient route to unsymmetric formamidines (A.R. Katritzky, M. Sutharchanadevi and L. Urogdi, J. Chem. Soc., Perkin Trans. 1, 1990, 1847)

$$BtCH(R')NC + HNR_2^2 \rightarrow BtCH(R')N{=}CHNR_2^2 \quad \mathbf{(26)}$$

(26) + R^3MgX → $R^3CH(R')\text{-}N{=}CHNR^2_2$ (Bt = benzotriazolyl)

Cyclopropenimines have been isolated from the [2+1] addition of isonitriles to strained cycloalkynes (A. Krebs and H. Kimling, Angew. Chem. Int. Ed., 1971, **10** 409). From the [3+1] addition of N-neopentylidene-*tert*.butylamine N-oxide to alkyl or arylisocyanides are obtained stable 4-imino-1,2-oxazetidines (D. Moderhack and M. Lorke, Angew. Chem. Int. Ed., 1980, **19**, 45). Isocyanides undergo [4+1] cycloaddition reactions with α,β unsaturated ketones activated by electron-withdrawing groups (E. Vilsmaier, R. Baumheier and M. Lemmert, Synthesis, 1990, 995)

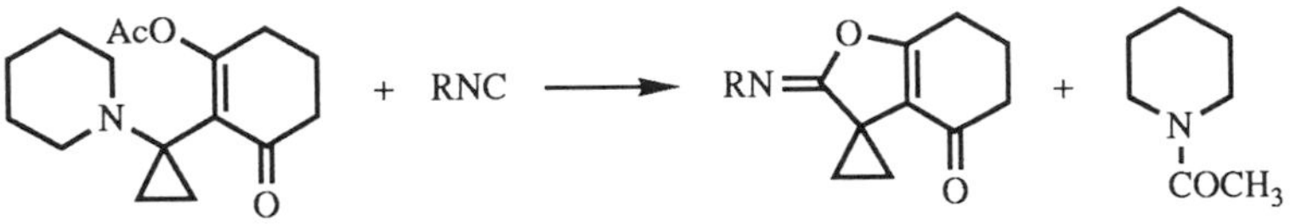

The use in the synthesis of natural products of "four-component" condensations of isonitriles with amines, carbonyl compounds and acid components has been reviewed (I. Ugi, Angew. Chem. Int. Ed., 1982, **21**, 810)

Preparation of a carbopenem intermediate **(27)** illustrates the particular suitability of these condensations in β-lactam synthesis.

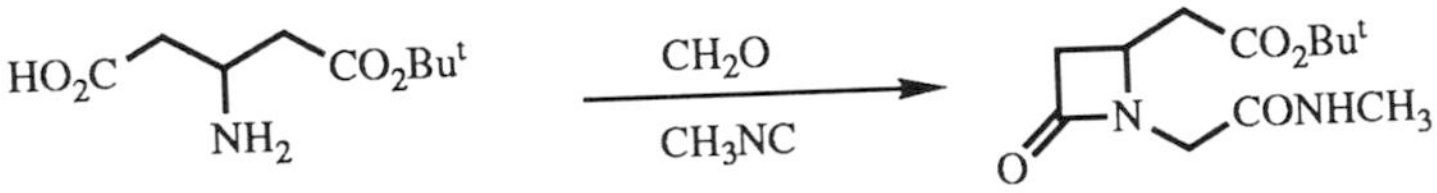

(27)

Moderhack (Synthesis, 1984, 1083) has produced a more general survey (1969-84) of the preparation of four-membered rings from isocyanides.

A recent review (C. Rüchardt *et al.*, Angew. Chem. Int. Ed., 1991, **30**, 893) discusses mechanistic and synthetic aspects of the rearrangement of isocyanides to cyanides. With flash pyrolysis side reactions are suppressed, allyl isocyanides rearrange without allyl isomerisation, and chiral isocyanides retain configuration. The transformation has been used in efficient synthesis of anti-inflammatory drugs.

Reactions with organometallic reagents. With isocyanides not possessing an α-hydrogen atom, organolithum or Grignard reagents undergo α-addition to form metalloaldimines (M.P. Periasamy and H.M. Walborsky, Org. Prep. Proceed. Int., 1979, **11**, 293 and cited references).

$$RNC + R'M \rightarrow RN{=}C(R')M$$

Readily available 1,1,3,3-tetramethylbutyl isocyanide (H.M. Walborsky and G.E. Niznik, J. Org. Chem., 1972, **37**, 187] is a convenient source of lithium aldimines, which act as masked acyl carbanions

$$RCOCO_2H \xleftarrow[\text{ii) } H_3O^+]{\text{i) } CO_2} Me_3CCH_2CMe_2N{=}C(R)Li \xrightarrow{D_3O^+} RCD{=}O$$

If the isocyanide possesses an α-hydrogen, organometallic reagents abstract a proton forming *α-metallated isocyanides*. The synthetic uses of these versatile intermediates has been extensively surveyed (U. Schollköpf Angew. Chem. Int. Ed., 1977, **16**, 339; P. Beak and D.B. Reitz, Chem. Rev. 1978, **78**, 275) and involve the usual reactions of carbanions with alkylating or acylating agents and with carbonyl compounds. Illustrative examples of the usefulness of these intermediates are their application to the chain elongation of primary amines (U. Schöllkopf *et al*, Justus Liebigs Ann. Chem., 1977, 40), the preparation of oxazoles (A.P. Kozikowski and A. Ames, J. Amer. Chem. Soc., 1980, **102**, 860) and the conversion of α-isocyanoacetic acid derivatives to α-branched amino acids such as α-methyldopa (U. Schőllkopf, D. Hoppe and R. Jentsch, Chem. Ber., 1975, **108**, 1580).

p-*Toluenesulphonylmethyl isocyanide (TosMic)*. This stable isocyanide has found application in a variety of syntheses. It is commercially available, or may be obtained by a Mannich reaction of *p*-toluenesulphonic acid, formaldehyde and formamide to given N-tosylmethylformamide, followed by dehydration with $POCI_3$(B. E. Hoogenboom, O. H. Oldenziel and A. M. van Leusen, Org.Synth., Coll Vol. **6**, 987). Treatment of the iscyanide with base yields the α carbanion which can be alkylated or acylated. Thus dihalides (**28**) may be transformed to diketones (**29**), (A. M. van Leusen *et al.*, Recl. Trav. Chim. Pays-Bas, 1988, **104**, 50), while ketones undergo homologation to α, β-enones (**30**) or keto-alcohols (**31**). The latter sequence has been applied to the synthesis of 21-hydroxy - 20-ketosteroids (D. van Leusen and A. M. van Leusen, Tetrahedron Lett., 1984, **25**, 2581; J Moskal and A. M. van Leusen *ibid*, 1984, **25**, 2585)

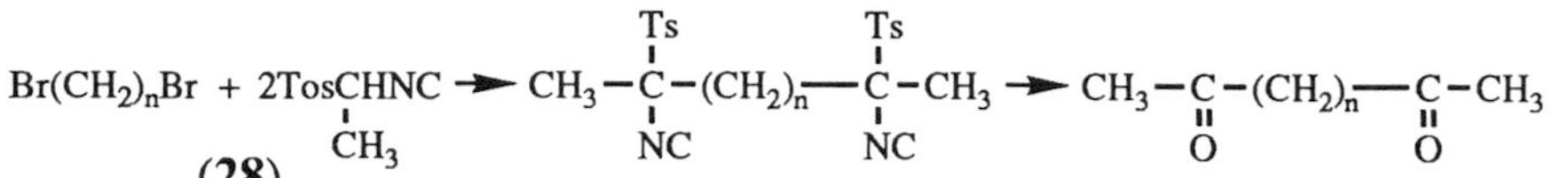

$$Br(CH_2)_nBr + 2TosCH(CH_3)NC \rightarrow CH_3{-}C(Ts)(NC){-}(CH_2)_n{-}C(Ts)(NC){-}CH_3 \rightarrow CH_3{-}C(=O){-}(CH_2)_n{-}C(=O){-}CH_3$$

(**28**) n = 3, 4. n = 3, 4. (**29**)

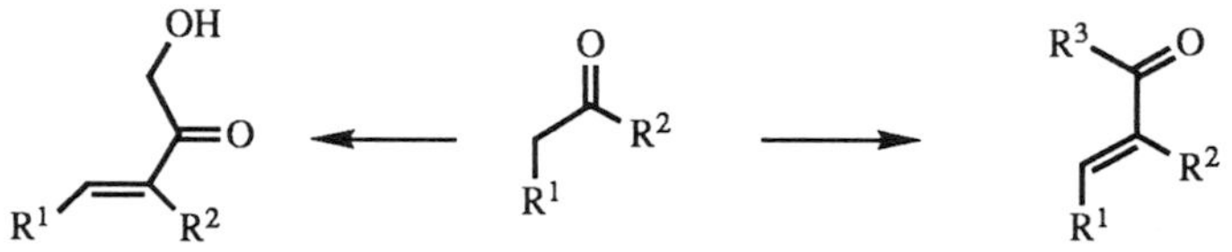

(**31**) (**30**)

TosMic converts ketones to nitriles in a single operation, with the nitrile carbon originating from the methylene of the reagent (O. H. Oldenziel, D. van Leusen and A. M. van Leusen, J. Org. Chem., 1977, **42**, 3114). The synthesis of heterocycles by the cycloaddition of tosylmethylisocyanide to substrates is exemplified by the preparation of imidazole (**32**) in which the 2-position is unsubstituted (A. M. van Leusen, J. Wilderman, and O. H. Oldenziel, J. Org. Chem., 1977, **42**, 1153)

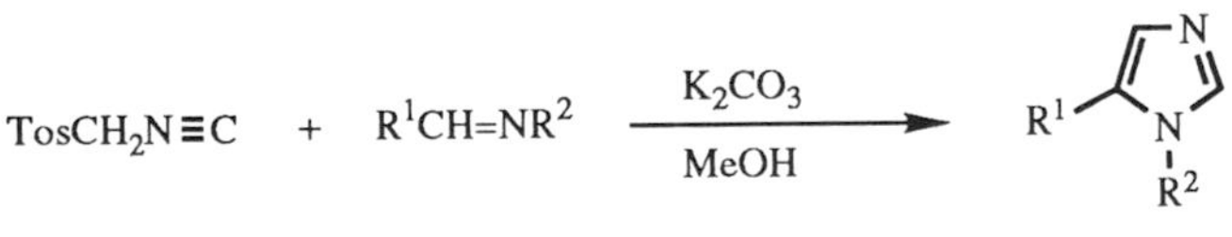

$$TosCH_2N{\equiv}C \quad + \quad R^1CH{=}NR^2 \xrightarrow[MeOH]{K_2CO_3}$$

(**32**)

5. Carbonic acid and derivatives

Carbon dioxide

1. Carboxylation of lithium trialkylalkynylborates constitutes a stereospecific synthesis of *(Z)*-α,β-unsaturated acids (M.-Z. Deng, Y.-T. Tang, and W.-H. Xu, Tetrahedron Lett., 1984, **25**, 1797)

$$[R'_3BC{\equiv}CR^2]Li \xrightarrow[\text{ii) AcOH}]{\text{i) } CO_2} R'CH{=}C(R^2)CO_2H.$$

2. Phosphine catalyses the one-step formation of oxazolidinones from propargylic alcohols, carbon dioxide and primary amines (J. Fournier, C. Bruneau and P.H. Dixneuf, Tetrahedron Lett., 1990, **31**, 1721)

$$HC{\equiv}CCMe_2OH + CO_2 + RNH_2 \rightarrow$$

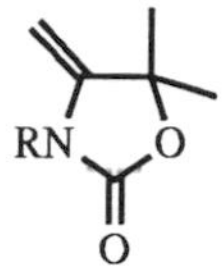

RN O
O

Halogenated oxazolidines may also be obtained from allylic amines, iodine and carbon dioxide (T. Toda and Y. Kitagawa, Angew. Chem. Int. Ed., 1987, **26**, 334)

3. Reaction of carbon dioxide with decamethylsilicocene gives spirocompound (**33**) (P. Jutzi and A. Mohrke, Angew. Chem. Int. Ed., 1989, **28**, 762)

$$(Me_5C_5)_2Si + CO_2 \rightarrow Me_2Si \quad SiMe_2$$

O O O O

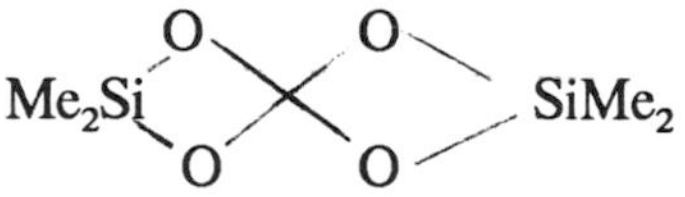

(**33**)

4. Primary amines condense with Dry Ice in the presence of diimides to form disubstituted ureas (H. Ogura *et al.*, Synthesis, 1978, 394) These may also be obtained by the reaction of carbon dioxide with primary amines in the presence of ruthenium complexes and terminal alkynes. (J. Fournier *et al.*, J. Org. Chem., 1991, **56**, 4456)

5. An electrochemical fixation of carbon dioxide into oxoglutaric acid. yielding isocitric acid, has been achieved using isocitrate dehydrogenase as electrocatalyst (K. Sugimura, S. Kuwabata and Y. Yoneyama, J. Amer. Chem. Soc., 1989, **111**, 2361)

6. The synthetic use of reactions of carbon dioxide catalysed by transition metals has been extensively studied in recent years (T. Tsuda *et al.*, J. Org. Chem., 1988, **53**, 3140, and references cited therein) Examples are given in the preparative sections on carbonates and carbamates.

Carbonic acid Although molecular orbital calculations suggest that there is appreciable activation energy (>160 kJ mol^{-1}) involved in the formation and dissociation of carbonic acid, it cannot exist in aqueous solution because of acid and base catalysis of its decomposition. However, thermolysis of ammonium hydrogencarbonate gives free carbonic acid, with a finite existence in a mass spectrometer. (J.K. Terlouw, C.B. Lebrilla and H. Schwarz, Angew. Chem. Int. Ed., 1987, **26**, 354)

Carbonic acid monoesters At high pH, alkyl monocarbonates, present in the ionised form, decarboxylate in a manner which is independent of pH, but which is strongly influenced by the nature of the alkyl group. (C.K. Sauers, W.P. Jencks and S. Groh, J. Amer. Chem. Soc., 1975, **97**, 5546)

$$RO\text{-}CO\text{-}O^- + HO^- \rightarrow ROH + CO_3^{2-}$$

A monocarbonate, stabilised by chelation with magnesium, has been alkylated and acylated (B.J. Whitlock and H.W. Whitlock, J. Org. Chem., 1974, **39**, 3144)

6. Carbonic acid diesters, carbonates

Preparation

1. Butadienylcarbonates are obtained by reaction of chloroformates with enolate ions of crotonaldehyde or tiglic aldehyde, the *(E)*-isomers being the major products (P.F. de Cusati and R.A. Olofson, Tetrahedron Lett., 1990, **31**, 1405)

$$\mathrm{CH_3CH{=}C(R^1)CHO} \; + \; \mathrm{Cl{-}\overset{O}{\overset{\|}{C}}{-}OR^2} \longrightarrow \mathrm{CH_2{=}CH{-}C(R^1){=}CH{-}OCO_2R^2}$$

Similarly, aldehydes treated with potassium fluoride and crown ethers yield enolates which can be trapped by fluoro- or chloroformates to afford vinylic carbonates (R.A. Olofson *et al.*, J. Org. Chem., 1990, **55**, 1)

2. Diphenyl thionocarbonate is converted to diphenyl carbonate by dinitrogen tetroxide; the reaction is thought to proceed *via* an *S*-nitroso intermediate (H.J. Kim and Y.H. Kim, Synthesis, 1986, 970)

$$\mathrm{PhO{-}\overset{S}{\overset{\|}{C}}{-}OPh} \rightarrow [\mathrm{PhO{-}\overset{S^{+}{-}NO}{\overset{\|}{C}}{-}OPh}] \rightarrow \mathrm{(PhO)_2CO}$$

The thiocarbonate-carbonate transformation is also effected with potassium superoxide/2-nitrobenzenesulphonyl chloride (Y.H. Kim, B.C. Chung and H.S. Chang, Tetrahedron Lett., 1985, **26**, 1079) and with benzeneselenic anhydride (D.H.R. Barton, N.J. Cussons and S.V. Ley, J. Chem. Soc., Chem. Commun., 1978, 393)

3. Alkyl oxiranes react with carbon dioxide to form cyclic carbonates.

$$\text{R-oxirane} \; + \; \mathrm{CO_2} \longrightarrow \text{cyclic carbonate (R, O, =O)}$$

Autoclave conditions are required if the reactions are catalysed by antimony (V) compounds (H. Matsuda, A. Ninagawa and R. Nomura, Chem. Lett., 1979, 1261) or by copper (1) cyanoacetate (T. Tsudo, Y. Chujo and T. Saigusa, J. Chem. Soc., Chem. Commun., 1976, 415), but reaction occurs at one atmosphere pressure in the presence of $\mathrm{MoCl_5}$-triphenylphosphine (M. Ratzenhofer and H. Kisch, Angew. Chem. Int. Ed., 1980, **19**, 317) Oxiranes are also converted to cyclic carbonates by treatment with β-butyrolactone and a phase-transfer catalyst (T. Nishikubo *et al.*, Tetrahedron Lett., 1986, **27**, 3741)

4. A synthesis of diphenyl carbonate, avoiding the use of phosgene, involves the reaction of phenyl dichloroacetate with sodium hydride; the reaction may proceed by formation of a dichloroketen as an intermediate (K.R. Fountain and M. Pierschbacher, J. Org. Chem, 1976, **41**, 2039)

5. Mixed carbonates can be obtained by the reaction of alcohols with dialkyl azodicarboxylates and hexamethylphosphorous triamide (G. Grynkiewicz, J. Jurczak and A. Zamojski, Tetrahedron, 1975, **31**, 1411)

$$ROH + R'O_2CN{=}NCO_2R' + (Me_2N)_3P \rightarrow ROCOR'$$

Alkylfuryl carbonates are prepared by reaction of 2-trimethylsiloxyfuran with methyl lithium to give lithium-2-furanolate, followed by quenching with alkyl chloroformates (J.H. Näsman, Synth. Commun., 1987, **17**, 1035)

6. Vicinal diols may be protected as cyclic carbonates by reaction with N,N'-carbonyldiimidazole (J.P. Kurney and A.H. Ratcliffe, Synth. Commun., 1975, **5**, 47) or with di-2-pyridyl carbonates (S. Kim and Y.K. Ko, Heterocycles, 1986, **24**, 1625)

Reactions

Carbonate esters undergo hydrolysis by a stepwise mechanism involving tetrahedral intermediates analogous to those for carboxylate esters (T.H. Fife and J.E.C. Hutchins, J. Amer. Chem. Soc., 1981, **103**, 4194); the nearly constant carbon isotope effect indicates that transition-state structures are similar for all substrates (J.F. Marlier and M.H. O'Leary, J. Amer. Chem. Soc., 1990, **112**, 5996) The hydrolysis of ethyl 2-hydroxy-5-nitrocarbonate shows a rate enhancement at high pH attributed to neighbouring group catalysis by the ionised phenoxy group (T.H. Fife and J.E.C. Hutchins, J. Amer. Chem. Soc., 1972, **94**. 2837)

A quantitative rearrangement of benzofuranone-derived enol carbonates (**34**) to the corresponding carbon-acylated isomers (**35**) is promoted by 4-dimethylaminopyridine (T.H. Black *et al*, J. Chem. Soc., Chem. Commun., 1986, 1524)

Ph CO$_2$R OCO$_2$R O O O

(**34**) (**35**)

Allyl alkyl and alkyl vinylic carbonates react with Pd(O) complexes to form π-allyl alkoxides or enolates, which undergo facile decarboxylation to intermediates such as (**36**); complex (**36**) transfers an allyl group to a variety of carbonucleophiles.

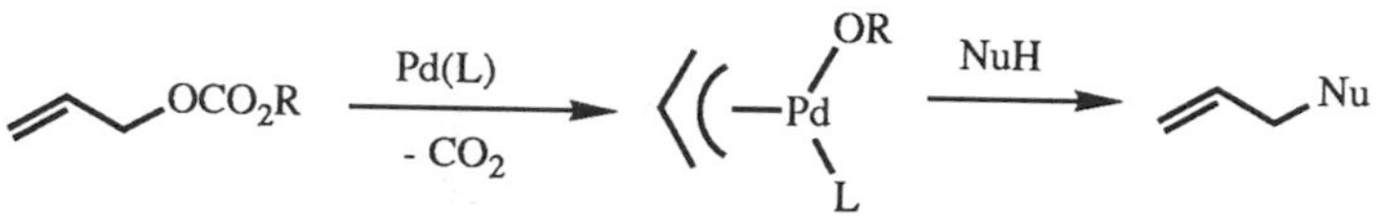

OCO$_2$R Pd(L) - CO$_2$ OR Pd L NuH Nu

(**36**)

Trimethylsilyl enol ethers can be converted to α-allyl ketones

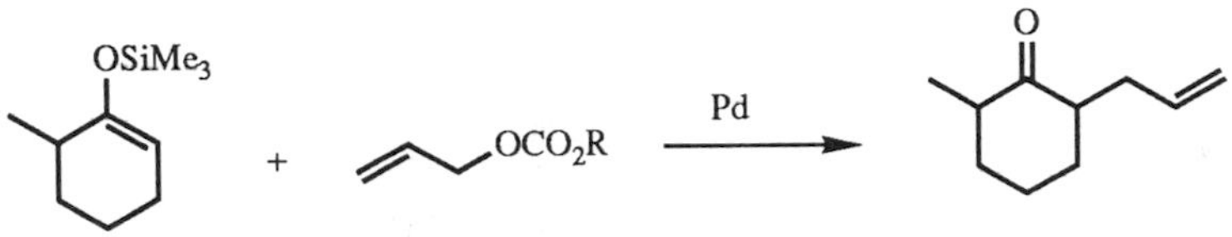

OSiMe$_3$ + OCO$_2$R Pd O

The scope of these reactions has been discussed (J. Tsuji and I. Minami, Acc. Chem. Res., 1987, **20**, 140). An interesting synthesis of 1,2-dien-4-ynes (**37**) employs a palladium-catalysed coupling of substituted 2-alkynyl carbonates with terminal acetylenes (T. Mandai *et al.*, Tetrahedron Lett., 1990, **31**, 7179)

R^1 R^2 C≡CR^3 OCO$_2$Me + HC≡CR^4 Pd(O) R^1 R^3 R^2 ≡R^4

(**37**)

Unsubstituted allenes are available by the hydrogenolysis (ammonium formate) of alkyl-2-ynyl carbonates (J. Tsuji *et al.*, J. Chem. Soc., Chem. Commun., 1986, 922)

Alcohols may be converted to allyl ethers under neutral conditions by Pd(O)-catalysed reactions of allyl ethyl carbonate; with sugars, anomeric hydroxyl groups react selectively (R. Lakhmiri, P. Lhoste and D. Simon, Tetrahedron Lett., 1989, **30**, 4669)

HOH_2C ... OH (furanose, OR OR) + $CH_2{=}CHCH_2OCO_2Et$ ⟶ HOH_2C ... $OCH_2CH{=}CH_2$ (furanose, OR OR)

Similarly, allylic aryl carbonates, obtained from phenols by sequential reaction with triphosgene and an allylic alcohol, undergo Pd(O)-mediated decarboxylation to give allylic aryl ethers; the method is particulary useful for sterically crowded phenols (R.L. Larock and N.H. Lee, Tetrahedron Lett., 1991, **32**, 6315)

Diaryl carbonates, by an SnAr mechanism, undergo a base-initiated thermolysis, leading to diaryl ethers (P. Jost *et al.*, Tetrahedron Lett., 1982, **23**, 43311)

Cyclic carbonates are regioselectively cleaved by bromotrimethylsilane, with concomitant decarboxylation (H.R. Kricheldorf, Angew. Chem. Int. Ed., 1979, **18**, 689)

(R-substituted cyclic carbonate) + $SiMe_3Br$ ⟶ $R{-}CH(OSiMe_3){-}CH_2Br$

In a reaction proceeding without loss of carbon dioxide, ethylene carbonate yields dialkyl carbonates $(RO)_2CO$ on treatment with $Bu_2Sn(OR)_2$ (H.D. Schorf and W. Kusters, Chem. Ber., 1972, **105**, 5640)

7. Chloroformic esters (carbonochloridates)

Chlorovinyl chloroformates are produced by reaction of phosgene and zinc dust with α-chloroaldehydes and ketones in which both other α-positions are occupied by halo or alkyl groups (M.P. Bowman *et al.*, J. Org. Chem., 1990. **55**, 5982)

Chloroformates on Ag^+-catalysed loss of chloride give carboxylium ions, which by decarboxylation provide access to unstable carbocations (P. Peak, Acc. Chem. Research, 1976, **9**, 935)

Complexes formed between tertiary alkylamines and trichloroethyl chloroformate dealkylate to form highly crystalline carbamates which can be cleaved by zinc to give a secondary amine (T.A. Montzka, J.D. Matiskella and R.A. Partyka, Tetrahedron Lett., 1974, 1325).

$$Cl_3CCH_2OC(O)Cl + R_3N \rightarrow Cl_3CCH_2OC(O)NR_3Cl \rightarrow$$

$$Cl_3CH_2OC(O)NR_2\ Cl_3CH_2OC(O)NR_2 \rightarrow RNH_2$$

The use of chloroformate esters to monodealkylate tertiary amines is claimed to rival the von Braun reaction, with vinyl and α-chloroethyl chloroformates being particularly efficient (J.H. Cooley and E.J. Evain, Synthesis, 1989, 1, and references cited therein)

Continuing attention has been given to the formation from chloroformates of mixed anhydrides to act as oxycarbonylation reagents. Isopropenyl chloroformate reacts with N-hydroxysuccinimides or hydroxybenzotriazoles to give stable mixed carbonates **(38)** which form active esters with carboxylic acid/dimethylaminopyridine (K. Takeda *et al.*, Synthesis, 1991, 689)

$$CH_2{=}C(CH_3)OCOCl + HONR_2 \rightarrow \underset{\mathbf{(38)}}{CH_2{=}C(CH_3)OC(O)NR_2}$$

$$\mathbf{(38)} + R'CO_2H \rightarrow R'C(O)ONR_2$$

Related mixed carbonates can be obtained by the treatment of primary alcohols with 1,1-bis[6-(trifluoromethyl)benzotriazolyl]carbonate (Takeda *et al.*, Synthesis, 1987, 557) Chloroformates react with thiazolidine-2-thione or 2-mercapto-5-methyl 1,3,4,thiadiazole to give carbamates which are very effective alkoxycarbonylation reagents towards amino alcohols and polyols (M. Allainmat *et al.*, Synthesis, 1990, 27)

Fluoroformic esters are usually prepared by treating the analogous chloroformates with potassium fluoride, activated by a little 18-crown-6 ether (J. Cuomo and R.A. Olofson, J. Org. Chem., 1979, **44**, 1016) Benzyl and *tert*-butyl fluoroformates are not available by this route, but may be obtained by fragmentation of the corresponding chloroalkyl carbonates (V. A. Dang *et al.*, J. Org. Chem., 1990, **55**, 1847)

$$R'CH(Cl)OC(O)R^2 + KF \rightarrow R'CHO + KCl + FC(O)OR^2$$

(R^2 = benzyl, *tert*-butyl)

The development of a practical route to carbonylchloridefluoride, ClCOF, by the treatment of fluorotrichloromethane with oleum, (G. Siegemund, Angew. Chem. Int. Ed., 1973, **12**, 918) has made its use in the formation of fluoroformates from alcohols more attractive (L. Warkerle and I. Ugi, Synthesis, 1975, 598) Unlike chloroformates, fluoroformates are stable in DMSO and DMF, enabling the *per*-carboalkoxylation of polar substrates such as glucose and thymidine (V.A. Dang and R.A. Olofson, J. Org. Chem., 1990, **55**, 1851)

8. Carbonic acid halides

Carbonyl chloride, phosgene

The use of phosgene to prepare a variety of five- and six-membered heterocycles has been reviewed (E. Kühle, Angew. Chem. Int., Ed., 1973, **12**, 630; H. Habed and A.G. Zeiler, Chem. Rev., 1973, **73**, 75) and a more general survey of phosgene chemistry has appeared (H. Eckert and B. Forster, Angew. Chem. Int. Ed., 1987, **26**, 894)

Phosgene reacts with α-azidocarboxylic acids to form 4-alkyliden-2,5-oxazolidindiones (F. Effenberger *et al.*, Synthesis, 1988, 218) With N-silylimine (**39**) phosgene gives an unusual dication (**40**) (A. Hamed, J.C. Jochins and M. Przybylski, Synthesis, 1989, 401).

$$(Me_2N)_2C{=}N\text{-}SiMe_3 + COCl_2 + SbCl_5$$

(39)

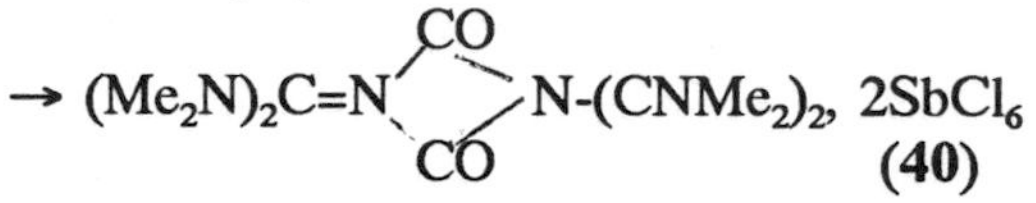

(40)

"Diphosgene", "triphosgene". Although it is now commercially available in toluene solution, the severe toxicity of phosgene has limited its use in the laboratory. Recently a liquid equivalent, trichloromethyl chloroformate, $Cl_3CC(O)Cl$, ("diphosgene") and a stable crystalline solid, bis-trichloromethyl carbonate, $(Cl_3CO)_2CO$ ("triphosgene") have been advocated as phosgene equivalents. Triphosgene, as a solid, is more convenient but is appreciably more expensive.

Both can replace phosgene in a wide variety of reactions. Diphosgene converts amines to isocyanates, aminoacids to isocyanoacyl chlorides and aminoalcohols to isocyanatochloroformates, ONCROCOCl, (K. Kurita, T. Matsumura and Y. Iwakura, J. Org. Chem., 1976, **41**, 2070); it is less efficient than phosgene in transforming aromatic diamines to diisocyanates (A. Efraty *et al.*, J. Org. Chem., 1980, **45**, 4059) but is superior in the dehydration of N-alkyl or N-aryl formamides to isocyanides (G. Skorna and I. Ugi, Angew. Chem. Int. Ed., 1977, **16**, 259)

In the presence of activated charcoal it is decomposed to liberate phosgene *in situ*, a reaction which has been used to convert α-aminoacids to N-carboxy-α-aminoacid anhydrides (R. Katakai and Y. Iizuka, J. Org. Chem., 1985, **50**, 715). Reduction of α-aminoacids to the corresponding aminoalcohols, followed by reaction with diphosgene constitutes a simple large-scale synthesis of chiral oxazolidines, valuable auxiliaries in directed lithiations (L.N. Pridgen *et al.*, J. Org. Chem., 1989, **54**, 3231)

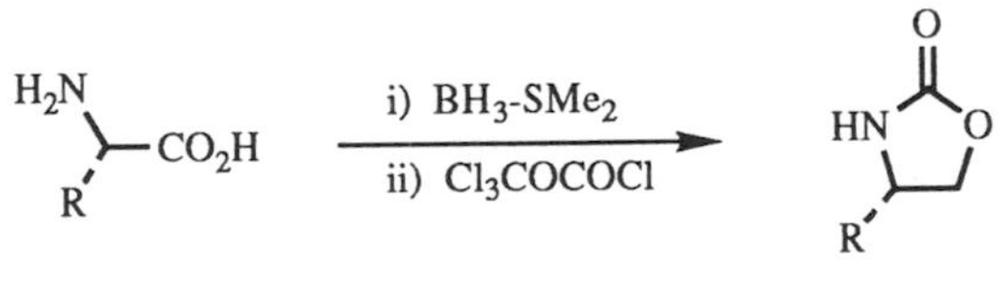

With *N*-hydroxysuccinimide it forms *N,N'*-disuccinimidyl carbonate, a useful substitute for carbodiimides in the preparation of active esters and peptides (H. Ogura *et al.*, Tetrahedron Lett., 1979, 4745)

Similarly triphosgene (1 mol) reacts with amines, alcohols and formamides (3 mol) to give isocyanates, chloroformates and isonitriles respectively (H. Eckert and B. Forster, Angew. Chem., 1987, **26**, 894), and with aminoacids to form N-carboxyanhydrides (W.H. Daly and D. Poche, Tetrahedron Lett., 1988, **29**, 5859)

Both diphosgene and triphosgene can be used with DMSO in the Swern oxidation of alcohols to carbonyl compounds (S. Takano *et al.*, Tetrahedron Lett., 1988, **29**, 6619; C. Palomo *et al.*, J. Org. Chem., 1991, **56**, 5948).

Carbonyl fluoride COF_2, has been used as a fluorinating agent, converting secondary amines to the corresponding N-F compound, triphenylmethane to trityl fluoride, and diphosphine $Ph_2PCH_2CH_2PPh_2$ to the tetrafluoride.

Dicarbonic acid diesters. These esters have received increasing attention as alkoxycarbonylating agents for amine protection in peptide syntheses. The most widely used, di-*tert*-butyl dicarbonate (**41**) is obtained from a tricarbonate diester (**42**) (B. M. Pope, Y. Yamamoto and D. S. Tarbill, Org.Synth. 1977, **57**; 15)

t-BuOK + $CO_2 \rightarrow t$-BuOCO_2K

t-BuOCO_2K + $COCl_2 \rightarrow (t$-Bu$OCO_2)_2CO$ (**42**)

(**42**) + diazobicyclooctane $\rightarrow (t$-BuOCO$)_2O$ (**41**)

Dicarbonate **41** forms *tert*-butyl esters from carboxylic acids (V. Pozdnev, Zh. Obshch. Khim., 1988, **58**, 670) or from benzyl carboxylates in the presence of triethylsilane and palladium (II) acetate, (M. Sakaitani *et al.*, Tetrahedron Lett., 1988, **29**, 2983). With hydroxylamine it gives t-BuO(CO)ONH_2, which acts as an *N*-carbo-*tert*-butoxylation agent at acid pH (R. B. Harris and I. B. Wilson, Tetrahedron Lett., 1983, **24**, 231). Reaction with formanide yields di-*tert*-butyliminocarbonate $[t$-BuOC(O)$]_2NH$, a useful substitute for phthalimide in the Gabriel synthesis of amines (L. Grehn and U. Ragnarrsson, Synthesis, 1987, 275). Reaction of the *trans*-pyrrolidine diester **(43)** with **(41)** gives an unusual mixed carbonic-carbamic anhydride **(44)** (D. S. Kempe and T. P. Curran, J. Org. Chem., 1988, **53**, 5729)

EtO_2C H H N CO_2Et H **(43)** + **(41)** —DMAP→ EtO_2C H H N CO_2Et C–O–C–OBu^t O O **(44)**

Diallyl dicarbonate introduces the alloxycarbonyl protecting group in peptide and nucleotide synthesis (G. Sennyer, G. Barcelo and L. P. Senet, Tetrahedron Lett., 1987, **28**, 5809)

9. Sulphur analogues of carbonic acid derivatives

Carbon oxysulphide reacts with the enolate ion of propiophenone to yield the β-betothiolic acid ester PhCOCH(Me)COSMe (E. Vedejs and B. Nader, J. Org. Chem., 1982, **47**, 3193). With phenol **(45)** the product of treatment with COS is cleanly determined by the reaction temperature (A. O. Fulton and M. Qutob, J. Chem. Soc., Perkin Trans. 1, 1972, 2660)

OH Bu^t Bu^t CH_2SOC N ←COS, 35°— N·CH_2 Bu^t OH Bu^t **(45)** —COS, 80°→ OH Bu^t Bu^t HOH_2C

Carbon disulphide. The chemistry of CS_2 and its derivatives has been the subject of a recent comprehensive monograph (Carbon Disulphide in Organic chemistry, A. D. Dunn and W. D. Rudolf, Ellis Harwood, 1989) and only selected reactions are presented here. Addition of CS_2 to alkynes, giving cyclic thiones **(46)** occurs at atmospheric pressure in the presence of bis-amine disulphides (F. M. Benitez and J. R. Grunwell, J. Org. Chem., 1978, **43**, 2917)

$$RC\equiv CR + CS_2 + \left(\begin{matrix}R^1\\ \ \ \ NS\\ R^2\end{matrix}\right)_2 \longrightarrow$$

(**46**)

With carbodiimides as activator, CS_2 condenses with substituted propargylamines to form isothiocyanates, precursors of thiazoles (G. Ferrand *et al.*, Eur. J. Med. Chem., 1976, **11**, 49). The reaction of CS_2 with ketones $RCOCH_3$ in basic conditions, followed by dialkylation gives keten dithioacetals, $RCOCH{=}C(SR^1)_2$, valuable three-carbon synthons (R. K. Dieter, Tetrahedron, 1986, **42**, 3029, and references cited therein). Hindered *N,N*-dialkylhydroxylamines react rapidly with CS_2 to give corresponding secondary amines; the rate-determining step is the relief of steric compression, and less crowded hydroxylamines undergo competing dealkylation (M. A. Schwartz, J. Gue and X. Hu, Tetrahedron Lett; 1992, **33**, 1687)

10. Dithiocarbonic O, S-acid, xanthic acid

Potassium *O*-alkylthiocarbonates, prepared from COS, potassium and an alcohol, undergo transesterification, and with chloroformates give dialkoxycarbonylsulphides (**47**) (M. A. Palominos and J. C. Vega, Synthesis, 1990 825).

$$ROC(O)SK + R^1OH \rightleftharpoons R^1OC(O)SK + ROH$$
$$R^1OC(O)SK + ClC(O)OR^2 \rightarrow ROC(O)SC(O)R^2 \quad (\mathbf{47})$$

Xanthate salts are alkylated by esters of sulphuric or sulphonic acids (I. Degani, R. Fochi and V Regondi, Synthesis, 1979, 1789; D. Trimnell *et al*, J. Org. Chem., 1975, **40**, 1337); phase transfer catalysis is useful and has been used in a one pot synthesis of hexose xanthates (P. Di Cesare and B. Gross, Synthesis , 1980, 714; A. W. M. Lee *et al.*, Synth. Commun., 1989, **19**, 547) Yields of alkenes from the pyrolysis of xanthates of tertiary alcohols are enhanced if the potassium salts are used (K. G. Rutherford, R. M. Ottenbritt and B. K. Tang, J. Chem. Soc (C), 1971, 582)

Primary, secondary or tertiary alcohols may be deoxygenated by reduction of their xanthate diesters, the Barton-McCombie reaction. The reaction proceeds by a radical chain mechanism, and may be induced by Bu_3SnH, R_3SiH/thiols, or $Ph_2SiH_2/Et_3B/O_2$ (D. H. R. Barton, D. O. Jang and J. C.

Jaszberenyi, Tetrahedron Lett., 1990, **31**, 4681; J. N. Kirwan, B. P. Roberts and C. R. Willis, *ibid*, 5093, and references cited therein). The deoxygenation of secondary alcohols in nucleosides may also be achieved by treatment with PhOC(S)Cl, and reaction of the resulting thiocarbonate ester, PhOC(S)OR with Bu_3SnH (M. J. Robins, J. S. Wilson and F. Hansske, J. Amer. Chem. Soc., 1983, **105**, 4059) Rearrangement of *O,S* diesters to *S, S*-diesters occurs smoothly and without alkene formation in the presence of trifluoracetic acid (M. W. Fichtner and N. F. Haley, J. Org. Chem., 1981, **46**, 3141).

11. Carbamic acid esters, carbamates

Preparation

1. Chloroalkyl carbamates are formed by treatment of diols with (dichloromethylene)ammonium chloride, the reaction proceeding *via* an immonium carbonate **(48)** (B. LeClef *et al.*, Angew. Chem. Int. Ed., 1973, **12**, 404)

$$HO(CH_2)_nOH + Cl_2C{=}NH_2Cl \longrightarrow (CH_2)_n\langle O,O\rangle C{=}NCl \longrightarrow Cl(CH_2)_nCH_2OCONH_2$$

(48)

Symmetric diols can also be selectively converted to monocarbamates by reaction with alkyl isocyanates, generated *in situ* from alkyl halides and potassium cyanate under phase-transfer catalysis (M. Prashad, J.C. Tomesch, and W.J. Houlihan, Synthesis, 1990, 477). Isocyanates may also be formed in the reaction medium by treatment of amides with lead tetraacetate (A. Brandstrom, B. Lamm and I. Palmertz, Acta Chem. Scand. (B), 1974, **28**, 699), or by oxidation of isonitriles with thallium (III) nitrate (F. Kienzle, Tetrahedron Lett., 1972, 1771)

The addition of alcohols to isocyanates is catalysed by bases or by Lewis acids (H. Irie *et al.*, J. Chem. Soc., Perkin Trans 1, 1989, 1209 and references cited therein)

2. Reactions of alcohols with trichloroacetyl isocyanate gives

trichloroacetyl carbamates which are hydrolysed to unsubstituted carbamates on filtration through neutral alumina; the method is versatile and tolerates a variety of labile functional groups (P. Kocovsky and I. Stieborova, J. Chem. Soc., Perkin Trans. 1, 1987, 1969)

$$Cl_3CCON{=}C{=}O + ROH \rightarrow Cl_3CCOHNC(O)OR \rightarrow NH_2C(O)OR$$

3. Chloromethyl carbonates, prepared by treatment of alcohols or phenols with chloromethyl chloroformate are converted to carbamates by reaction with primary or secondary alkylamines (T. Patonay, E. Patonay-Peli and R. Mogyorodi, Synth. Commun., 1990, **20**, 2863). Similarly di-2-pyridyl-carbonate (available from triphosgene and 2-hydroxypyridine) reacts with alcohols to form mixed carbonates which give high yields of carbamates under mild conditions on the addition of amines (A.K. Ghosh, T.T. Duong and S.P. McKee, Tetrahedron Lett., 1991, **32**, 4251)

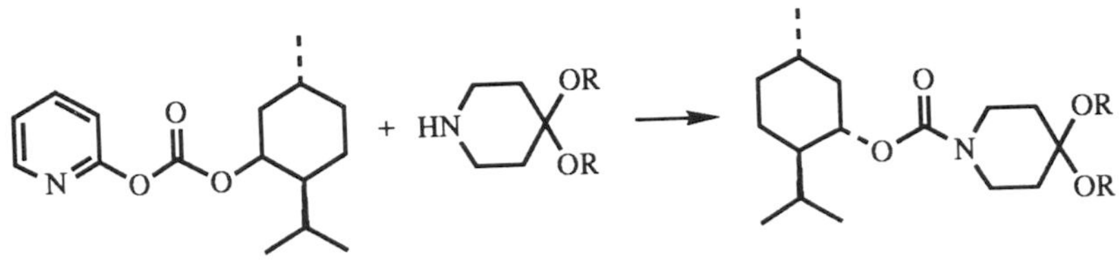

O N O O + HN OR OR → O O N OR OR

4. The reductive carbonylation of aromatic nitro compounds to the corresponding methyl carbamates is catalysed by ruthenium carbonyls in the presence of tetraethylammonium chloride (S. Cenini *et al.*, J. Org. Chem., 1988, **53**, 1243)

$$ArNO_2 + 3CO + MeOH \rightarrow ArNHCO_2Me + 2CO_2$$

5. Ruthenium complexes also catalyse the one-step synthesis of vinyl carbamates from secondary amines, carbon dioxide and terminal alkynes (R. Mahé *et al.*, J. Org. Chem., 1989, **54**, 1518)

$$RC{\equiv}CH + CO_2 + HNR'_2 \rightarrow RCH{=}CHOC(O)NR'_2$$

6. Carbon dioxide, trapped by (tetraphenylporphinato)aluminium acetate, is activated towards reaction with secondary amines and epoxide, yielding carbamates. (F. Kojima, T. Aida and S. Inoue, J. Amer. Chem. Soc., 1986,

108, 391).

$$Et_2NH + CO_2 + CH_3\overset{O}{CH-CH_2} \rightarrow Et_2NCO_2CH_2CH(OH)CH_3$$

7. Amines undergo oxidative alkoxycarbonylation in the presence of an alkali metal halide and platinum group metal (S. Fukuoka, M. Chono and M. Kohno, J. Org. Chem., 1984, **49**, 1458)

$$RNH_2 + CO + R'OH + \frac{1}{2}O_2 \rightarrow RNHCO_2R' + H_2O$$

8. Primary aliphatic or aromatic carboxamides on treatment with *N*-bromosuccinimide or 1,3-dibromo-5,5-dimethylhydantoin in the presence of mercuric acetate undergo a quantitative Hofmann rearrangement to carbamates of amines having one less carbon (S-S Jew *et al.*, Tetrahedron Lett., 1990, **31**, 1559, and references therein for related procedures). Carbamates are also available from a modified Curtius rearrangement, using diphenylphosphonyl azide (T. Shiori, K. Ninomiya and S. Yamada, J. Amer. Chem. Soc., 1972, **94**, 6203)

$$RCO_2H + R'OH + N_3P(O)(OPh)_2 \rightarrow RNHCO_2R'$$

9. Silyl carbamates, $RNHCO_2SiR'_3$, can be prepared by the reaction of trialkylsilanols with isocyanates (F.T. Chiu *et al.*, J. Pharm. Sci., 1982, **71**, 542) and by treatment of N-*tert*-butoxycarbonyl or N-benzyloxycarbonyl amines with *tert*-butyldimethylsilyl triflate or *tert*-butyldimethylsilane/palladium acetate respectively (M. Sakaitani and Y. Ohfune, J. Org. Chem., 1990, **55**, 870; J. Amer. Chem. Soc., 1990, **112**, 1150. They afford cyclic carbamates by intramolecular trapping of an electrophile in the presence of fluoride.

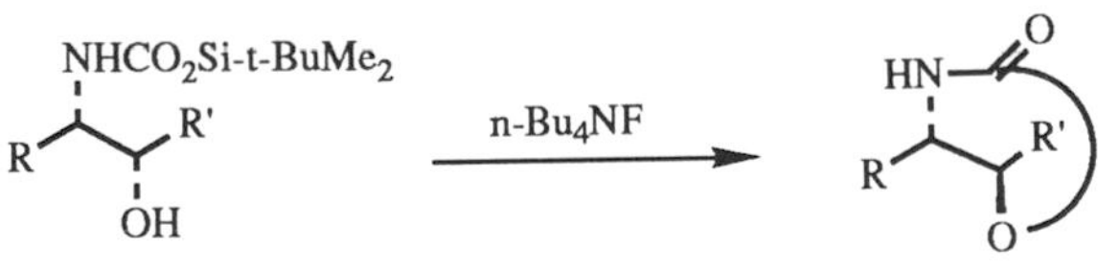

10. An unusual allylic amination, induced by a sulphur diimide, produces carbamates as intermediates (G. Kresze and H. Munstërer, J. Org. Chem.,

1983, **48**, 3561)

$MeO_2CN{=}S{=}NCO_2Me + Me_2C{=}CHCH_3$
$\rightarrow MeCH{=}C(CH_3)CH_2N(CO_2Me)SNHCO_2Me$
$\xrightarrow{KOH} MeCH{=}C(CH_3)CH_2NHCO_2Me$

Reactions

Trialkylboranes react with N-chloro-N-sodiocarbamates to form N-alkyl carbamates (N. Wachter-Jurcsak and F.E. Scully, Tetrahedron Lett., 1990, **37**, 5261)

The *tert*-butylcarbamate of O-benzylhydroxylamine can be converted to N-alkylhydroxylamines by alkylation and subsequent hydrogenolysis (R. Sulsky and J.P. Demers, Tetrahedron Lett., 1989, **30**, 31)

$t\text{-}BuOCONHOCH_2Ph \rightarrow t\text{-}BuOCON(R)OH$

Allyl carbamates transfer an allyl group to carbon nucleophiles under palladium salt catalysis (I. Minimi *et al*, Tetrahedron Lett., 1985, 26, 2449). Extension of the reaction to 2-butenylene dicarbamates results in an efficient synthesis of 4-vinyl-2-oxazolidines (T. Hayashi, A. Yamamoto and Y. Ito, Tetrahedron Letts., 1987, **28**, 4837)

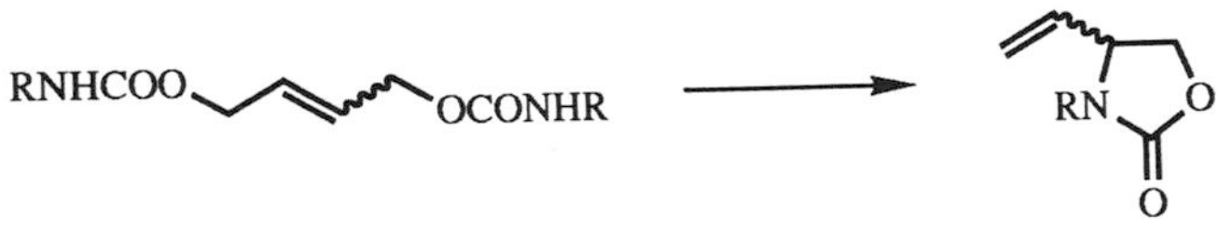

RNHCOO OCONHR → RN O O

An ingenious synthesis of α-amino acids involves the reaction of Grignard reagents with the adduct between chloral and ethyl carbamate, followed by hydrolysis of trichloromethyl and carbamate groups (C. Kashima, Y. Aoki and Y. Omote, J. Chem. Soc., Perkin Trans., 1, 1975, 2511)

$Cl_3CCH{=}NCOOEt + RMgX \rightarrow Cl_3CH(R)NHCO_2Et \rightarrow HO_2CCH(R)NH_2$

Carbamates of N-hydroxypyridine-2-thione are a convenient source of dialkylaminyl radicals, protonated in weak acids to radical cations. These are useful synthetic intermediates. Related monoalkylaminium cations can be similarly obtained from 3-hydroxy-4-methylthiazole-2(3H)-thione carbamates (M. Newcomb and K.A. Weber, J. Org. Chem., 1991, **56**, 1309)

Irradiation of benzyl N-bromo-N-methylcarbamates produces carbamyl radicals which cyclise and trap bromine to afford spirocyclic oxazolidinones (P.F.Dicks *et al*, J. Chem. Soc., Perkin Trans I, 1987, 1243)

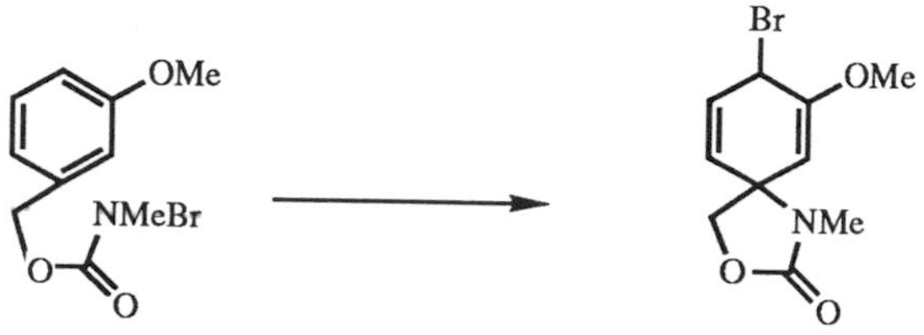

The anion of an alcohol-derived benzoyl carbamate may be used to deliver an amino nitrogen to an electrophilic carbon, and allows the conversion of sugars to amino sugars with inversion of configuration (S. Knapp *et al.*, J. Org. Chem., 1990, **55**, 5700)

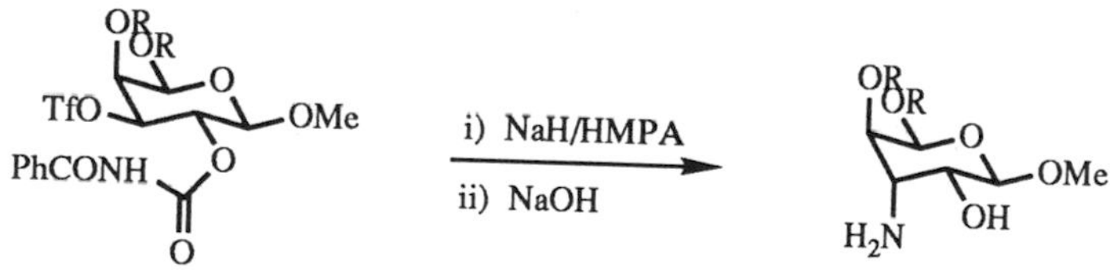

i) NaH/HMPA
ii) NaOH

Derivatives of 2-amino alcohols can also be obtained by iodination of 2-allenic O-carbamates; the *syn* isomer predominates in the product mixture (R.W. Friesen, Tetrahedron Lett., 1990, **30**, 4249)

Treatment of α-methoxycarbamates with LDA gives N-vinylcarbamates which react with enolate anions of alkyl acetates to afford β-amino acid derivatives. (T. Shono *et al.*, Tetrahedron Letters, 1989, **30**, 1253)

The ability of O-carbamates to act as directors in the ortho-lithiation of aromatic compounds has been comprehensively surveyed (V. Snieckus, Chem. Rev., 1990, **90**, 879). The organolithium intermediate may be quenched by electrophile or may undergo the anionic equivalent of an *ortho*-Fries rearrangement.

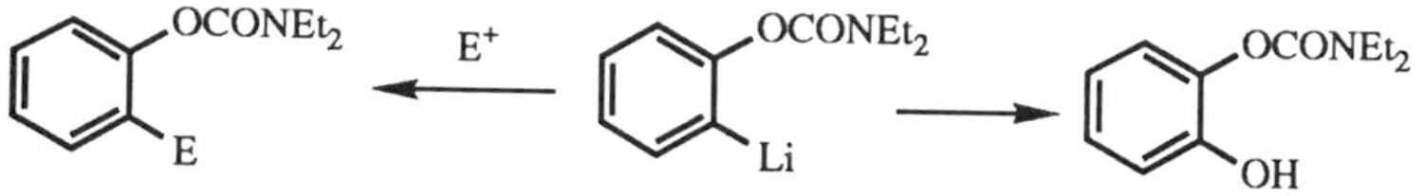

Enol carbamates, prepared from vinyl chloroformate and trimethylsilylamides also undergo α lithiation (S. Sengupta and V. Snieckus. J. Org. Chem., 1990, **55**, 5680) *N-tert*-butoxycarbonyl-pyrrole or indole lithiate in the 2-position (I. Hasan *et al.*, J. Org., Chem., 1981, **46**, 157). When allyl alcohols are converted to N,N-dialkylcarbamates they readily form homoenolate anions from which 4-oxoalkanoates may be obtained (D. Hoppe, R. Hanko, and A. Brönneke, Angew. Chem. Int. Ed., 1980, **19**, 625)

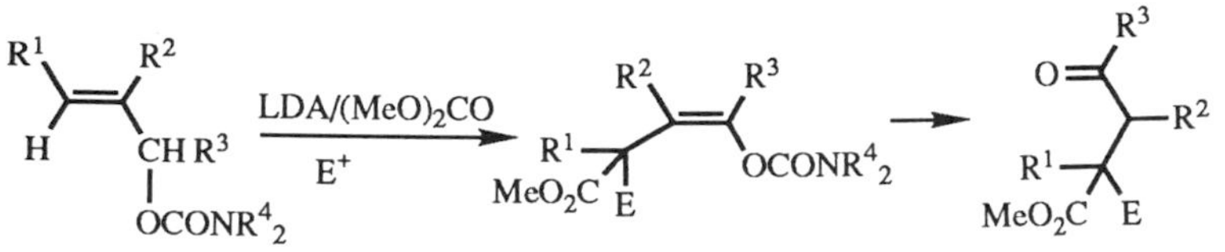

In a useful variant of carbamate-directed lithiation, lithium salts of primary or secondary aromatic amines are carboxylated; the resulting carbamate salt protects the amine from electrophiles and controls subsequent *ortho*-metalation (A.R. Katritzky, M. Black and W-Q Fan, J. Org. Chem., 1991,

56, 5045)

NHMe → i) n-BuLi ii) CO_2 → Me-N-CO_2Li → i) t-BuLi ii) E^+ iii) H_3O^+ → NHMe, E

Carbamates as blocking groups Because of their low tendency to racemise, and their ease of formation and removal, carbamates continue to enjoy popularity as protective groups for the amine function, particularly in peptide and protein synthesis. Their use has been the subject of a comprehensive treatise (Protective Groups in Organic Synthesis, 2nd Edition, T.W. Greene and P.G.M. Wuts, Wiley, 1991, p315 *ff*) in which 89 carbamates are listed. Many are variants of *tert*-butyl and benzyl carbamates, differing in the strength of acid required for their removal; e.g. 2,4-dichlorobenzyl carbamates are 80 times more stable to acid than are unsubstituted benzyl carbamates (Y.S. Klausner and M. Chorev, J. Chem. Soc., Perkin Trans. 1, 1977, 627). In reactions of amino sugars, O-nitrobenzyl carbamates are useful amino protecting groups as they may be removed by photolysis (B. Amit, V. Zehari and A. Patchornik, J. Org. Chem., 1974, **39**, 192)

The acid stable, base labile fluorenylmethyl (FMOC) carbamates have become increasingly used, and the practical aspects of their application to peptide synthesis has been discussed (Solid Phase Peptide Synthesis, E. Atherton and R.C. Sheppard, Oxford University Press, 1989) FMOC carbamates are deblocked on treatment by amines, pyridine being adequate for the removal of 2,7-dibromo-FMOC derivates (L.A. Carpino, J. Org. Chem., 1980, **45**, 4250)

12. Orthocarbonic acid esters

Preparations and reactions of orthocarbonates have been reviewed (W. Kanthlehuer *et al.*, Synthesis, 1977, 73) Mixed orthocarbonates are intermediates in the oxidation of acetals of long chain aldehydes with m-chloroperbenzoic acid (W.F. Bailey, M. Shih, J. Amer. Chem. Soc., 1982, **104**, 1769)

$R_2C(OEt)_2 \rightarrow [(RO)_2C(OEt)_2] \rightarrow (RO)_2C{=}O$

Lower tetraalkyl orthocarbonates can be obtained from reaction of sodium alkoxides with $SnCl_4$ to give disodium hexaalkoxystannates, followed by treatment with carbon disulphide (S. Sakai *et al.*, Synthesis, 1984, 233) The reaction of sodium 4-methylphenoxide with carbon tetrachloride in an acetonitrile solution containing copper (I) chloride results in a quantitative formation of the corresponding tetraaryl orthocarbonate (T.H. Chan, J.F. Harrod and P. van Gheluwe, Tetrahedron Lett., 1974, 4409)

Tetramethylorthocarbonate undergoes facile transacetylation with alkanediols in xylene, using p-toluenesuphonic acid as catalyst to afford symmetrical spiro orthocarbonates (49) in good yields (T.Endo and M.Okawars, Synthesis, 1984, 837)

$$(MeO)_4C + 2HO(CH_2)_nOH \rightarrow (CH_2)_n\langle {}^{O}_{O} \rangle C \langle {}^{O}_{O} \rangle (CH_2)_n$$

(49)

Spiroorthocarbonates polymerise with expansion to form poly(ether-carbonates) (T. Endo, H. Katsuki and W.J. Bailey, Makromol. Chem., 1976, **177**, 3231) while carbonate (**49**, n = 2) has been used as an acetalisation reagent (D.H.R. Barton, C.C. Davies and P.D. Magnus, J. Chem. Soc., Chem. Commun., 1975, 432)

Second Supplements to the 2nd Edition of Rodd's Chemistry of Carbon Compounds, Vol. 1C, edited by M. Sainsbury
© 1992 Elsevier Science Publishers B.V., Amsterdam

Chapter 11

CARBAMATES AND THEIR ALLIES

D. W. ANDERSON

1. Carbamates

Preparations. (1) The synthesis of carbamates from dialkyl-azodicarboxylates has been described (Barneis, Z., Broeir, Y. and Bittner, S., Chem. Ind., London, 526–527, (1976)).
(2) A number of crown ether and cryptand systems have been synthesised which contain urethane units (Vogtle, F. et al., Justus Liebigs Ann. Chem., 1586–1591, (1978)). Conformational flexibility and complexation properties of these compounds have been studied.
(3) N–Acyl–N–formylcarbamates may be prepared by singlet oxygen oxidation of 5–unsubstituted 4–alkoxyoxazoles (Graziano, M. L., Iesce, M. R. and Scarpati, R., J. Heterocycl. Chem., **16**(1), 129–131, (1979)). They are photo– and thermo–stable but sensitive to hydrolysis under very mild conditions.
(4) *Tert*–butyl percarbamates have been prepared by the reaction of amines with *tert*–butylimidazolylcarboxylate (Maillard, B. et al., Tetrahedron, **38**(24), 3569–3577, (1982)).
(5) Carbamate esters labelled at the carbonyl with ^{14}C are synthesised from readily available [^{14}C]–phosgene which is first converted to an isolable [^{14}C]–labelled alkyl or aryl chloroformate and subsequently reacted with the appropriate amine. (Nasser, M. N., Agha, B. J. and Digensis, G. A., J. Label Compound Radiopharm., **23**(6), 667–676, (1986)).
(6) N–(Alkoxymethylene)carbamates are synthesised from imidate hydrochlorides or imidates and formates in the presence of amine bases (Allmann, R., Krestel, M., Kupfer, R. and Wurthwein, E. U., Chem. Ber., **119**(8), 2444–2457, (1986)).
(7) Fluoroisopropenyl carbamates are prepared using 1,3–dihalogenoprop–2–yl chloroformate (Shimizu, M., Tanaka, E. and Yoshioka, H., J. Chem. Soc., Chem. Commun., 136–137, (1987)).
(8) A convenient synthesis of *tert*–butyl N–(2–bromoethyl)

carbamate has been reported (Beylin, V. G. and Goel, C. P., Org. Prep. Procedure Int., **19**(1), 78–80, (1987)).
(9) Monomeric beryllium bis(dialkylamides) react with CO_2 to give bis(dialkylcarbamoyloxy)beryllium compounds (Noth, H. and Schlosser, D., Chem. Ber., **121**(10), 1715–1717, (1988)). Insertion of CO_2 into the BeN bonds of $[Be(NPr^i{}_2)_2]_2$ and $[Be(NMe_2)_2]_3$ occurs only at the terminal BeN bonds.
(10) Carbamates may be prepared by simply stirring the substrate, sodium or potassium cyanate, and Cl_3CCO_2H in DCM overnight at room temperature (Depreux, P., Bethegnies, G. and Marcincal, A., Synth. Commun., **19**(15), 2737–2740, (1989)).
(11) Carbamates are obtained in high yield from the reaction of CO_2 with aliphatic amines and ortho esters (Ishii, S., Nakayama, H., Yoshida, Y. and Yamashita, T., Bull. Chem. Soc. Jpn., **62**(2), 455–458, (1989)).
(12) Trialkylboranes react with N-chloro-N-sodiocarbamates to form N-alkylcarbamates (Wachterjurcsak, N. and Scully, F. E., Tetrahedron Lett., **31**(37), 5261–5264, (1990)). The reaction is best carried out without isolating the intermediate N-chlorocarbamate salt.
(13) The stereoselective formation of cyclic carbamates has been achieved by intramolecular trapping of a *tert*-butyldimethylsilyloxycarbonyl group with allylic esters upon activation with fluoride and catalysis by Pd^0 (Spears, G. W., Nakanishi, K. and Ohfune, Y., Tetrahedron Lett., **31**(37), 5339–5342, (1990)).
(14) An efficient and general synthesis of optically active ene carbamates has been reported (Montgomery, J., Wieber, G. M. and Hegedus, L. S., J. Amer. Chem. Soc., **112**(17), 6255–6263, (1990)). Subsequent Pd^{II}-assisted carboacylation proceeds with complete control of stereochemistry.

from isocyanates. (1) Carbamates may be prepared by the condensation of alcohols with isocyanates in the presence of the Lewis acid catalysts $AlCl_3$ or $BF_3.OEt_2$ (Ibuka, T., Chu, G.-N., Aoyagi, T., Kitada, K., Tsukida, T. and Yoneda, F., Chem. Pharm. Bull., **33**, 451, (1985)). Bases such as dimethylaminoethanol (Pirkle, W. H. and Adams, P. E., J. Org. Chem., **44**(13), 2169–2175, (1979)), pyridine and Et_3N (Roush, W. R. and Adam, M. A., J. Org. Chem., **50**(20), 3752–3752, (1985); Hauser, F. M., Rhee, R. P. and Ellenberger, S. R., J. Org. Chem., **49**(12), 2236–2240, (1984)), lithium alkoxides (Bailey, W. J. and Griffith, J. R., J. Org. Chem., **43**(13), 2690–2692, (1978)) and tris(dimethylamino)-N-methylphosphine imide (Kuhlmeyer, R., Schwesinger, R. and Prinzbach, H., Tetrahedron Lett., **25**, 3429,

(1987)) have also been used to catalyse the reaction.
(2) Selective monocarbamoylation of diols may be achieved by reaction with alkyl isocyanates, generated *in situ* from alkyl halides and potassium cyanate under phase-transfer catalysis (Prashad, M., Tomesch, J. C. and Houlihan, W. J., Synthesis, 477–480, (1990)). Under these conditions the ease of reaction is *cis*-diols > *trans*-diols and diols > monoalcohols. An increase in the number of methylene groups between the two hydroxyl groups leads to a decrease in product yields.
(3) The direct addition of tertiary alcohols to isocyanates usually gives no reaction at low temperatures and produces olefins on being heated. Catalysts such as lithium alkoxides and dibutyltin diacetate allow the synthesis of tertiary alkyl N-phenyl- (Bailey, W. J. and Griffith, J. R., J. Org. Chem., **43**(13), 2690–2692, (1978)) and other carbamates (Francis, T. and Thorne, M. P., Can. J. Chem., **54**(1), 24–30, (1976); Nikiforov, A., Jirovetz, L. and Buchbauer, G., Liebigs Ann. Chem., 489–491, (1989)). 2-Oxazolidinones are obtained from cyclisation of the carbamates of unsaturated tertiary alcohols.

from alkyl halides. Reaction of CO_2, aliphatic amines, and alkyl halides gives alkyl carbamate esters (Yoshida, Y., Ishii, S. and Yamashita, T., Chem. Lett., 1571–1572, (1984); Yoshida, Y., Ishii, S., Watanabe, M. and Yamashita, T., Bull. Chem. Soc. Jap., **62**(5), 1534–1538, (1989)). Reactions with secondary alkyl bromides gave higher yields than those using primary or tertiary alkyl bromides. Copper-promoted (an intermediate Cu^{I} carbamato-complex has been isolated and characterised) (Saegusa, T. et al., J. Chem. Soc., Chem. Commun., 815–816, (1978)) and phase-transfer catalysed (Gomezparra, V., Sanchez, F. and Torres, T., Synthesis, 282–285, (1985)) conditions have been reported.

from epoxides. (1) Carbamic esters are obtained directly by the reaction of CO_2, epoxides and primary or secondary aliphatic amines (Yoshida, Y. and Inoue, S., J. Chem. Soc., Perkin Trans. I, 3146–3150, (1979); Saito, N., Hatakeda, K., Asand, T., Toda, T. and Ito, S., J. Chem. Soc. Jap., Chem. Ind. Chem., 1196–1201, (1986)) or under catalysis with (5,10,15,20-tetraphenylporphinato)aluminium acetate [Al(TPP)O_2CCH_3] (Kojima, F., Aida, T. and Inoue, S., J. Amer. Chem. Soc., **108**(3), 391–395, (1986)).
(2) Alkylcarbamic esters of 1,2-diols or hemiacetals may be synthesised by reaction of CO_2, primary or secondary aliphatic amines, and epoxides or vinyl ethers, respectively. An

analogous reaction using aromatic amines does not occur. Arylcarbamic esters may be synthesised however from CO_2 and an aromatic amine *via* a zinc carbamate (Yoshida, Y. et al., Bull. Chem. Soc. Jpn., **61**, 2913–2916, (1988)).

Reductive carbonylation. The reductive carbonylation of nitroarenes to urethane is catalysed by platinum^{II} – tin^{IV} (Watanabe, Y., Tsuji, Y. and Suzuki, N., Chem. Lett., 105–106, (1982); Watanabe, Y., Tsuji, Y., Takeuchi, R. and Suzuki, N., Bull. Chem. Soc. Jap., **56**(11), 3343–3348, (1983)) or $Ru_3(CO)_{12}$ and $Ru(CO)_3(PPh_3)_2$ (Cenini, S. et al., J. Chem. Soc., Chem. Commun., 1286–187, (1984)). The reaction is enhanced by addition of a tertiary amine.

$$ArNO_2 + 3CO + EtOH \xrightarrow[Et_3N]{PtCl_2(PPh_3)_2SnCl_4} ArNHCOOEt + 2CO_2$$

Oxidative Carbonylation. Carbamates have been prepared from amines, alcohols, CO and O_2 in the presence of a catalyst system comprising platinum or palladium metal and iodide (Fukuoka, A. S., Chono, M. and Kohno, M., J. Org. Chem., **49**(8), 1458–1460, (1984); ibid., J. Chem. Soc., Chem. Commun., 399–400, (1984)).

$$R^1NH_2 + CO + R^2OH + \tfrac{1}{2}O_2 \longrightarrow R^1NHCO_2R^2 + H_2O$$

Spectroscopy. (1) An NMR study of the conformations of alkyl and aryl N-(alkylsulphonylmethyl)-N-methylcarbamates and aryl N-(arylsulphonylmethyl)-N-methylcarbamates has been used to investigate hindered internal rotation in carbamates (Vaneststammer, R. and Engberts, J. B. F. N., Rec. Trav. Chim.-J. Roy. Neth. Chem., **90**(12), 1307–1319, (1971)).
(2) Activation parameters for hindered rotation about the C–N double bond have been obtained for some trimethylgermyl, trimethylplumbyl, and *tert*-butyl esters of N,N-dimethylcarbamic, N,N-dimethylmonothiocarbamic, and N,N-dimethyldithiocarbamic acids (Lemire, A. E. and Thompson, J. C., Can. J. Chem., **53**(24), 3732–3738, (1975)).
(3) The structure of N-methylenecarbamates has been investigated by crystal-structure analysis, spectroscopy and

quantum-mechanical calculations (Wurthwein, E. U., Kupfer, R., Meier, S., Krestel, M. and Allmann, R., Chem. Ber., **121**(4), 591–596, (1988)).
(4) Chemical ionisation mass spectra have been reported for a number of carbamates of general structure $RNHCO_2C_2H_5$ (Wright, A. D., Bowen, R. D. and Jennings, K. R., J. Chem. Soc., Perkin Trans. II, 1521–1528, (1989)). The mechanism of formation of the observed fragment ions, and their analytical utility, are discussed.
(5) The structure and intramolecular dynamics of bis(diisobutylselenocarbamoyl)triselenide have been studied in solution by ^{77}Se NMR spectroscopy (Mazaki, Y. and Kobayashi, K., Tetrahedron Lett., **30**(21), 2813–2816, (1989)). A novel four-coordinated structure was identified in the solution at low temperature which undergoes rapid interconversion to the chain structure on raising the temperature.

Reactions. (1) N,N–Disubstituted carbamates give alkenes, amine and CO_2 upon heating (Daly, N. J. and Ziolkowski, F., Aust. J. Chem., **24**, 2541, (1971); Daly, N. J. and Ziolkowski, F., J. Chem. Soc., Chem. Commun., 911, (1972); Daly, N. J., Heweston, G. M. and Ziolkowski, F., Aust. J. Chem., **26**, 1259, (1973); Atkinson, R. F. et al., J. Org. Chem., **46**(13), 2804–2806, (1981)).
(2) The anodic oxidation of various carbamates, oxazolidones and imidazolidones in methanol has been studied (Shono, T. et al., J. Chem. Soc. Jap., Chem. Ind. Chem., 1782–1787, (1984)). Methyl N,N–dialkylcarbamate yields three types of product, α–methoxylated compound, enamine-type product and dealkylated carbamate (Shono, T., Hamaguchi, H. and Matsumura, Y., J. Amer. Chem. Soc., **97**(15), 4264–4268, (1975)).
(3) A mild method for the conversion of carbamates to carbinols involves a Cl_3SiH–induced cleavage reaction (Pirkle, W. H. and Hauske, J. R., J. Org. Chem., **42**(16), 2781–2782, (1977)).
(4) The mechanism of cleavage of carbamate anions (Ewing, S. P., Lockshon, D. and Jencks, W. P., J. Amer. Chem. Soc., **102**(9), 3072–3084, (1980)), the solvolysis of phenyl N,N-dimethylcarbamates in alkaline and acid media (Vontor, T., Drobilic, V., Socha, J. and Vecera, M., Collect. Czech. Chem. Commun., **39**(1), 281–286, (1974)) and the effects of hydroxy-functionalised micelles on the basic hydrolysis of carbamates (Broxton, T. J. and Chong, R. P. T., J. Org. Chem., **51**(16), 3112–3115, (1986)) have been investigated. The presence of an azole group (pyrrole, indole, carbazole) in carbamate derivatives produces a remarkable increase in reactivity

towards basic hydrolysis of up to 35,000 times faster than corresponding methylamino derivatives (Savelli, G. et al., Gazz. Chim. Ital., **121**(4), 205–208, (1991)).
(5) Aminoalkylphosphonates are synthesised from carbamates in which a new carbon–phosphorus bond forming reaction, the reaction of α–methoxyurethanes with trialkyl phosphites in the presence of Lewis acid catalysts is used (Shono, T. et al., J. Amer. Chem. Soc., **97**(15), 4264–4268, (1975); Shono, T., Matsumura, Y. and Tsubata, K., Tetrahedron Lett., **22**(34), 3249–3252, (1981)). $TiCl_4$ or $BF_3.Et_2O$ are effective catalysts.
(6) Aza–1 bicyclobutanes are readily obtained by photolysis of carbamates (Bartnik, R., Cebulska, Z. and Laurent, A., Tetrahedron Lett., **24**(39), 4197–4198, (1983)).
(7) N–Formyl, –acetyl, and –benzoyl groups can be removed from secondary amides with amines under mild conditions after *t*–butoxycarbonylation, giving acid–labile *t*–butyl carbamates (Grehn, L., Gunnarsson, K. and Ragnarsson, U., J. Chem. Soc., Chem. Commun., 1317–1318, (1985)).
(8) Carbamates are N–alkylated in good yields by electro–reduction in the presence of alkyl halides (Shono, T., Kashimura, S. and Nogusa, H., Chem. Lett., 425–428, (1986)).
(9) β–Phenylseleno carbamates are synthesised by reaction of olefins with phenylselenenyl chloride and carbamates in the presence of $AgBF_4$ (Salazar, J. A. et al., Tetrahedron Lett., **27**(22), 2513–2516, (1986)). The reaction constitutes a useful method for the conversion of olefins to β–functionalised protected amines.
(10) N,N–Disubstituted β,γ–unsaturated urethanes and bromine react in DCM at room temperature. Crotyl urethanes and bromine give mixtures of corresponding saturated urethanes (dibromine adducts), oxazolidin–2–ones and tetrahydro–2H–1,3–oxazine–2–ones. Reaction of γ,γ–dimethylallyl urethanes with bromine gives similar results. A useful synthetic route to N–substituted tetrahydro–2H–1,3–oxazin–2–ones is demonstrated by the reaction of cinnamyl urethanes with bromine (Muhlstadt, M., Olk, B. and Widera, R., J. Prakt. Chem., **328**(2), 163–172, (1986)).
(11) N–Cinnamyl urethanes are cyclised to 3,5–disubstituted oxazolidin–2–ones by aromatic sulphenyl chlorides in DCM (Muhlstadt, M., Olk, B. and Widera, R., J. Prakt. Chem., **328**(2), 173–180, (1986)).
(12) N–Acyliminium ions add γ–oxygenated allyltins in the presence of Lewis acids to give carbamates bearing both an oxygen function and a terminal double bond (Yamamoto, Y. and Schmid, M., J. Chem. Soc., Chem. Commun., 1310–1312, (1989)).

α-Ethoxycarbamates are used as precursors for the N-acyliminium ions ("activated imines") (Speckamp, W. N. et al., Tetrahedron Lett., **26**(26), 3151-3154, (1985)). $BF_3.OEt_2$ or $TiCl_4$ can be used as an activator, but $SnCl_4$ is not effective. The stereochemistry of the adducts is determined by conversion to 2-oxazolidone derivatives and in certain cases very high diastereoselectivity is achieved.

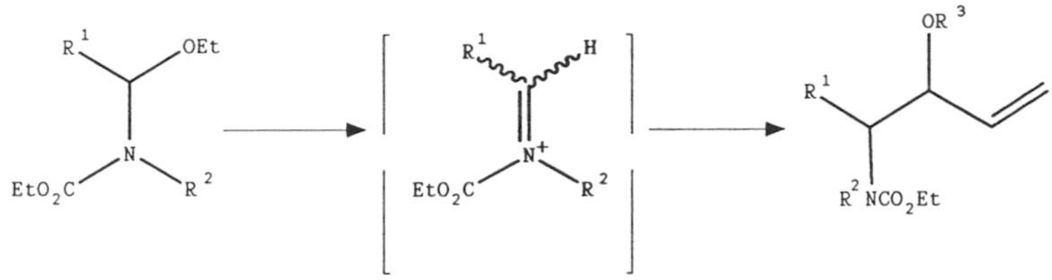

(13) Carbamates react with sodium under acyloin conditions to give substituted oxalate diamides (Crumrine, D. S., Dieschbourg, T. A., O'Toole, J. G., Tassone, B. A. and Vandeburg, S. C., J. Org. Chem., **52**(16), 3699-3701, (1987)). The best yields are obtained with phenyl-substituted systems.

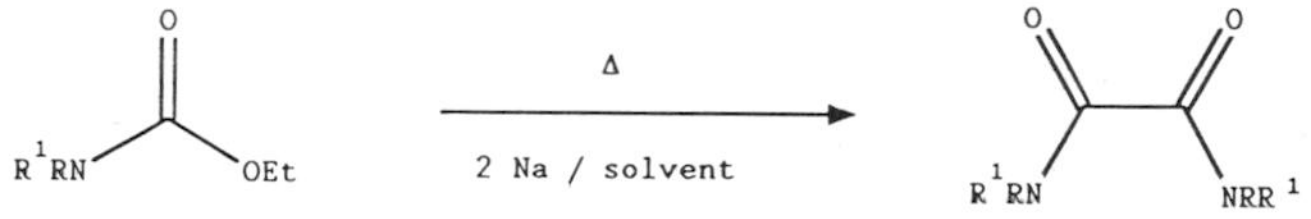

(14) Methyl carbamates react smoothly with sodium hydrogen telluride in DMF to give amines (Zhou, X. J. and Huang, Z. Z., Synth. Commun., **19**(7-8), 1347-1349, (1989)).

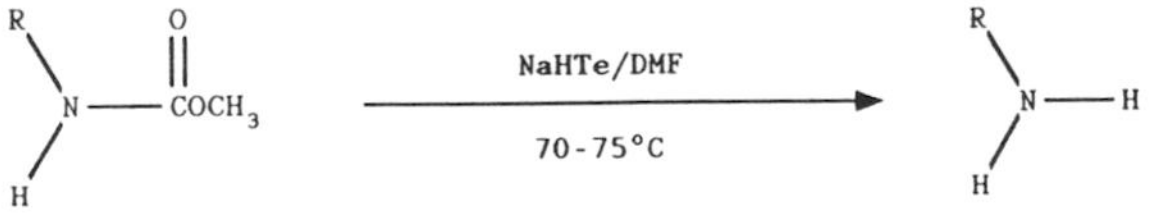

(15) Phenyltellurinyl acetate or trifluoroacetate in combination with ethyl carbamate effects regio- and

stereoselective aminotellurinylation of olefins in the presence of $BF_3.Et_2O$ in $CHCl_3$ under reflux to give ethyl[(2-phenyltellurinyl)alkyl] carbamates (Ogura, F. et al., J. Org. Chem., **54**(18), 4398–4404, (1989)). Phenyltellurinyl trifluoromethanesulphonate may be used at lower temperatures and without Lewis acid. In refluxing 1,2-dichloroethane, 2-oxazolidinone is obtained.

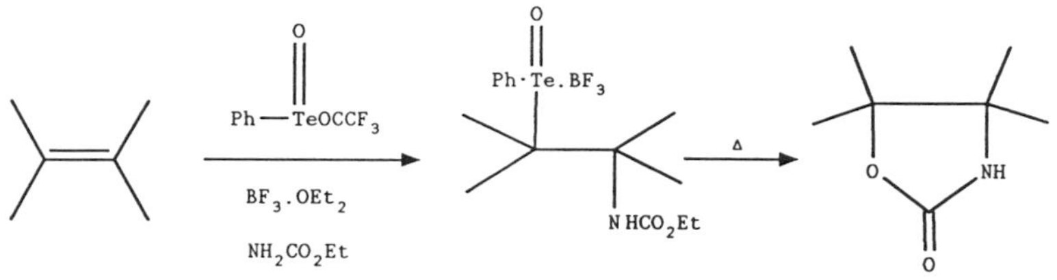

(16) Ene reactions of N-sulphinylcarbamates derived from chiral alcohols 8-phenylmenthol and trans-2-phenylcyclohexanol have been shown to proceed with high levels (>95%) of asymmetric induction (Whitesell, J. K. et al., J. Amer. Chem. Soc., 112(21), 7653–7659, (1990)). Where regioisomeric products are possible, as with unsymmetrical alkenes, generally only the adduct with the more stable double bond is formed.
(17) Alkyl α-thiocyano and α-isothiocyano ethyl carbamates are obtained by the reaction of the corresponding alkyl α-chloroethyl carbamate with KSCN or NH_4SCN (Caubere, P. et al., Tetrahedron Lett., 27(50), 6067–6070, (1986)).

with organometallics. (1) Symmetrical ketones have been prepared through the addition of organolithium and Grignard reagents to N,N-dialkylurethanes (Michael, U. and Hornfeldt, A.-B., Tetrahedron Lett., 5219–5222, (1970); Scilly, N. F., Synthesis, 160, (1973)).
(2) Treatment of methylenebis(N,N-dimethyldithiocarbamate) with n-BuLi produces the lithiomethylene derivative which functions as a formyl anion equivalent. Reaction with alkyl halides followed by hydrolysis with mercuric ion gives the corresponding aldehydes (Nakai, T. and Okawara, M., Chem. Lett., 731–732, (1974)).
(3) Symmetrical and unsymmetrical ketones are obtained *via* methoxyamides generated *in situ* from organometallic addition reactions to N-methoxy-ureas and urethanes (Hlasta, D. J. and Court, J. J., Tetrahedron Lett., **30**(14), 1773–1776, (1989)).

$$X-C(=O)-N(CH_3)-OCH_3 \xrightarrow[2.\ R^2M]{1.\ R^1M} R^1-C(=O)-R^2$$

(4) Dianions from simple N-*t*-butyl benzylic-type carbamates are readily formed with alkyllithium bases and undergo alkylation with a variety of electrophiles. Both secondary and tertiary α-oxo carbanions are easily accessible. DIBAL cleavage of the carbamate provides a high yield, general synthesis of alkylated benzylic alcohols (Barner, B. A. and Mani, R. S., Tetrahedron Lett., **30**(40), 5413-5416, (1989)).
(5) α-Metalated enol carbamates are conveniently generated (*sec*-BuLi/TMEDA/THF/-78°C) and well-behaved acyl anion equivalents (Sengupta, S. and Snieckus, V., J. Org. Chem., **55**(22), 5680-5683, (1990); Kocienski, P. and Dixon, N. J., Synlett., 52, (1989)). α-Lithiation of a thioenol carbamate has been demonstrated (Hoppe, D., Beckmann, L. and Follmann, R., Angew. Chem., Int. Ed. Engl., **19**(4), 303-304, (1980)).
(6) The *t*-Boc group activates the α'-lithiation of piperidinyl and related carbamates to give lithium reagents which add to electrophiles to provide α'-elaborated carbamates (Beak, P. and Lee, W.-K., Tetrahedron Lett., **30**(10), 1197-1200, (1989)).
Other carbamates have been used by various workers (Armande, J. C. and Pandit, U. K., Tetrahedron Lett., 897-898, (1977); Hassel, T. and Seebach, D., Angew. Chem. Int. Ed. Engl., **17**, 274, (1978); Macdonald, T. L., J. Org. Chem., **45**, 193-194, (1980); Beak, P., Zajdel, W. J. and Reitz, D. B., Chem. Rev., **84**, 471-523, (1984); Hassel, T. and Seebach, D., Helv. Chim. Acta., **61**, 2237, (1987); Comins, D. L. and Weglarz, M. A., J. Org. Chem., **53**(19), 4437-4442, (1988); Ahlbrecht, H. and Kornetzky, D., Synthesis, 775-777, (1988)).

Arylcarbamates. (1) A facile synthesis of alkyl N-arylcarbamates has been reported (Leardini, R. and Zanardi, G., Synthesis, 225-227, (1982)).

$$ArHN-\overset{O}{\overset{\|}{C}}H \xrightarrow{Pb(OAc)_4} ArN{=}C{=}O \xrightarrow{ROH} ArHN-\overset{O}{\overset{\|}{C}}-OR$$

(2) Arylhydroxymethylcarbamates are prepared from aryl carbamates by reaction with paraformaldehyde in THF and a

catalytic amount of acid (Fahmy, M. A. H. and Fukuto, T. R., J. Agr. Food Chem., **20**(1), 168–169, (1972)).
(3) A base-catalysed thermal decomposition of aryl urethanes has been reported (Blahak, J., Justus Liebigs Ann. Chem., 1353–1356, (1978)).
(4) α–Sulphenyl–α–aminonitrile is converted to alkyl N–methyl–N–phenylcarbamates by concurrent autoxidation and substitution with alkoxide ions (Chuang, T. H. et al., Synlett., 733–734, (1990)).

CN
PhS
NMePh
ROLi / O_2 / CuI
THF
O
RO
NMePh

Allylic urethanes. (1) Halonium–initiated cyclisation of allylic urethanes has been studied (Parker, K. A. and Ofee, R., J. Amer. Chem. Soc., **105**(3), 654–655, (1983)).
(2) α–Anions of allyl carbamates are useful synthetic intermediates (Kramer, T., Schwark, J.–R. and Hoppe, D., Tetrahedron Lett., **30**(50), 7037–7040, (1989)) and have been reviewed (Hoppe, D., Angew. Chem., Int. Ed. Engl., **23**, 932–948, (1984)).
(3) The allylation of carbonucleophiles by allylic carbamates under neutral conditions has been reported (Tsuji, J. et al., Tetrahedron Lett., **26**(20), 2449–2452, (1985)).

Vinylogous urethanes.

Preparations. (1) Various methods have been reported for the synthesis of open–chain and cyclic, vinylogous urethanes containing intramolecular hydrogen bonds (Walter, W. and Fleck, T., Justus Liebigs Ann. Chem., 670–681, (1976)).
(2) N–Alkyl lactams are converted to vinylogous urethanes *via* (methylthio)alkylideniminium salts (Gugelchuk, M. M., Hart, D. J. and Tsai, Y. M., J. Org. Chem., **46**(18), 3671–3675, (1981)).
(3) Vinyl carbamates may be obtained *via* interaction of alkylidenecarbenes with isocyanates (Stang, P. J. and Anderson, G. H., J. Org. Chem., **46**(22), 4585–4586, (1981)). Other methods include dehydrohalogenation of α–chloroalkyl carbamates or reaction of vinyloxycarbonyl chloride with amines (Olofson, R. A., Yamamoto, Y. S. and Wancowicz, D. J., Tetrahedron Lett., 1563–1566, (1977); Olofson, R. A. et al., ibid., 1567–1570,

(1977); Olofson, R. A. and Schnur, R. C., ibid., 1571–1574, (1977)). The latter method has been used for amino group protection in peptide synthesis.

$$(CH_3)_2C{=}C(OSO_2CF_3)Si(CH_3)_3 \xrightarrow[\text{glyme } -20°C]{n\text{-}Bu_4NF} (CH_3)_2C{=}C:$$

$$\xrightarrow[2.\ H_2O]{1.\ RN{=}C{=}O} RNHCOOCH{=}C(CH_3)_2$$

(4) Treatment of ω–azido–β–dicarbonyl derivatives with Ph_3P leads to a transient phosphinimine (Staudinger reaction), which cyclises to vinylogous urethanes and amides *via* an intramolecular aza–Wittig reaction (Lambert, P. H., Vaultier, M. and Carrie, R., J. Org. Chem., **50**, 5352–5356, (1985). Reaction of 2–[(triphenylphosphoranylidene)amino]benzyl alcohol with CO_2 and CS_2 gives benzoxazin–ones and –thiones respectively (Molina, P., Arques, A. and Molina, A., Synthesis, 21–23, (1991)).

(5) Sulphide contraction (Eschenmoser, A. et al., Helv. Chim. Acta, **54**, 710, (1971)), or the use of lactam derivatives (Maitte, P. et al., J. Org. Chem., **44**(17), 3089, (1979); Celerier, J. P., Lhommet, G. and Maitte, P., Tetrahedron Lett., 22, 963–964, (1981); Celerier, J. P., Richaud, M. G. and Lhommet, G., Synthesis, 195–197, (1983); Maitte, P. et al., Synthesis, 130–133, (1981)) has been used to obtain vinylogous urethanes.

(6) Terminal alkynes react with secondary amines and CO_2 in the presence of ruthenium catalysts such as $Ru_3(CO)_{12}$ and $RuCl_3.3H_2O$ to afford vinyl carbamates (Sasaki, Y. and Dixneuf, P. H., J. Chem. Soc., Chem. Commun., 790–791, (1986); Mahe, R., Dixneuf, P. H. and Lecolier, S., Tetrahedron Lett., **27**(52), 6333–6336, (1986); Sasaki, Y. and Dixneuf, P. H., J. Org. Chem., **52**(2), 314–315, (1987)).

$$R{-}C{\equiv}CH \xrightarrow[Ru_3(CO)_{12}]{Et_2NH,\ CO_2} (R)(H)C{=}C(H){-}O{-}C(O){-}NEt_2 + (R)(H)C{=}C(H){-}O{-}C(O){-}NEt_2 + (H)(H)C{=}C(R){-}O{-}C(O){-}NEt_2$$

Reactions. (1) Vinylogous urethanes have been used for the synthesis of alkaloids such as pyrrolizidines (Pinnick, H. W. and Chang, Y. H., J. Org. Chem., **43**(24), 4662–4663, (1978)), indolizidines (Howard, A. S., Gerrans, G. C. and Michael, J. P., J. Org. Chem., **45**(9), 1713–1715, (1980)) and quinolizidines (Gerrans, G. C., Howard, A. S. and Orlek, B. S., Tetrahedron Lett., 4171–4172, (1975)).
(2) The synthesis and use of enantio- and erythro-selective lithium enolates derived from vinylogous urethanes has been described (Schlessinger, R. H. et al., J. Org. Chem., **51**(15), 3068–3070, (1986); Schlessinger, R. H., Iwanowicz, E. J. and Springer, J. P., J. Org. Chem., **51**(15), 3070–3073, (1986)).

Halourethanes. (1) Reactions of iodine isocyanate (INCO), N,N-dichlorourethane (Cl_2NCO_2Et) and N-chlorourethane ($ClNHCO_2Et$) with olefins has been reviewed (Swern, D., Amer. Chem. Soc., Div. Petrol. Chem., Prep. **15**(2), E39–E51, (1970)).
(2) Carbalkoxynitrenes may be generated from N,N-dichlorocarbamates by electrolysis (Fuchigami, T. and Nonaka, T., Chem. Lett., 1087–1090, (1977)).
(3) The reactivity of N,N-dichlorourethanes towards alkenes has been studied (Balon, Y. G. and Paranyuk, V. E., Zh. Org. Khim., **16**(3), 556–563, (1980)).
(4) Alkyl, aryl and oxime carbamates react with thionyl chloride to give the corresponding N-chlorosulphinyl derivatives (Fukuto, T. R. and Fahmy, M. A. H., J. Agr. Food Chem., **29**(3), 567–572, (1981)).
(5) The structure and chemical properties of N-monobromocarbamates have been investigated (Mochalin, V. B., Maksimova, T. N., Filenko, N. I., Bakova, O. V. and Khenkina, T. V., Zh. Org. Khim., **18**(6), 1202–1205, (1982)).
(6) The irradiation of benzyl N-bromo-N-methylcarbamates affords products due to Ar_{1-5} cyclisation. The intramolecular aromatic cyclisation is regiospecific and not strongly influenced by electronic effects (Goosen, A. et al., J. Chem. Soc., Perkin Trans. I 1243–1245, (1987)).

Silyl urethanes.

Preparations. (1) Organosilicon adamantyl-containing urethanes have been synthesised (Ushchenko, V. P., Kim, A. D., Khardin, A. P. and Brel, V. K., Zh. Obshch. Khim., **46**(9), 2157–2158, (1976)).
(2) Silicon derivatives of oxycarbamic acid have been used to prepare O-silylurethanes (Sheludyakov, V. D., Dmitrieva, A. B.,

Gusev, A. I., Apalkova, G. M. and Kirilin, A. D., Zh. Obshch. Khim., **54**(10), 2298–2301, (1984)).

$$[(CH_3)_3SiO][(CH_3)_3Si]N{-}C(=O){-}OSi(CH_3)_3 \xrightarrow{H_2NRNH_2} (CH_3)_3SiO{-}C(=O){-}NH{-}R{-}NH{-}C(=O){-}OSi(CH_3)_3$$

(3) Alkyl- or aryl-bis(trimethylsilyl)amines react with chlorothionoformates to give N-alkyl or N-aryl-N-trimethylsilylthionurethanes (Walter, W. and Akram, M., Phosphor. Sulfur Relat. Elem., **21**(3), 291–300, (1985)).
(4) Trimethylsilyl N-aryl carbamates are prepared by the reaction of the respective aniline derivatives, hexamethyldisilazane and CO_2. Two mol-% of anhydrous $CoCl_2$ is an effective catalyst for the process (Knausz, D., Koos, Z., Rohonczy, J. and Ujszaszy, K., Acta Chim. Hung., **120**(2), 167–170, (1985)).
(5) Reaction of $SiCl_4$ with R_2NH and CO_2 in a hydrocarbon solvent at atmospheric pressure and room temperature gives N,N-dialkyl-silylcarbamates $Si(O_2CNR_2)_4$ (Calderazzo, F. et al., Gazz. Chim. Ital., **120**(12), 819–820, (1990)).

$$Si[O{-}C(=O){-}NR_2]_4$$

Reactions. (1) A kinetic study of the hydrolysis of trimethylsilylurethanes of the type $Me_3Si(para\text{-}XC_6H_4)NCO_2Et$ has been reported (Grusseruyken, H., Hanig, K., Wagner, S., Ruhlmann, K. and Schlapa, J., J. Organometal. Chem., **260**(1), 51–67, (1984)).

(2) N,O-Bis(trimethylsilyl)carbamate ($Me_3SiNHCO_2SiMe_3$) is extremely useful for the silylation of alcohols, phenols and carboxylic acids (Birkofer, L. and Sommer, P., J. Organometal. Chem., **99**(1), C1-C4, (1975)). The only by-products are CO_2 and NH_3.

Amino protection in Peptide Synthesis. (1) The introduction and removal of urethane-type protecting groups for amines has been reviewed (Green, T. W., in "Protective Groups in Organic Synthesis". Wiley, New York, (1981); Geiger, R. and Konig, W., in "The Peptides" Gross, E. and Meienhofer, J. (Eds.) Academic Press: New York, **3**, chapter 1, (1981); Bodanszky, M., in "Principles of Peptide Synthesis". Springer-Verlag, p90-102, (1984); Carpino, L. A., Accounts of Chemical Research, 20, 401-407, (1987); Jones, J., in "The Chemical Synthesis of Peptides" Clarendon Press: Oxford, p17-32, (1991)).
(2) The chemistry of the dithiasuccinoyl (DTS) function, an amino protecting group removable by reduction, has been studied (Barany, G. and Merrifield, R. B., J. Amer. Chem. Soc., **99**(22), 7363-7365, (1977)). The dithiasuccinoyl (Dts) amino protecting group was developed for application to orthogonal schemes of peptide synthesis. The group is stable under the acidolytic conditions used to remove *tert*-butyl and benzyl-based protecting groups and resistant to the photolytic conditions used to cleave the acid-stable *o*-nitrobenzyl and α-methylphenacyl esters. It is rapidly and quantitatively removed in the presence of these groups by mild and specific treatment with thiols, borohydrides and trialkylphosphines. The kinetics and mechanism of the thiolytic removal of the Dts group have been reported (Barany, G. and Merrifield, R. B., J. Amer. Chem. Soc., **102**(9), 3084-3095, (1980)).
(3) The 2,2,2-trichloro-*tert*-butyloxycarbonyl (TCBOC) group is an acid and base stable amino protecting group (Eckert, H., Listl, M. and Ugi, I., Angew. Chem., **90**(5), 388-389, (1978)).

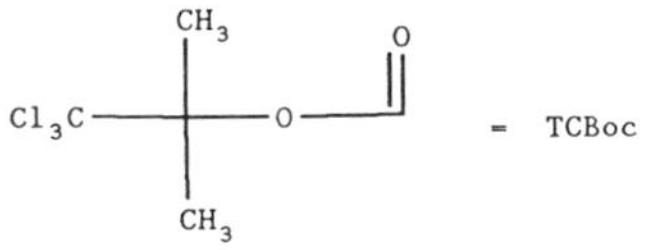

$Cl_3C-C(CH_3)_2-O-C(=O)-$ = TCBoc

(4) Base-labile Nα-fluorenylmethoxycarbonyl (Fmoc) amino acids (Carpino, L. A. and Han, G. Y., J. Amer. Chem. Soc., **92**, 5748-5749, (1970); *ibid.*, J. Org. Chem., **37**, 3404, (1970)) are well established in solid phase peptide synthesis (Fields, G. B.,

and Noble, R. L., Int. J. Pept. Protein Res., **35**, 161–214, (1990); Atherton, E. and Sheppard, R. C., in "The Peptides, Analysis, Synthesis, Biology" (Udenfreind, S. and Meienhofer, J., Eds. Academic Press, New York **9**, chapter 1, (1987); Atherton, E. and Sheppard, R. C., "Solid Phase Peptide Synthesis: A Practical Approach" Oxford University Press, Oxford, (1989)). (+) and (−)-1-(9-fluorenyl)ethyl chloroformate (FLEC) is a useful reagent for the resolution of chiral amino acids and amines via diastereomer formation and HPLC.

(5) The effect of structure on the ease of solvolytic deblocking of an array of α-halo-*tert*-alkyl carbamates has been studied in a search for acid-stable, solvolytically-deblocked amino protecting groups (Carpino, L. A. et al., J. Org. Chem., **49**(5), 836–842, (1984)).

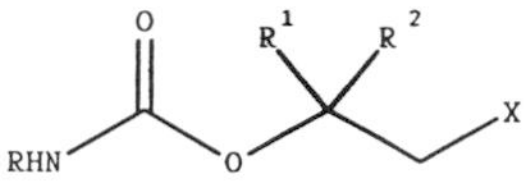

The corresponding thiocarbamates undergo isomerisation and other reactions due to participation of the sulphur atom. 1,1,1,3,3,3-hexachloro-2-(bromomethyl)-2-propyl carbamates are relatively unreactive towards solvolytic deblocking. The α-bromo-*tert*-butyloxycarbonyl (α-Br-*t*-Boc) group undergoes self cleavage upon warming in methanol or ethanol. Cleavage does not take place in nonpolar solvents such as CHCl3, DCM or benzene. A deficiency of this group is its moderate sensitivity toward acidic reagents. The 1,3-dibromo-2-methyl-2-propyloxycarbonyl group (DB-*t*-Boc) is easily beblocked by warming in EtOH or MeOH and is much more stable towards acids.

(6) Catalytic transfer hydrogenation has been used for the cleavage of the benzyloxycarbonyl group (Jackson, A. E. and Johnstone, R. A. W., Synthesis, 685, (1976); Anantharamaiah, G. M. and Sivanandaiah, K. M., J. Chem. Soc. Perkin Trans. I, 490–491, (1977)). An electrosynthetic procedure for the cleavage of the benzyloxycarbonyl group from protected amino acids and peptides has also been described (Casadei, M. A. and Pletcher, D., Synthesis, 1118–1119, (1987)).

$$R^1R^2N{-}CO_2CH_2C_6H_5 \xrightarrow[\text{MeOH/AcOH/NaClO}_4\ \ 90\text{-}99\%]{\text{Pd/C cathode 15mA cm}^{-2}} R^1R^2NH + CO_2 + C_6H_5CH_3$$

O-2-(Trimethylsilyl)ethyl and O-*tert*-butyl carbamates are converted into the corresponding O-benzyl carbamates using benzyl trichloroacetimidate (Barrett, A. G. M. and Pilipauskas, D., J. Org. Chem., **55**(17), 5170-5173, (1990)).

(7) 1-[2-(Trimethylsilyl)ethoxycarbonyloxy]benzotriazole (Teoc-OBt) and 1-[2-(trimethylsilyl)ethoxycarbonyloxy]pyridin-2,5-dione (Teoc-OSu) are superior to other reagents for the introduction of the Teoc group into amino acids (Shute, R. E. and Rich, D. H., Synthesis, 346-349, (1987)). The group is stable under basic and hydrogenolytic conditions, and cleaved by tetra-alkylammonium fluorides, strong acid and Lewis acids.

(8) The (allyloxy)carbonyl function is an N-blocking group which is removable by hydrogenolysis (Tsuji, J. and Yanakawa, T., Tetrahedron Lett., 613-616, (1979); Hutchins, R. O., Learn, K. and Fulton, R. P., Tetrahedron Lett., 27-30, (1980)). It is also subject to C-O bond scission by organocuprates (Anderson, R. J., Hendrick, C. A. and Siddall, J. B., J. Amer. Chem. Soc., **92**, 735-737, (1970); Ho, T., Synth. Commun., 15-17, (1978)), nickel carbonyl (Corey, E. J. and Suggs, W. J., J. Org. Chem., **38**, 3223-3224, (1973)) and homogeneous Palladium^0 - catalysed exchange deprotection (Jeffrey, P. D. and McCombie, S. W., J. Org. Chem., **47**, 587-590, (1982)).

(9) The *tert*-butoxycarbonyl group can be attached to amines, as free bases or their salts, under mild conditions using ultrasonic irradiation (Einhorn, J., Einhorn, C. and Luche, J. L., Synlett, 37-38, (1991)).

(10) After protection of amines with the 4-azidomethylenoxy-benzyloxycarbonyl (AZ) group it may be cleaved in the presence of Z and methyl ester groups and independent removal of AZ and Boc groups is possible (Loubinoux, B. and Gerardin, P., Tetrahedron Lett., **32**(3),351-354, (1991)).

(11) A comparison has been made of the coupling efficiencies in solid phase peptide synthesis using Mpc-2-[4-(methyl-sulphonyl)phenylsulphonyl]ethoxycarbonyl or Fmoc-amino acids. The former were found to be superior and to give a more homogeneous product (Schielen, W. J. G. et al., Int. J. Peptide Protein Res., **37**, 341-346, (1991)).

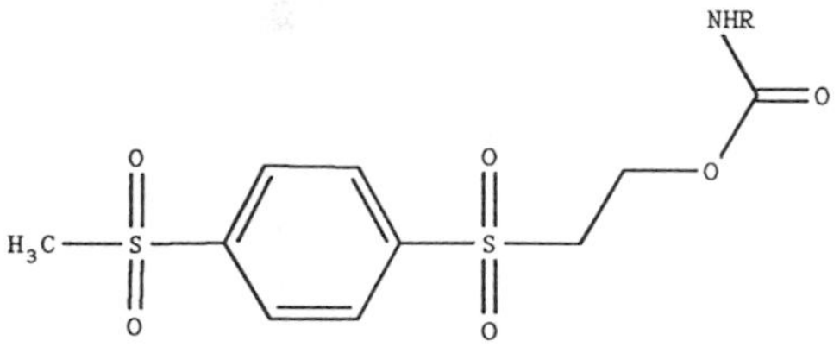

Mpc

Coupling Reagents in Peptide Synthesis. The mixed anhydride producing compounds 1-ethoxycarbonyl-2-ethoxy-1,2-dihydroquinoline (EEDQ) (Belleau, B. and Malek, G., J. Amer. Chem. Soc., **90**, 1651, (1968)) and 1-isobutoxycarbonyl-2-isobutoxy-1,2-dihydroquinoline (IIDQ) (Kiso, Y., Kai, Y. and Yajima, H., Chem. Pharm. Bull., **21**, 3507, (1973)) are useful coupling reagents which are readily prepared, easily stored and cause little racemisation or other side reactions.

Thiocarbamates

Preparations. (1) Aminolysis of primary and secondary alkyl xanthate esters yields thiourethanes while dithiourethanes are formed, equally exclusively, by the aminolysis of tertiary alkyl xanthate esters (Barrett, G. C. and Martins, C. M. O. A., J. Chem. Soc., Chem. Commun., 638-639, (1972)).
(2) The reaction of N,N-disubstituted carbamoyl chlorides with potassium alkyl or benzyl dithiocarbonates yields disubstituted thiocarbamates (Damico, J. J. and Schafer, T., Phosphor. Sulfur Relat. Elem., **8**(3), 301-304, (1980)).

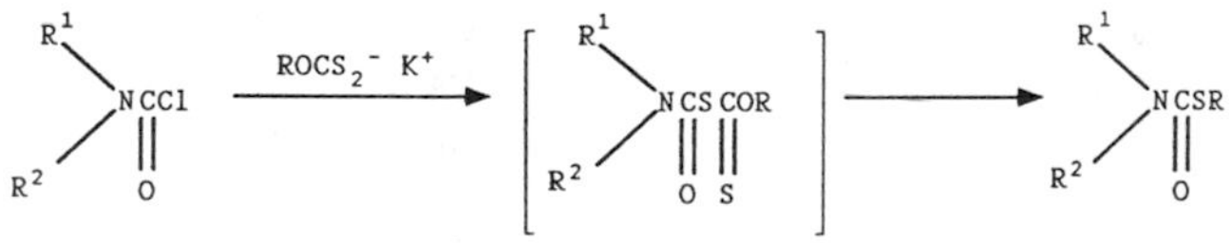

$$R^1R^2NC(O)Cl \xrightarrow{ROCS_2^- \ K^+} [R^1R^2NC(O)SC(S)OR] \longrightarrow R^1R^2NC(O)SR$$

(3) The phase-transfer catalysed preparation of S-alkyl thiocarbamates has been reported (Wang, C. H., Synthesis, 622-623, (1981)).
(4) In connection with a study of thio- and selenocarbonic

acids, bis(N,N-dialkylselenocarbamoyl)selenides and bis(N,N-dialkyldiselenocarbamato)selenium^{II} have been prepared (Henriksen, L., Synthesis, 771-773, (1982)).
(5) O,O-Diethylphosphono and phosphonothioyl N,N-dialkyldithiocarbamates have been prepared (Sodhi, G. S. and Kaushik, N. K., J. Indian Chem. Soc., **60**(8), 806-808, (1983)).
(6) A number of thiocarbamic S-esters have been synthesised in good yield by the reaction of thiol with substituted N-methyl-N-nitrosoureas in anhydrous CH_3CN (Yoshida, K., Isobe, M., Yano, K. and Nagamatsu, K., Bull. Chem. Soc Jap., **58**(7), 2143-2144, (1985)).

$$R^1NH_2 \xrightarrow[2.\ NaNO_2/HCO_2H]{1.\ CH_3NCO} R^1HN{-}C({=}O){-}N(NO)CH_3 \xrightarrow[CH_3CN]{R^2SH} R^1HN{-}C({=}O){-}SR^2$$

(7) The insertion of CS_2 into the BeN bonds of bis(diisopropylamino)beryllium yields bis(diisopropyldithiocarbamato)beryllium (Noth, H. and Schlosser, D., Chem. Ber., **121**(10), 1711-1713, (1988)). In the reaction of trimeric $Be(NMe_2)_2$ with CS_2 all Me_2N bridge bonds remain intact: only the two terminal Me_2N groups add CS_2.
(8) A facile method for the synthesis of bis(dithiocarbamates) and bis(dithiocarbonimidates) has been reported (Garin, J., Melendez, E., Merchan, F. L., Merino, P. and Tejero, T., Bull. Soc. Chim. Belg., **97**(10), 791-792, (1988)).
(9) The selenium-catalysed synthesis of S-alkyl thiocarbamates from amines, CO, sulphur and alkyl halides has been reported (Sonoda, N., Mizuno, T., Murakami, S., Kondo, K., Ogawa, A., Ryo, I. and Kambe, N., Angew. Chem. Int. Ed. Engl., **28**(4), 452-453, (1989)).
(10) Alkoxythiocarbonyl(ethoxycarbonyl)sulphides are used for converting amines to thiourethanes (Martin, A. A., Zeuner, F. and Barnikow, G., Z. Chem., **30**(3), 90-91, (1991)).
(11) Treatment of Viehe's salt ($Me_2N^+{=}CCl_2.Cl^-$) with one equivalent of butanethiol followed by hydrolysis gives the thiocarbamate, whereas addition of the reaction mixture to a solution of sodium hydrogen selenide in EtOH yields the selenothiocarbamate $Me_2NC({=}Se)SBu$ (Stick, R. V. et al., Aust. J. Chem., **41**, 549-561, (1988)). A similar result was obtained with thiophenol.

Reactions. (1) Pyrolysis of N-aryldithiocarbamates, prepared

from arylamines, affords isothiocyanates (Ottenbrite, R. M., J. Chem. Soc., Perkin I, 88–90, (1972)).
(2) The decomposition of O-ethyl-N-lithio-N-substituted thiocarbamates is accelerated by addition of excess CS_2 or SO_2, and isothiocyanates are obtained in moderate yields even at room temperature (Fujinami, T., Ashida, M. and Sakai, S., J. Chem. Soc. Jap., Chem. Ind. Chem., 773–774, (1978)).
(3) Free enthalpies of activation of the restricted rotation at the partial C–N double bond in a number of thiocarbamic acid esters have been determined and the results compared with earlier studies (Kleinpeter, E., Widers, R. and Mulhstadt, M., J. Prakt. Chem., **320**(2), 279–282, (1978)).
(4) The reactivity of dithiocarbamic esters and methods for the preparation of 3,5-substituted 2-thiohydantoins has been examined (Blotny, G., Synthesis, 391–392, (1983)).
(5) Activation of dithiocarbamate salts with 2-halo-3-alkyl-4-phenylthiazolium salt and subsequent one-pot nucleophilic reaction with N, S, and O nucleophiles gives substituted thioureas, dithiocarbamates and thiocarbamates respectively, under very mild conditions (Sugimoto, H., Makino, I. and Hirai, K., J. Org. Chem., **53**(10), 2263–2267, (1988)).

Oxazolidinones.

Preparations. (1) $Pb(OAc)_4$ in pyridine effects rapid, high-yield, Hofmann-like rearrangement of β-hydroxy primary amides to 2-oxazolidinones (Simons, S. S., J. Org. Chem., **38**(3), 414–416, (1973)). The products in turn give the corresponding β-hydroxy amines.

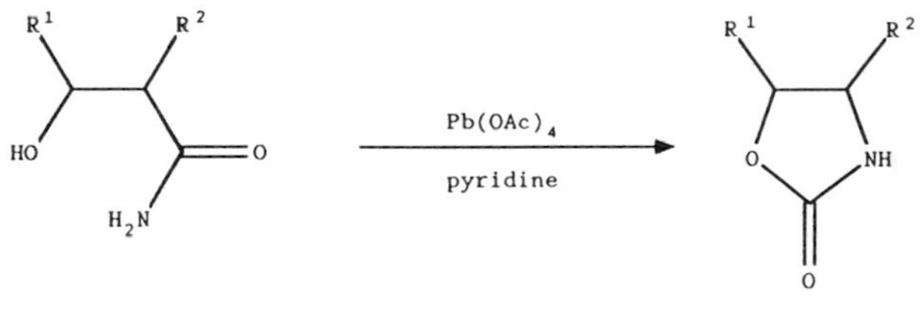

(2) Ph_4SbI is a general and versatile catalyst for the selective formation of unusual cycloadducts, 3,4-disubstituted oxazolidinones, in the reaction of oxiranes with isocyanates (Baba, A., Fujiwara, M. and Matsuda, H., Tetrahedron Lett., **27**(1), 77–80, (1986)).

$$R^1-CH(O)CH_2 + R^2-N=C=O \xrightarrow{Ph_4SbI} \text{oxazolidin-2-ones (4-}R^1\text{ and 5-}R^1\text{, N-}R^2)$$

(3) The synthesis of fatty 2-oxazolones from epoxy fatty acids has been reported (Farocqi, J. A., Chem. Ind.,- London, 245-246, (1986)).

$$R-CH(O)CH-R^1 \xrightarrow[\text{DMF, LiCl}]{PhNCO} \text{3-phenyl-4-R-5-}R^1\text{-oxazolidin-2-one}$$

(4) An efficient synthesis of the Evans' chiral auxiliary, (4S)-4-isopropyl-2-oxazolidinone, has been described (Wuts, P. G. M. and Pruitt, L. E., Synthesis, 622-623, (1989)). Schotten-Baumann acylation of valine with phenyl carbonochloridate, followed by borane reduction and cyclisation affords the product in 81% overall yield. Since Evans' initial reports (Evans, D. A., Aldrichim. Acta 15(2), 23, (1982); Evans, D. A., Ennis, M. D. and Mathre, D. J., J. Amer. Chem. Soc., 104, 1737, (1982)) a variety of oxazolidinones have found widespread application in natural product synthesis.

$$\text{valine} \xrightarrow[\text{pH 8.5}]{ClCO_2Ph} \text{N-phenoxycarbonyl valine} \xrightarrow[\text{THF}]{BH_3} \text{N-phenoxycarbonyl valinol} \xrightarrow[\text{THF}]{t\text{-BuOK}} \text{(4S)-4-isopropyl-2-oxazolidinone}$$

(5) A single-pot reductive conversion of amino acids to their respective 2-oxazolidinones uses trichloromethyl chloroformate as the acylating agent (Pridgen, L. N., Prol, J., Alexander, B. and Gillyard, L., J. Org. Chem., **54**(13), 3231–3233, (1989)).
(6) Trans 4,5-disubstituted oxazolidin-2-ones are formed with high diastereoselectivity *via* a photoinitiated radical alkylation using 5-(phenylthio)oxazolidin-2-ones derived from (S)-α-amino acids (Kano, S., Yokomatsu, T. and Shibuya, S., J. Org. Chem., **54**(3), 513–515, (1989)).
(7) The reaction of CO_2, propargyl alcohol derivatives and primary amines, catalysed by phosphine, gives N-substituted 4-methylene-2-oxazolidinones (Fournier, J., Bruneau, C. and Dixneuf, P. H., Tetrahedron Lett., **31**(12), 1721–1722, (1990)).
(8) O-Propargyl carbamates undergo an intramolecular nucleophilic addition to the acetylenic bond to furnish 4-methylene-2-oxazolidinones (Tamaru, Y. et al., Tetrahedron Lett., **31**(34), 4887–4890, (1990)).
(9) 4-Alkylidene-3-tosyloxazolidin-2-ones may be obtained from propargyl alcohols and p-toluenesulphonyl isocyanate using a CuI/Et_3N catalyst (Ohe, K., Ishihara, T., Chatani, N., Kawasaki, Y. and Murai, S., J. Org. Chem., **56**(6), 2267–2268, (1991)).

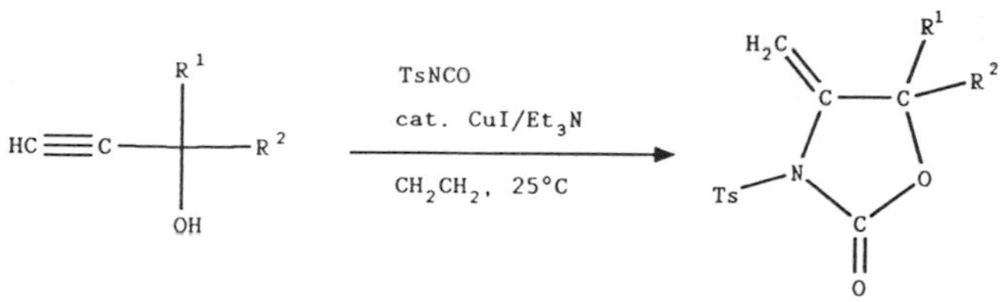

TsNCO
cat. CuI/Et_3N
CH_2CH_2, 25°C

Reactions. (1) Mild ring cleavage of N-*tert*-butoxycarbonyl-2-oxazolidinones, allows recovery of β-amino alcohols (Jommi, G., Ripa, A., Ripa, G. and Sisti, M., Gazz. Chim. Ital., **118**(1), 75–76, (1988)).
(2) Optically-active α-alkylsuccinates have been obtained from stereoselective alkylation of chiral oxazolidinones and subsequent removal of the chiral auxiliary (Fadel, A. and Salaun, J., Tetrahedron Lett., **29**(48), 6257–6260, (1988)).
(3) The 2-oxazolone moiety has synthetic potential as a unique leaving group in carboxyl (Kunieda, T. et al., J. Org. Chem., **47**(22), 4291–4297, (1982); Kunieda, T. et al., Tetrahedron, **39**(20), 3253–3260, (1983)) and phosphoryl (Nagamatsu, T. and Kunieda, T., Tetrahedron Lett., **28**(21), 2375–2378, (1987)) activating processes. Reagents have been developed for the formation of β-lactam compounds from β-amino acids (Kunieda,

T. et al., Tetrahedron Lett., **29**(18), 2203–2206 (1988)). Compounds used include diphenyl 2-oxo-3-oxazolinylphosphonate, p-chlorophenyl bis(2-oxo-3-oxazolinyl)phosphinate and tris(2-oxo-3-oxazolinyl)phosphine oxide. These reagents are readily obtainable from 2-oxazolone and the corresponding halogenophosphorus compounds.
(4) 3,5-Dioxo-4-methyl-1,2,4-oxadiazolidine is a good leaving group efficiently used for carboxylic acid activation in peptide synthesis and ester formation (Grenouillat, D., Senet, J. P., and Sennyey, G., Tetrahedron Lett., **28**(47), 5827–5828, (1987)). 2-Acyl-3,5-dioxo-1,2,4-oxadiazolidine (acyl-MODD), arising from the addition of the stable 2,2′-carbonyl bis(3,5-dioxo-4-methyl-1,2,4-oxadiazolidine (COMODD) to the carboxylic acid, is not isolated but reacted *in situ*. N-protected α-amino acids and O-protected α-hydroxy acids activated with COMODD provide a stable activated intermediate suitable for the synthesis of β-keto esters (Jouin, P. et al., Tetrahedron Lett., **29**(22), 2661–2664, (1988)).

Thiazolidinones. (1) The preparation, reactions, spectroscopy and biological activity of thiazolidinones has been reviewed (Stenberg, V. I. et al., Chem. Rev., **81**, 175–203, (1981)).
(2) Sulphur dichloride reacts with thiocarbamic acid esters to form aminoacyltrisulphides (Muhlstadt, M. and Widera, R., J. Prakt. Chem., **320**(1), 123–127, (1978)). Reaction with unsaturated thiocarbamic acid esters yields heterocycles of the 2-thiazolidinone and 3-dithiazinone type.
(3) 2-Thiazolidinones may be synthesised stereospecifically from *vic*-iodoalkylcarbamates (Brunet, E., Carreno, M. C. and Ruano, J. L. G., Heterocycles, **23**(5), 1181–1195, (1985)).
(4) The pharmacological activities of thiazolinones, thiazolidinediones and thiazoles have been reported (Heindel, N. D. and Hoko, C. C., J. Heterocycl. Chem., 1007, (1970); Heindel, N. D., Reid, J. R. and Willis, J. E., J. Med. Chem., **14**, 453, (1971); Chaudhary, M., Parmar, S. S., Chaudhry, S. K., Chaturvedi, A. K. and Ramsastry, B. V., J. Pharm. Sci., **65**, 443, (1976); Chaudhary, S. K., Verma, M., Chaturvedi, A. K. and Parmar, S. S., J. Pharm. Sci., **64**, 614, (1975)).
(5) Thiazolidinone and thiazole derivatives of long-chain fatty acids have been prepared (Mustafa, J., Ahmad, M. S. and Osman, S. M., J. Chem. Research, 220–221, (1989)).

Thiazolidinethiones. (1) An infrared study has been made of N-methyl-1,3-thiazolidine-2-thione and -2-selenone (Devillanova, F. A., Sathyanarayana, C. N. and Verani, G., J. Heterocycl.

Chem., **15**(6), 945–947, (1978)) and 1,3-thiazolidin-2-one, -2-thione, -2-selenone and their 1-oxa-analogues (Cristiani, F., Devillanova, F. A. and Verani, G., J. Chem. Soc., Perkin Trans. II, 324–327, (1977)). In each series, the zwitterionic form increases on passing from oxygen to selenium.

(2) Certain thiazolidine-2-thiones are effective sulphur-transfer agents for the conversion of oxiranes into the corresponding thiiranes (Woodgate, P. D. et al., J. Chem. Soc., Perkin Trans. I, 52–57, (1981)).

Oxazolidineselenones. (1) 2-Oxazolidineselenones are synthesised by $HgCl_2$-assisted reactions of 1,2-aminoalcohols and CSe_2 (Kjaer, A. and Skrydstrup, I., Heterocycles, **28**(1), 269–273, (1989)). Prior to this study, N-methyl-2-oxazolidineselenone, prepared from N-methylethanolamine and CSe_2, was the only known monocyclic 2-oxazolidineselenone (Devillanova, F. A. and Verani, G., J. Heterocycl. Chem., 17, 571, (1980)). The parent compound and a series of C-substituted derivatives have been produced as crystalline, colourless, light-sensitive compounds which may be stored indefinitely at -25°C in the dark.

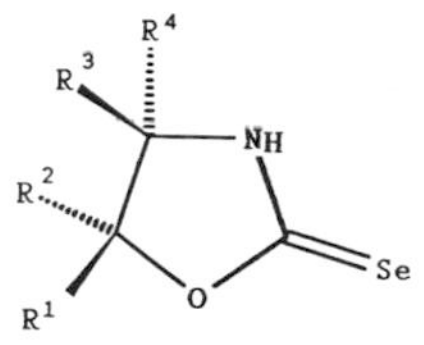

2. Urea and its derivatives

Reviews. (1) The chemistry and biochemistry of urea and related biosynthetic intermediates (Kennedy, J., Urea Cycle, [Proc. Symp.], 39–56, (1975), Grisolia, S., Baguena, R. and Mayor, F. (eds.) Wiley: New York)) and the synthesis and properties of N–sulphenyl acylureas (Anderson, M., Brinnand, A. G., Camilleri, P., Langner, E. J. and Weaver, R. C., Spec. Publ. – R. Soc. Chem. **79**, 184–205, (1990)) have been reported.
(2) Inclusion compounds of urea, thiourea and selenourea have been reviewed (Takemoto, K. and Sonoda, N., Inclusion Compd., **2**, 47–67, (1984), Atwood, J. L., Davies, J. E. D. and MacNicol, D. D. (eds.) Academic: London, U. K.)).
(3) The preparation, properties and applications of N–alkylated ureas including tetramethylurea (TMU), tetraethylurea (TEU), and cyclic analogues such as 1,3–dimethyl–2–imidazolidinone ("dimethylethylene–urea", DMEU) and 1,3–dimethyl–3,4,5,6–tetrahydro–2(1H)–pyrimidinone ("dimethylpropyleneurea", DMPU) have been described (Barker, B. J., Rosenfarb, J. and Caruso, J. A., Angew. Chem. Int. Ed. Engl., **18**, 503–507, (1979)).

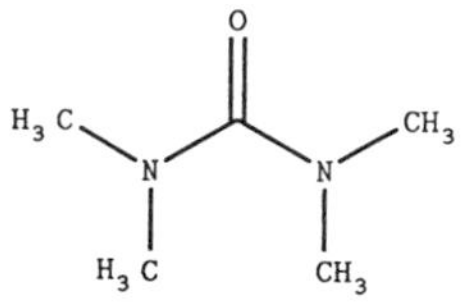

TMU

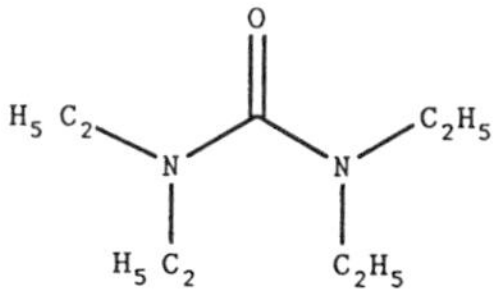

TEU

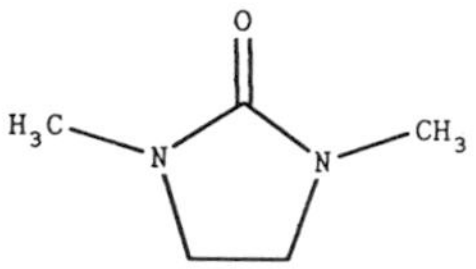

DMEU

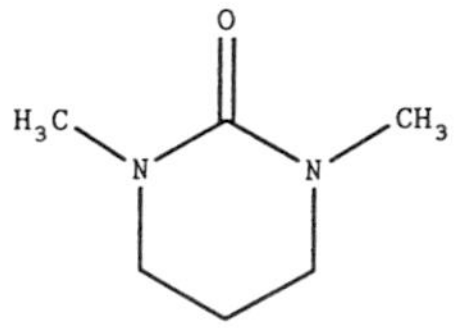

DMPU

Preparations. (1) The reaction of amines or diamines with CO in the presence of selenium yields ureas (Sonoda, N. et al., J. Amer. Chem. Soc., **93**(23), 6344, (1971)) and cyclic ureas (Yoshida, T., Kambe, N., Ogawa, A. and Sonoda, N., Phosphor. Sulfur Relat. Elem., **38**(1/2), 137–148, (1988)).

(2) Methods for the synthesis of ureas include the reaction of amines with isocyanates (Stachell, R. A. and Stachell, D. P. N., Chem. Soc. Rev., **4**, 231, (1975)), CO_2 (Ogura, H., Takeda, K., Tokue, R. and Kobayashi, T., Synthesis, 394–396, (1978)) and with carbonyl equivalents such as phosgene (Hassel, T. and Seebach, D., Helv. Chim. Acta, **61**(6), 2237–2240, (1978)), carbonyldiimidazole or disuccinimido carbonate (Takeda, K. and Ogura, H., Synth. Commun., **12**, 213, (1982); Ogura, H. et al., Tetrahedron Lett., **24**(42), 4569–4572, (1983)) and carbonyl selenide (Se=C=O) (Sonoda, N. et al., Angew. Chem. Int. Ed. Engl., **18**(9), 692, (1979)).

(3) Simple aminolysis of alkoxycarbamates does not proceed readily under mild conditions (Lyon, P. A. and Reese, C. B., J. Chem. Soc. Perkin I, 131–137, (1978); Chheda, G. B. and Hong, C. I., J. Med. Chem., **14**, 748, (1971)). However, this reaction can be accomplished when the OR group is a better leaving group such as a phenoxy group. (Adamiak, R. W. and Stawinski, J., Tetrahedron Lett., 1935–1936, (1977)). If an amine is first treated with a Grignard reagent to form the magnesium salt, it subsequently reacts efficiently to displace an alkoxy group of a carbamate providing the corresponding urea in high yield.

(4) Ureas may be prepared by the thermolysis of allylic pseudoureas (Tsuboi, S., Stromquist, P. and Overman, L. E., Tetrahedron Lett., 1145–1148, (1976)).

(5) A general synthesis of highly substituted ureas, isoureas and guanidines involves the preparation of metal complexes of cyanamides and their alkylation to cyanamidium salts (Jochims, J. C., Abuelhalawa, R., Zsolnai, L. and Huttner, G., Chem. Ber., **117**(3), 1161–1177, (1984)).

(6) β-Phenylseleno ureas are quantitatively prepared by mild acid treatment of β-phenylseleno cyanamides (Salazar, J. A. et al., J. Chem. Soc., Chem. Commun., 312–314, (1987)).

(7) The synthesis of substituted urea compounds has been reported (Henklein, P., Jahrling, R., Teubner, H., Tietze, H. and Ott, T., Pharmazie, **44**(3), 225–226, (1989)).

(8) N-chlorination and subsequent treatment with alkali of suitably substituted benzamidines yields ureas with phenyl or alkyl migration. Under the same conditions, 2-iminopyrrolidine is converted into 2-oxohexahydropyrimidine with ring enlargement (Sonnenschein, H. and Schmitz, E., Synthesis, 443–444, (1989)).

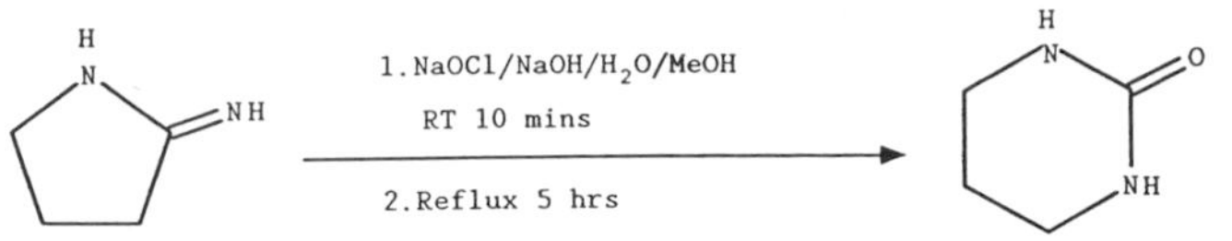

1. $NaOCl/NaOH/H_2O/MeOH$ RT 10 mins
2. Reflux 5 hrs

(9) The synthesis of cryptands containing urea and thiourea moieties (Bogatsky, A. V. et al., Synthesis, 137, (1984)) and spherands composed of cyclic urea and anisyl units (Cram, D. J. et al., J. Amer. Chem. Soc., **106**(23), 7150–7167, (1984)) has been reported.

Reactions. (1) A method for the N–methylation of ureas has been reported (Auerbach, J., Zamore, M. and Weinree, S. M., J. Org. Chem., **41**(4), 725–726, (1976)).
(2) Ureas are dehydrated with dichlorocarbene in a phase-transfer reaction (Schroth, W., Kluge, H., Matthias, M., Krieg, R. and Schadler, H. D., Z. Chem., **21**(1), 25–27, (1981)).
(3) N-*t*-Butyl substituted derivatives of urea have been used for the tert-butylation of aromatic compounds (Nevrekar, N. B. et al., Chem. Ind.,-London, 206–207, (1983)).
(4) 1-Hydroxyxanthones have been prepared from *ortho*-hydroxybenzoyl ureas (Mucheli, M. V. R. and Kudav, N. A., Chem. Ind.,-London, 31–32, (1985)).
(5) The reaction of allylic ureas with phenylselenenyl chloride (PhSeCl) in the presence of silica gel affords 2-oxazolines; 2-oxazolines are also stereospecifically obtained from β-phenylseleno ureas by chemoselective alkylation of the selenium atom followed by basic treatment (Salazar, J. A. et al., J. Chem. Soc., Chem. Commun., 450–452, (1989)).
(6) The successful intramolecular Diels-Alder reactions of indolylureas have been contrasted with the failure of the analogous carbamates to cyclise (Kraus, G. A. et al., J. Org. Chem., **54**(10), 2425–2428, (1989)).

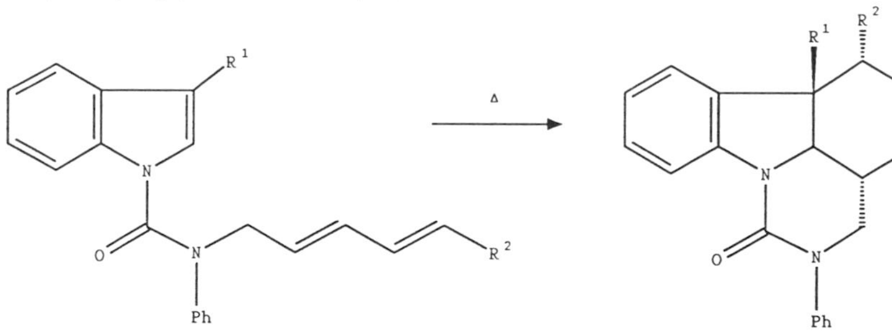

Imidazolidinones. (1) N-2-Propenyl-, N-3-butenyl, N-4-pentenyl, and N-5-hexenylureas undergo aminocarbonylation (0.1-0.01 equiv. $PdCl_2$, 3.0 equiv. $CuCl_2$, 1 atm. CO, MeOH, 0°C - ambient temp.) to yield 4-[(methoxycarbonyl)methyl]-2-imidazolidinones, 4-[(methoxycarbonyl)methyl]-3,4,5,6-tetrahydro-2(1H)-pyrimidinones, 1,3-diazabicyclo[4.3.0]nonane-2,4-diones and 1,3-diazabicyclo[4.4.0]decane-2,4-diones respectively (Yoshida, Z. et al., J. Amer. Chem. Soc., **110**(12), 3994-4002, (1988)). (2) Olefinic ureas may be cyclised to imidazolidinones using N-bromosuccinimide in CCl_4 (Balko, T. W., Brinkmeyer, R. S. and Terando, N. H., Tetrahedron Lett., **30**(16), 2045-2048, (1989)).

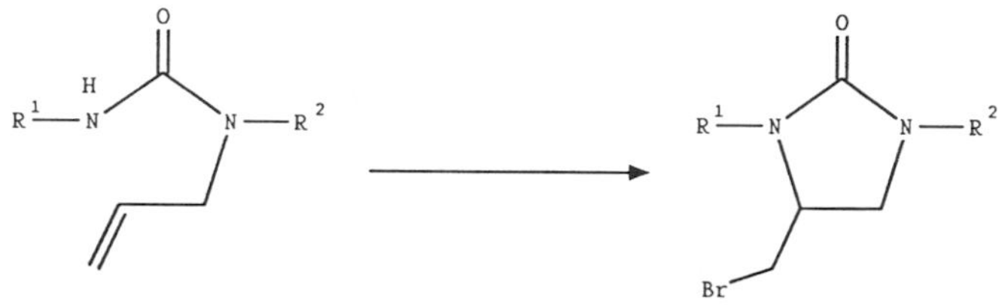

(3) N-Benzyloxyureas are alkylated with 1,2- and 1,3-dibromoalkanes and subsequently deprotected to provide 1-hydroxy-3-alkyl or 3-aryl imidazolidinones and tetrahydropyrimidinones (Sulsky, R. and Demers, J. P., Synth. Commun., **19**(11-12), 1871-1874, (1989)).

Disubstituted ureas.

Preparations. (1) N,N'-Disubstituted ureas may be prepared from the decomposition of aryl azides (Iqbal, A. F. M., Helv. Chim. Acta, **59**(2), 655-660, (1976)) or reaction of nitro-compounds with bromomagnesium alkyl- or aryl-amides and iron pentacarbonyl (Yamashita, M. et al., J. Chem. Soc., Chem. Commun., 670, (1976)).

$$R^1NHMgBr + Fe(CO)_5 \xrightarrow[2.\ H^+]{1.\ R^2NO_2} R^1NHCONHR^2$$

(2) N,N'-Bisalkylideneureas have been prepared (Fetyukhin, V. N., Vovk, M. V. and Samarai, L. I., Zh. Org. Khim., **17**(7), 1420-1429, (1981)).

$$R^1R^2C(Cl)-N=C=O + HN=CR^3R^4 \longrightarrow \left[R^1R^2C(Cl)-N-C(=O)-N=CR^3R^4 \right] \longrightarrow R^1R^2C=N-C(=O)-N=CR^3R^4$$

(3) A convenient synthesis of cyclic N,N'-dialkylureas has been reported (Bogatsky, A. V., Lukyanenko, N. G. and Kirichenko, T. I., Synthesis, 464–465, (1982)).

(4) N-Substituted trichloroacetamides in alkaline medium can be used as *in situ* isocyanate generating reagents for the synthesis of acylureas, sulphonylureas and symmetrical ureas (Atanassova, I. A., Petrov, J. S. and Mollov, N. M., Synthesis, 734–736, (1987); Atanassova, I. A., Petrov, J. S. and Mollov, N. M., Synth. Commun., **19**(1&2), 147–153, (1989)). This non-phosgene method has been applied to the synthesis of N,N'-di-3,4-methylenedioxybenzylurea in high yield.

(5) A simple synthesis of N,N'-disubstituted ureas from carbamates involves displacement of an alkoxy group by the magnesium salt of an amine generated *in situ* by treatment with ethylmagnesium bromide (Basha, A., Tetrahedron Lett., **29**(21), 2525–2526, (1988).

$$R^1\text{-}NHCO_2Et + NH_2\text{-}R^2 \xrightarrow[\text{DCM}]{\text{EtMgBr}} R^1HN-C(=O)-NHR^2$$

(6) A number of methods are known for the synthesis of symmetrical ureas most of which make use of phosgene. Symmetrical dialkylureas can be prepared in high yields by a convenient non-phosgene method starting from alkyl chlorides and sodium isocyanate.

(7) A non-phosgene route for the synthesis of *sym*-N,N'-diethyldiphenylurea has been proposed (Ayyangar, N. R. et al.,

Chem. Ind., London, 599–600, (1988)).
(8) Symmetrically and unsymmetrically disubstituted ureas are obtained by aminolysis of bis(4-nitrophenyl)carbonate (Izdebski, J. and Pawlak, C., Synthesis, 423–425, (1989)).
(9) Difluorophosphine-substituted ureas are prepared by the reaction of PF_2Cl with N,N'-bis(trimethylsilyl)ureas (Kruger, W., Schmutzler, R., Schiebel, H. M. and Wray, V., Polyhedron, **8**(3), 293–300, (1989)).
(10) 1,2-Diisocyanatocubane has been prepared and is a stable crystalline solid which, on hydrolysis with HCl, gives 1,2-diaminocubane bishydrochloride. Reaction in dilute solution in acetone with the minimum amount of water produces "cubanourea", a hetero[3.2.2]propellane (Eaton, P. E. and Pramod, K., J. Org. Chem., **55**(22), 5746–5750, (1990)).

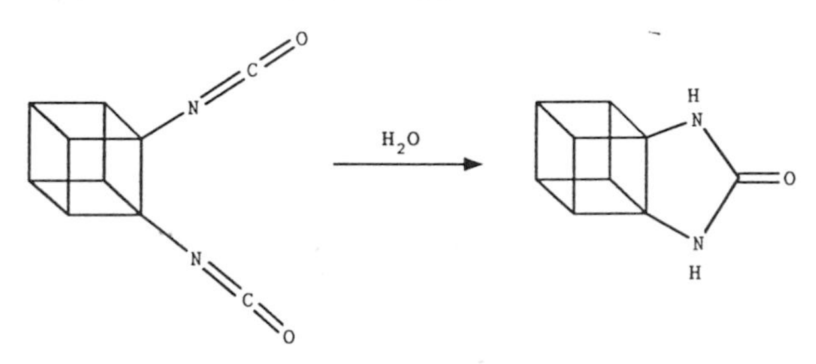

Reactions. (1) N,N'-Bis(trimethylsilyl)urea is a useful silylating agent for alcohols and carboxylic acids (Verboom, W., Visser, G. W. and Reinhoudt, D. N., Synthesis, 807–809, (1981)).
(2) Hexachlorocyclotriphosphazene reacts with N,N'-dimethylurea to yield spirotris-1,1,3,3,5,5-N,N'-dimethylureidocyclotriphosphazene (Prakash, H., Indian J. Chem., Sect. A, **20**(2), 170–171, (1981)).
(3) A wide variety of N,N-disubstituted ureas have been dehydrated under mild conditions using a $CHCl_3$/NaOH/TEA catalytic two-phase system (Schroth, W., Kluge, H., Frach, R., Hodek, W. and Schadler, H. D., J. Prakt. Chem., **325**(5), 787–802, (1983)).
(4) An acid-catalysed rearrangement of N-alkyl-N'-phenoxyureas allows the synthesis of 2-alkylaminophenols from phenols (Endo, Y., Shudo, K. and Okamoto,, T., Synthesis, 471–472, (1983)).
(5) In the presence of F_3CCO_2H N-phenyl-N'-phenoxyurea rearranges to N-(4'-hydroxy-2-biphenylyl)urea and N-carbamoyl-2-hydroxydiphenylamine (Endo, Y., Terashima, T. and Shudo, K., Tetrahedron Lett., **25**(48), 5537–5540, (1984)).
(6) A 1,3-dimethylurea-formaldehyde condensate, bis(1,3-

dimethylureido)methane, has been prepared and an X-ray analysis performed (Ebisuno, T., Takimoto, M., Takahashi, M. and Shiba, R., Bull. Chem. Soc. Jpn., **61**(12), 4441–4442, (1988)). The compound forms complexes with alkali metal ions.

Trisubstituted ureas. (1) Trisubstituted ureas and thioureas are obtained from hexaalkylphosphorus triamides, primary amines, and CO_2 or CS_2 (Yamazaki, N., Tomioka, T. and Higashi, F., Synthesis, 384–385, (1975)).

$$P(NR^1{}_2)_3 + CO_2 + R^2NH_2 \xrightarrow[60°C\ 6hrs]{pyridine} NR^1{}_2CONHR^2$$

(2) 1,1,3-Trisubstituted ureas are converted with phosphoryl chloride to hexasubstituted carbamimidic acid anhydrides (Torizuka, M., Kikugawa, Y. and Nagashima, H., Synthesis, 226–228, (1986)).

Tetrasubstituted ureas. (1) Tetrasubstituted ureas are produced in radical reactions of bis-N,N-dialkylaminomethoxymethane and 2-N,N-dimethylamino-3-alkyl-1,3-oxazolidines (Kurbanov, D., Pastushenko, E. V., Zlotskii, S. S. and Rakhmankulov, D. L., Zh. Org. Khim., **21**(4), 899–902, (1985)).

O NR NMe$_2$ → Me_2N–C(=O)–N(Et)(R)

(2) The reaction of Et_2NH with CO_2 and $PdCl_2(MeCN)_2$ as a catalyst gives tetraethylurea or diethylformamide selectively under mild reaction conditions by employing the $PPh_3/CCl_4/MeCN$ and the HCOONa/methyl cellusolve system, respectively (Fujiwara, Y. et al., Tetrahedron Lett., **27**(16), 1809–1810, (1986)).

$$Et_2NH + CO_2 \xrightarrow{PdCl_2(MeCN)_2} Et_2N\text{–}C(=O)\text{–}NEt_2 + Et_2N\text{–}CHO$$

(3) N,N,N',N'-Tetrasubstituted ureas have been obtained by reaction of lithium aliphatic amides in THF with CO followed by O_2 prior to work up (Nudelman, N. S., Lewkowicz, E. S. and Perez, D. G., Synthesis, 917–920, (1990)). N,N,N',N'-tetrasubstituted oxomalonamides (oxopropanediamides) can be obtained under similar reaction conditions by carrying out the reaction in the presence of amines.

Arylureas. (1) Arylureas may be synthesised by the reaction of aromatic amines with CO using selenium and Et_3N as co-catalyst (Kondo, K., Sonoda, N. and Tsutsumi, S., J. Chem. Soc., Chem. Commun., 307–308, (1972)) or with CO_2 in the presence of ethylene chlorophosphite in pyridine (Chiriac, C., Rev. Roum. Chim., **32**(1), 29–31, (1987)).
(2) Aromatic cyclic ureas have been synthesised from *o*-amino or *o*-aminoalkyl substituted aromatic amines by reaction with CO and selenium (Sonoda, N. et al., Tetrahedron Lett., **27**(26), 3037–3040, (1986)).

NH_2 NH_2 — Se, CO → NH_2 HN O SeH — O_2 → HN O NH

Diarylureas. (1) Aromatic amines, aromatic nitro compounds, and CO react in the presence of catalytic amounts of Pd^{II} salts, Pt^{II} salts, organic phosphines, tertiary amine, and tetraethylammonium chloride to form N,N'-diarylureas (Dieck, H. A., Laine, R. M. and Heck, R. F., J. Org. Chem., **40**(19), 2819–2822, (1975); Tsuji, Y., Takeuchi, R. and Watanabe, Y., J. Organometal. Chem., **290**(2), 249–255, (1985)).
(2) N,N'-Diarylureas have been synthesised by the reaction of N-aryl substituted formamides with aminoarenes in the presence of a catalytic amount of dichlorotris(triphenylphosphine)-ruthenium [$RuCl_2(PPh_3)_3$] (Watanabe, Y. et al., J. Chem. Soc., Chem. Commun., 549–550, (1990)).

Hydroxyureas. (1) Routes to 1,3-dihydroxyureas have been described (Schmidt, J. and Zinner, G., Arch. Pharm., **312**(12), 1019–1026, (1979)).
(2) ^{14}C-Hydroxyurea has been prepared using an immobilised ^{14}C-cyanate (Ajami, A. M., J. Label. Compound Radiopharm., **20**(2), 167–171, (1983)).
(3) N-Benzyloxyureas and orthogonally protected N-hydroxycarbamates may be alkylated in high yields and subsequently deprotected to give N-alkylhydroxyureas and hydroxyamines (Sulsky, R. and Demers, J. P., Tetrahedron Lett., **30**(1), 31–34, (1989)).

Spectroscopy. The conformation of 1,3-disubstituted 1-hydroxyureas and 1-hydroxythioureas, mostly with aromatic substituents, has been investigated by means of dipole moments in dioxane solution and by infrared spectroscopy (Mollin, J., Fiedler, P., Jehlicka, V. and Exner, O., Collect. Czech. Chem. Commun., **44**(3), 895–907, (1979)).

Acylureas. (1) The acetylation of phenylurea and phenylthiourea has been studied (Cremlyn, R. J. and Badami, N., Chem. Ind.,-London, 691–692, (1976)).
(2) The reaction of aryl and alkyl thioamides with acyl isocyanates is a convenient source of substituted acylthioacylureas (Cohen, V. I., Helv. Chim. Acta, **59**(1), 350–352, (1976)).
(3) A synthesis of 1-aryl-3-acylureas has been reported (Richter, M., Richter, W. and Pallas, M., Z. Chem., **29**(6), 204–205, (1989)).
(4) Photooxygenation of imidazolin-2-ones yields diacylureas as the only products isolated at room temperature (Chawla, H. M. and Pathak, M., Tetrahedron, **46**(4), 1331–1342, (1990)). The reaction involves the formation of zwitterionic perepoxides leading to dioxetanes which decompose to diacylureas.
(5) Reaction of isocyanates with 2,3-diethoxypropionamide gives diethoxyacylureas which may be cyclised to 1-substituted uracils (Naim, A. and Shevlin, P. B., Syn. Commun., **20**(22), 3439–3442, (1990)).

Sulphonylureas. (1) The preparation of substituted ureas, acylureas and sulphonylureas has been reported (Anatol, J. and Berecoechea, J., Synthesis, 111–113, (1975)).
(2) Reaction of sulphonamides with isocyanates under Lewis acid catalysis affords sulphonylureas (Irie, H. et al., J. Chem. Soc. Perkin Trans. I, 1209–1210, (1989)).

$$R^1C_6H_4SO_2NH_2 \xrightarrow[R^2NCO]{BF_3.Et_2O} R^1C_6H_4SO_2NHCONHR^2$$

Nitrosoureas. (1) As part of an investigation into the chemistry of nitrosoureas, the synthesis and decomposition of deuterated 1,3-bis(2-chloroethyl)-1-nitrosourea has been examined (Brundrett, R. B. et al., J. Med. Chem., **19**(7), 958-961, (1976)).
(2) 1,2,2,2-Tetrachloroethyl carbamates are prepared by reaction of 1,2,2,2-tetrachloroethyl carbonochloridate with amines. Nitrosation and reaction with amines yields N-nitrosoureas (Barcelo, G., Senet, J.-P. and Sennyey, G., Synthesis, 1027-1029, (1987)).

$$Cl_3C\text{-}CH(Cl)\text{-}O\text{-}CO\text{-}Cl \xrightarrow[Dioxan]{R^1NH_2} Cl_3C\text{-}CH(Cl)\text{-}O\text{-}CO\text{-}NHR^1 \xrightarrow[CCl_4]{N_2O_4/NaOAc}$$

$$Cl_3C\text{-}CH(Cl)\text{-}O\text{-}CO\text{-}NR^1(NO) \xrightarrow[K_2CO_3/H_2O/THF]{R^2NH_2} R^2HN\text{-}CO\text{-}NR^1(NO)$$

(3) The thermolysis of trialkylnitrosoureas has been carried out neat, in protic and in aprotic solvents (Singer, S. S., J. Org. Chem., **47**(20), 3839-3844, (1982)). When run neat or in aprotic solvents, 1,3,3-triethyl-1-nitrosourea gave 70% of N,N-diethylalanine ethyl ester. Tetramethylurea was a major product from the thermolysis of 1,1,3-trimethyl-1-nitrosourea and 1-ethyl-3,3-dimethyl-1-nitrosourea.

Isoureas.

Preparations. (1) N-Hydroxyisoureas are obtained by the reaction of N-substituted hydroxylamines with cyanates (Neitzel, M. and Zinner, G., Arch. Pharm., **314**(1), 2-9, (1981)).
(2) O-Acetyl-N,N'-dimethylisourea has been synthesised and phosphorylated with PCl_5 (Dmitrichenko, M. Y., Donskikh, V. I., Rozinov, V. G., Rybkina, V. V. and Sergienko, L. M., Zh. Obshch. Khim., **55**(8), 1881-1882, (1985)).

Reactions. (1) N,N′-Diisopropyl-O-arylisoureas rearrange intramolecularly to the corresponding N,N′-diisopropyl-N′-arylurea in alkaline solution (Suttle, N. A. and Williams, A., J. Chem. Soc., Perkin Trans. II, 1369–1372, (1983)).

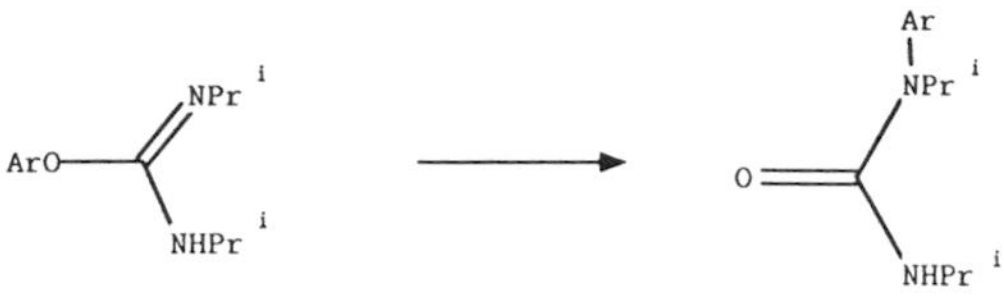

(2) Imidazolines are prepared by treatment of allylic O-methyl isoureas with phenylselenenyl chloride (PhSeCl) in the presence of silica gel; use of phenylselenenyltrifluoromethanesulphonate ($PhSeOSO_2CF_3$) and trifluoromethanesulphonic acid yields oxazines (Salazar, J. A. et al., J. Chem. Soc., Chem. Commun., 452–454, (1989).

Thioureas.

Reviews. (1) The use of thioureas in the synthesis of heterocycles (Griffin, T. S., Woods, T. S. and Klayman, D. L., Adv. Heterocycl. Chem., **18**, 99–158, (1975)) and radical reactions of thiourea and related compounds (Kandror, I. I., Kopylova, B. V. and Preidlina, R. Kh., Sulfur Rep., **3**(8), 289–320, (1984)) have been reviewed.
(2) Thioureas and their selenium analogues are included in wide-ranging reviews of thio- and selenocarbonyl compounds (Walter, W. and Voss, J., Org. Compd. Sulphur, Selenium, Tellurium, **4**, 141–185, (1977); Walter, W. and Voss, J., Org. Compd. Sulphur, Selenium, Tellurium, **5**, 139–186, (1979)).
(3) Phosphorus, arsenic, silicon and metal derivatives of thiourea have been reported (Friedman, H. A., Org. Prep. Proced. Int., **9**(3-4-5), 209–256, (1977)).
(4) The biological activities of thioureas have been described (Pandeya, S. N., Ram, P. and Shankar, V., J. Scient. Ind. Res., **40**(7), 458–466, (1981)).

Preparations. (1) Substituted thioureas and Schiff bases are obtained from organic isothiocyanates in DMSO (Chattopadhyaya, J. B. and Ramarao, A. V., Synthesis, 289–290, (1974)).
(2) N,N,N′,N′-Tetracarboxymethylthiourea (methylthiourea tetraacetate) has been synthesised (Carmonaguzman, E.,

Gonzalezvilchez, F. and Gonzalezgarcia, F., An. Quim., **72**(2), 134–138, (1976)).
(3) 1–Alkyl–1–cyanothioureas, 3–alkyl–2–thiobiurets, and 3–alkyl–2,4–dithiobiurets have been prepared (Benders, P. H. and Vanerkelens, P. A. E., Synthesis, 775–777, (1978)).
(4) Thioureas are produced by the three–component reaction of amines, CCl_4 and sulphide ions in the presence of a phase–transfer catalyst (Broda, W. and Dehmlow, E. V., Liebigs Ann. Chem., 1839–1843, (1983)).

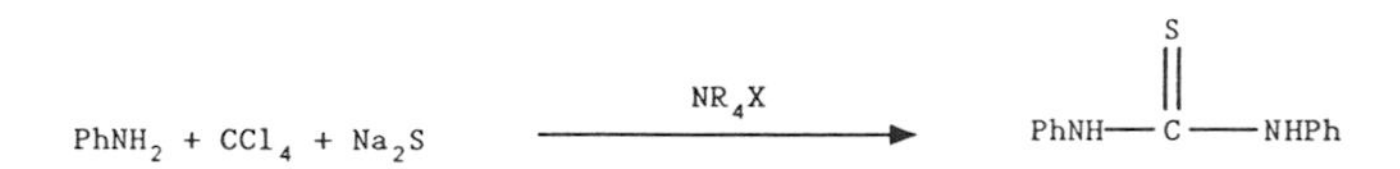

(5) A simple synthesis of thiourea derivatives involves the reaction of Ni^{II}–coordinated N–acyl thioureas with acid chlorides (Beyer, L., Hartung, J. and Widera, R., Tetrahedron, **40**(2), 405–412, (1984)).
(6) N–Alkylated thioureas, including those of macroheterocycles (Bogatsky, A. V., et al., Synthesis, 136, (1984)) and macrocyclic polyhydroxyethylenethioureas with structurally rigid fragments (Lukyanenko, N. G., Kirichenko, T. I., Dvorkin, A. A., Simonov, Y. A., Pastushok, V. N. and Malinovskii, T. I., Zh. Obshch. Khim. SSSR, **60**(2), 405–410, (1990)) have been prepared.
(7) The reaction of benzophenonazines with sodium in dry dioxane followed by treatment with CS_2 affords substituted thioureas (Singh, M. S. and Mehrotra, K. N., Indian J. Chem., Sect. B, **24**(2), 208–209, (1985)).
(8) Improved procedures for the preparation of cycloalkyl–, arylalkyl–, and arylthioureas have been reported (Rasmussen, C. R. et al., Synthesis, 456–459, (1988)).
(9) ^{13}C–, ^{14}C– and ^{15}N–labelled ethylenethioureas have been synthesised (Doerge, D. R., Cooray, N. M., Yee, A. B. K. and Niemczura, W. P., J. Label. Compound Radiopharm., **28**(6), 739–742, (1990)).

Spectroscopy. (1) The temperature–dependent spectra of several mono–, di–, and trialkylthioureas have been reported and free energy barriers to internal rotation about the C–N bonds have been calculated (Sullivan, R. H. and Price, E., Org. Magn. Resonance, **7**(3), 143–150, (1975)).
(2) ^{15}N NMR evidence has been presented for the formation of thionitrosyl compounds in the nitrosation of thioureas (Lown, J. W. and Chauhan, S. M. S., J. Chem. Soc., Chem. Commun., 675–

676, (1981)).
(3) ^{1}H and ^{13}C NMR studies of the conformations of 1,3-dipyridylthioureas have been reported (Sudha, L. V. and Sathyanarayana, D. N., J. Chem. Soc., Perkin Trans. II, 1647-1650, (1986)).

Reactions. (1) Dicyandiamide and aryldicyandiamides react with thiourea and N-arylthioureas in the presence of concentrated HCl to yield hexahydro-4,6-diimino-1-aryl-1,3,5-triazine-2-thiones and amidino/arylamidinothioureas. Condensation of amidinothiourea with arylcyanamides produces the same 1,3,5-triazine derivatives (Joshua, C. P. and Rajan, V. P., Aust. J. Chem., **28**(2), 427-432, (1975)).
(2) N,N′-Dilithio- or bis(bromomagnesio)-thioureas, prepared *in situ* from thioureas substituted by bulky groups and butyllithium or bromoethylmagnesium, decompose thermally to afford dialkyl- or diaryl-carbodiimides. The decomposition of N,N′-dilithiothioureas is accelerated by the addition of CS_2, and the carbodiimides are formed below room temperature (Sakai, S., Fujinami, T., Otani, N. and Aizawa, T., Chem. Lett., 811-814, (1976)).
(3) The reaction of ethylene thiourea with CS_2 has been investigated (Yokoyama, M. et al., Tetrahedron Lett., 3823-3826, (1978)).

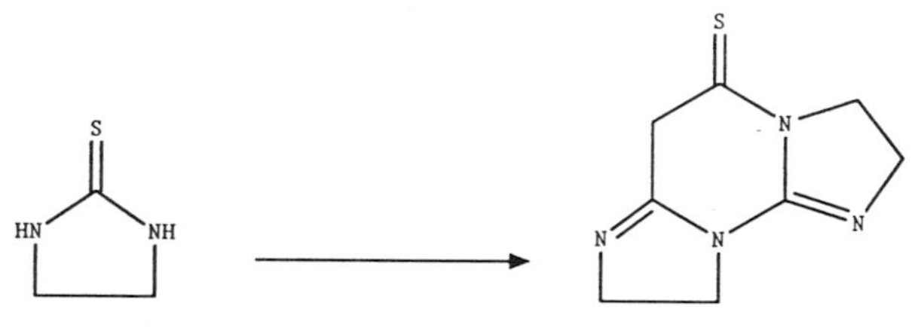

(4) The action of thiophosgene on thioureas and thioamides is a convenient method for their desulphurisation to ureas and amides (Abuzar, S., Sharma, S. and Iyer, R. N., Indian J. Chem., Sect. B, **19**(3), 211-212, (1980)).
(5) A new type of compound was obtained from the reaction of thiourea with CS_2 followed by methylation (Yokoyama, M. and Takeshima, T., Tetrahedron Lett., **21**(7), 635-636, (1980)).

S
H_2N NH_2
1. BuLi
2. CS_2
3. MeI
S
NCS NMe_2

(6) The interaction of thiourea and substituted thioureas with trifluoromethanesulphonic anhydride in DCM results in stable dication disulphide salts $(R_2N)_2^+CSSC^+(NR_2)_2.2CF_3SO_3^-$ (Maas, G. and Stang, P. J., J. Org. Chem., **46**(8), 1606–1610, (1981)).
(7) The ring cleavage of cyclic thioureas has been employed in a synthesis of N-(alkylthiocarbonyl)-N'-(alkylthiothiocarbonyl) alkanediamines and N,N'-bis(alkylthiothiocarbonyl)alkanediamines (Yokoyama, M., Hosi, K. and Imamoto, T., Synthesis, 908–910, (1981)).
(8) 1,3,5,6-Tetrasubstituted-iminohexahydro-1,3,5-triazine-2,4-dithiones have been synthesised from thioureas (Joshua, C. P. and Thomas, S. K., Synthesis, 1070–1071, (1982)).
(9) The reactions of N-monosubstituted thioureas with β-haloacyl halides in 5% NaOH–CH_2Cl_2 in the presence of a phase-transfer catalyst affords N-thioamido-β-lactams, thioureido acids and 5-hydroxy-2-thiohexahydropyrimidin-4-ones (Okawara, T., Nakayama, K. and Furukawa, M., Chem. Lett., 1791–1794, (1982); Okawara, T., Nakayama, K., Yamasaki, T. and Furukawa, M., J. Chem. Res. Synop., 188–189, (1985); *ibid.*, J. Chem. Res., 2215–2228, (1985)).
(10) Tetrahydropyrimidine-2-thiones are prepared from tertiary acetylenic alcohols and thiourea (Tarasova, D. A., Amosova, S. V., Ryskieva, G. K., Tsoi, L. A. and Sigalov, M. V., Zh. Org. Khim., **19**(11), 2452–2453, (1983)).
(11) Thiazolidinones and their hydrochlorides are synthesised by the reaction of cyclic thioureas with monochloroacetic acid and ethyl chloroacetate, respectively (Dehuri, S. N. and Nayak, A., J. Indian Chem. Soc., **60**(10), 970–974, (1983)).
(12) Thioureas undergo intramolecular Knoevenagel condensation under basic conditions whereby the amide carbonyl and the active methylene group react to form a double bond (Viski, P., Zueuvics, Z., Toldy, L., Sohar, P. and Dvortsak, P., Acta Chim. Hung., **112**(3), 323–334, (1983)).
(13) A method for the conversion of thiourea and monosubstituted thioureas into cyanamides involves desulphurisation with superoxide ion (Crank, G. and Makin, M. I. H., J. Chem. Soc., Chem. Commun., 53–54, (1984)).
(14) The reaction of thioureas with 3-bromomethyl-2-cyanocinnamonitrile gives hydrobromides of 2-aminothiazoles or 2-imino-3-thiazolines (Liebscher, J. and Mitzner, E., Tetrahedron Lett., **26**(15), 1835–1838, (1985)).
(15) Sulphides are conveniently prepared by treatment of aryl iodides with thioureas in the presence of a Ni^0 catalyst, then with CaO, and finally with a variety of electrophiles such as alkyl, acyl, silyl, aryl or alkenyl halides (Takagi, K., Chem.

Lett., 1379–1380, (1986)).
(16) N–Cyanocinnamamidines, prepared from thiourea and 3–arylpropenylnitriles, react with different benzylidene compounds to provide a simple method for the synthesis of 4,6–diaryl–2–cyanoiminopiperidines (Lorente, A. et al., Synthesis, 739–742, (1988)).

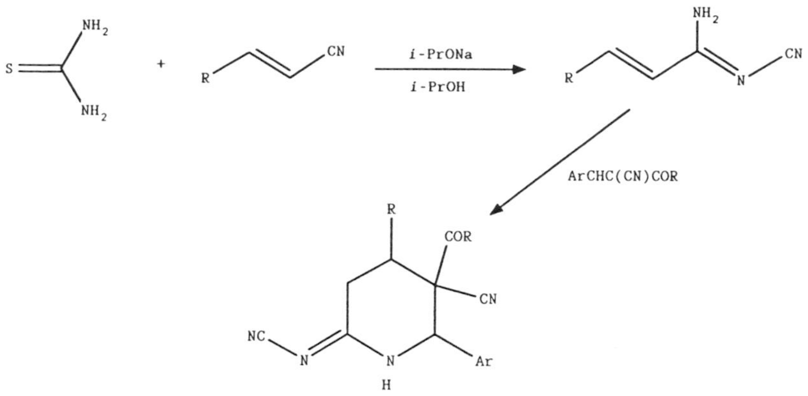

(17) Iodocyclisation of N–allyl or S–allyl thioureas yields dihydrothiazoles and dihydroimidazoles; homologous thioureas give dihydrothiazines (Creeke, P. I. and Mellor, J. M., Tetrahedron Lett., **30**(33), 4435–4438, (1989)).
(18) Thiocarbamates and thioureas are converted to the corresponding carbamates and ureas by baker's yeast (Kamal, A., Rao, M. V. and Rao, A. B., Chem. Lett., 655–656, (1990)).

Disubstituted thioureas.

Preparations. (1) Amines are converted into 1,1–disubstituted thioureas by using a combined reagent of Ph_3P and thiocyanogen (Tamura, Y., Adachi, M., Kawasaki, T. and Kita, Y., Tetrahedron Lett., 1753–1754, (1978)).
(2) N,N'–Disubstituted thioureas are prepared by treating 2–chloropyridinium salts with sodium trithiocarbonate and subsequently adding amines (Takikawa, Y., Inoue, N., Sato, R. and Takizawa, S., Chem. Lett., 641–642, (1982)).
(3) N,N'–Bis(aminomethylidene)thioureas are prepared by bis–iminoformylation of thiourea with formamide acetals (Knoll, A. and Liebscher, J., Synthesis, 51–53, (1984)).

$$S{=}C(NH_2)_2 \xrightarrow[MeOH]{NR_2CH(OCH_3)_2} S{=}C(N{=}CH{-}NR_2)_2$$

(4) Certain unsymmetrical N,N'-dialkylthioureas can be prepared by heating a sodium N-alkyldithiocarbamate, an alkylamine and a catalytic amount of NaOH in a two-phase solvent system (Robbins, J. D. and Neal, J. R., Synth. Commun, **16**(8), 891-897, (1986)).

Reactions. (1) Unsymmetrically 1,3-disubstituted thioureas without an intramolecular hydrogen bond react with diiodomethane in the presence of Et_3N to give thiazetidine isomers (Ried, W. and Mosinger, O., Chem. Ber., **111**(1), 155-167, (1978)).

$$RHN{-}C(=S){-}NHR^1 \xrightarrow{CH_2I_2} \text{RN=C(-N(R}^1\text{)-CH}_2\text{-S-)} + \text{R}^1\text{N=C(-N(R)-CH}_2\text{-S-)}$$

(2) The reaction of 1,3-disubstituted thioureas with α-haloacyl halides (Okawara, T., Nakayama, K. and Furukawa, M., Heterocycles, **19**(9), 1571-1574, (1982)) and α,ω-dibromoacyl chlorides (Okawara, T., Nakayama, K., Yamasaki, T. and Furukawa, M., Chem. Pharm. Bull., **34**(1), 380-384, (1986)) affords thiazolidin-5-ones and 5-bromoalkyl-2-iminothiazolidin-4-ones respectively. The latter compounds may be converted to the corresponding spiro derivatives under phase-transfer conditions or with NaOEt.

(3) Desulphurisation of N,N'-diarylthioureas by $Pb(OAc)_4$ oxidation in refluxing DCM and pyridine gives the corresponding N,N'-diarylureas (Debroy, A. Mazumdar, S. N., Barua, P. D. and Mahajan, M. P., Bull. Chem. Soc. Jap., **57**(1), 315-316, (1984)) whilst certain 1-substituted and 1,3-disubstituted thioureas are converted into the related mono- and 1,3-disubstituted formamidines using Raney nickel catalyst (Ali, M. U., Meshram, H. M. and Paranjpe, M. G., J. Indian Chem. Soc., **62**(9), 666-669, (1985)).

$$RHN{-}C(=S){-}NHR \longrightarrow RHN{-}C(SH){=}NR \xrightarrow[C_6H_6]{\text{Raney Ni}} RHN{-}CH{=}NR$$

(4) N, N′-Disubstituted thioureas may be converted into O-methyl-N,N′-disubstituted pseudoureas (Chern, J. W., Wise, D. S. and Townsend, L. B., Heterocycles, **23**(9), 2197–2200, (1985)).
(5) 2,6-Disubstituted-5,6-dihydro-4H-1,3-thiazin-4-ones are readily prepared by the $BF_3.OEt_2$-catalysed cyclisation of N-substituted-N′-(3-substituted propenoyl)thioureas (Dzurilla, A. M., Kutschy, P. and Kristian, P., Synthesis, 933–934, (1985)).
(6) Treatment of N-(2-hydroxyphenyl)-N′-phenylthioureas with the superoxide radical anion ($O_2^{-\cdot}$) at 20°C in CH_3CN, THF or DMSO gives 2-substituted aminobenzoxazoles (Chang, H. S., Yon, G. H. and Kim, Y. H., Chem. Lett., 1291–1294, (1986)).

Isothioureas.

Preparations. (1) The preparation and conformation of vinylogous thioureas and isothioureas has been studied (Goerdeler, J., Laqua, A. and Lindner, C., Chem. Ber., **107**(11), 3518–3532, (1974)).
(2) 1-Substituted 3-benzoyl-2-methylisothioureas and 1,3-disubstituted 2-benzoylguanidines are synthesised from dimethyl N-benzoylcarboimidodithioates and amines (Fukada, N., Hayashi, M. and Suzuki, Y., Bull. Chem. Soc. Jap., **58**(11), 3379–3380, (1985)). Imidazolidine and oxazolidine derivatives are obtained from ethylenediamine and 2-aminoethanol respectively.
(3) Extensive work has been done on the reaction of thiourea with alkyl halides to give isothiourea derivatives. However little information is available concerning such reactions with heterocyclic compounds. The reaction of thioureas with chloromethyl derivatives of heterocyclic compounds such as pyridine, quinoline and imidazole gives isothiourea dihydrochlorides (Horiuchi, J. et al., Chem. Pharm. Bull., **37**(4), 1080–1084, (1989)).

Reactions. (1) The main product from the pyrolysis at 155°C of

N,N-dimethyl-N'-(4-methyl-2-thiazolyl)-S-methylisothiourea is N,N-dimethyl-N'-(3,4-dimethyl-4-thiazolin-2-ylidene)thiourea (Yoda, R., Yamamoto, Y. and Matsushima, Y., Chem. Pharm. Bull., **32**(6), 2224–2229, (1984)).
(2) Pyrolysis of N-methyl-N'-(4-methyl-2-thiazolyl)-S-methylisothiourea at 155°C for 14 hours gives N-methyl-N',N''-bis(4-methyl-2-thiazolyl)guanidine (Yoda, R., Yamamoto, Y., Okada, T. and Matsushima, Y., Heterocycles, **23**(9), 2339–2345, (1985)).
(3) The stereochemistry of the rearrangement of S-phosphoryl-isothioureas into N-phosphorylthioureas has been investigated (Mikolajczyk, M., Kielbasinski, P. and Sut, A., Tetrahedron, **42**(16), 4591–4601, (1986)). The rearrangement proceeds with full retention of configuration at phosphorus *via* a mechanism that involves pseudorotation of a pentacovalent phosphorus intermediate.

Selenoureas and Telluroureas.

Preparations. (1) The preparation of selenoureas has been reviewed (Shine, R. J., Organic Selenium Compounds: Their Chemistry and Biology, 281, (1973), Klayman, D. L. and Gunther, W. H. H., (Eds.) Wiley; New York).
(2) Mono-, N,N'-di- and trisubstituted selenoureas may be obtained from methyl carbamimidothioates (S-methylpseudo-thioureas) (Cohen, V. I., Synthesis, 60–63, (1980)).
(3) Tetrasubstituted selenoureas and pentasubstituted selenosemicarbazides are prepared from CSe_2 (Henriksen, L., Synthesis, 773–776, (1982)).
(4) Treatment of N,N-diphenylamine or N-methylaniline with Viehe's salt ($Me_2N^+{=}CCl_2.Cl^-$) and subsequent addition to sodium hydrogen selenide gives selenoureas in excellent yields (Stick, R. V. et al., Aust. J. Chem., **41**, 549–561, (1988)). With aliphatic secondary amines only low to modest yields are obtained and attempts to prepare telluroureas were unsuccessful.
(5) A tellurourea has been synthesised using an electron-rich alkene and tellurium metal (Lappert, M. F., Martin, T. R. and McLaughlin, G. M., J. Chem. Soc., Chem. Commun., 635, (1980)).

Spectroscopy. Hindered rotation of the N-C(X)-bond in substituted selenourea and 4-selenazolones has been studied (Behrendt, S., Borsdorf, R., Kleinpeter, E., Grundel, D. and Hantschmann, A., Z. Chem., **16**(10), 405–406, (1976)).

Reactions. The oxidation of selenoureas in protic medium has been investigated (Treppendahl, S., Acta Chem. Scand., Ser. B, **29**(3), 385–388, (1975)). The reaction products are substituted benzoselenazolylguanidines in the case of N-methyl-N'-phenyl- and N,N'-diphenylselenourea. A cyclic trimer and a cyclic dimer of the corresponding carbodiimides are formed as by-products.

Hydantoins. (1) The chemical reactivity and pharmacological properties of hydantoins have been reviewed (Lopez, C. A. and Trigo, G. G., Adv. Heterocycl. Chem., **38**, 177, (1985)). However, much less is known about the sulphur dioxide equivalent of the hydantoins, 3-oxo-1,2,5-thiadiazolidine 1,1-dioxides. A general route for the preparation of these compounds, as well as the corresponding 3-imino analogues, has been presented (Lee, C.- H., Korp, J. D. and Kohn, H., J. Org. Chem., **54**(13), 3077–3083, (1989)).

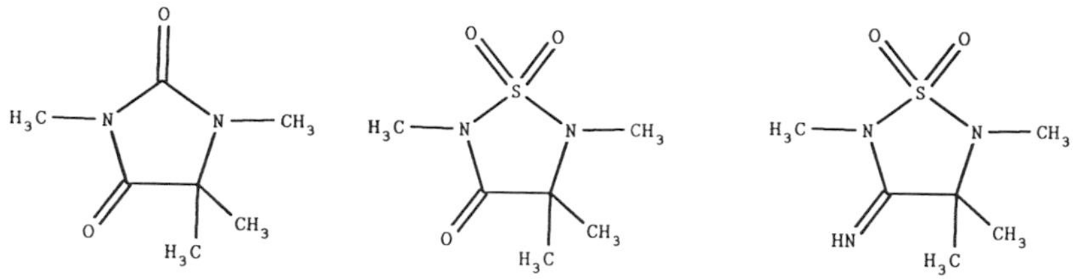

(2) The reaction of trifluoropyruvic acid hydrate with urea derivatives affords 5-hydroxy-5-trifluoromethylhydantoin derivatives in good yield (Mustafa, M. E., Takaoka, A. and Ishikawa, N., J. Fluorine Chem., **30**(4), 463–468, (1986)).
(3) A disubstituted urea, N-α-carbamylisopropyl-N'-α-isobutyronitrile urea, reacts at room temperature in water to yield an N-substituted hydantoin, specifically, 3-(α-carbamylisopropyl)-5,5-dimethylhydantoin (Uhrich, K., Olson, E. and Worman, J., Synth. Commun., **16**(11), 1387–1392, (1986)).
(4) An efficient and simple two-step synthesis of 1,3-dibenzyl-5-methylenehydantoin from cystine has been reported (Ravindranathan, T. et al., Synthesis, 38–39, (1989)).
(5) A detailed mechanistic study of the reaction of benzil with thioureas has been presented (Butler, A.R. et al., J. Chem. Soc. Perkin Trans. II, 731–740, (1989)). Benzil reacts under alkaline conditions with 1,3-dimethylthiourea to form 4,5-dihydroxy-1,3-dimethyl-4,5-diphenyltetrahydroimidazole-2-thione; with 1-methylthiourea to form 3-methyl-5,5-diphenyl-2-thiohydantoin; and with thiourea to form both the corresponding

hydantoin and 3a,7a-diphenyltetrahydroimidazo[4,5-d]imidazole-2,5-dithione.

(6) Kinetic studies of the acid and base hydrolysis of 1-acyl-2-thiohydantoins have been carried out (Congdon, W. I. and Edward, J. T., Can. J. Chem., **50**(23), 3767–79, 3780–3788, 3921–3923, (1972)).

3. Guanidine and its derivatives

Reviews. (1) The pharmacological and medicinal properties of guanidines have been extensively reviewed (Durant, G. J., Roe, A. M. and Green, A. L., Progr. Med. Chem., **7**, 124–213, (1970); Davidoff, F., N. Engl. J. Med., **289**(3), 141–146, (1973); Gerrard, S., J. Hosp. Pharm., **33**(1), 15, 17–18, 21, (1975); Greenhill, J. V., Dev. Drugs Mod. Med., 49–62, (1986), Gorrod, J. W., Gibson, G. G. and Mitchard, M. (eds.) Horwood: Chichester, U. K.) including the effects of guanidine on neuromuscular transmission (Debecker, J., Int. Encycl. Pharmacol. Ther., **14**(2), 503–513, (1972), Cheymol, J. (ed.) Pergamon: Oxford, U. K.), guanidine derivatives acting at histaminergic receptors (Durant, G. J., Chem. Soc. Rev., **14**(4), 375–398, (1985)) and the biochemistry and neurotoxicology of guanidino compounds (Mori, A., Pavlovian J. Biol. Sci., **22**(3), 85–94, (1987)).
(2) Metabolic effects of guanidines and biguanides (Schaefer, G., Pharmacol. Ther., **8**(2), 275–295, (1980)) and their interaction with biological membranes (Schaefer, G., Int. Encycl. Pharmacol. Ther., **107**, 165–185, (1981)) have been reviewed.
(3) Antiviral (Tershak, D. R., Yin, F. H. and Korant, B. D., Handb. Exp. Pharmacol., **61**, 343–375, (1982); Schrom, M. and Caliguiri, L. A., Dev. Mol. Virol., **4**, 271–283, (1984)), antifungal and antibacterial (Hudson, H. R., Ojo, I. A. O. and Pianka, M., Int. Pest Control, **28**(6), 148–149, 152–155, (1986)) activities of guanidines have been reported.
(4) The presence of guanidines in various marine natural products has been described (Shimizu, Y., Mar. Nat. Prod.: Chem. Biol. Perspect., **1**, 1–42, (1978), Scheuer, P. J. (ed.) Academic Press: New York); Chevolot, L., Mar. Nat. Prod.: Chem. Biol. Perspect., **4**, 53–91, (1981), Scheuer, P. J. (ed.) Academic Press: New York); Christophersen, C., Alkaloids, **24**, 25–111, (1985), Academic Press: New York)).

Preparations. (1) Isocyanide–mercuric chloride complexes react with primary and secondary amines to give guanidines and metallic mercury in high yields through a redox decomposition reaction. In the presence of Et_3N, the complexes react with an equimolar amount of primary amine to give carbodiimides and metallic mercury (Sawai, H. and Takizawa, T., J. Organometal. Chem., **94**(3), 333–343, (1975)).
(2) Guanidino–N–alkyl derivatives of arginine which retain the charge and have different hydrophobicity are useful in studying

the importance of the guanidino groups of arginine in peptides. The reaction between polysaccharides and cyanogen bromide yields cyanoesters (ROC≡N), which upon reaction with amines give polymeric pseudoureas ($POC(=NH)NR_1R_2$). These polymeric pseudoureas or N-substituted isoureas can be used for the synthesis of unsymmetrical or symmetrical guanidines (Pundak, S. and Wilchek, M., J. Org. Chem., **46**(4), 808–809, (1981)).
(3) Guanidines are conveniently synthesised from thioureas (Maryanoff, C. A. et al., J. Org. Chem., **51**(10), 1882–1884, (1986)) and sterically hindered guanidines have been prepared from chloroformamidinium or (dichloromethane)iminium salts and primary amines (Wieland, G. and Simchen, G., Liebigs Ann. Chem., 2178–2193, (1985)).

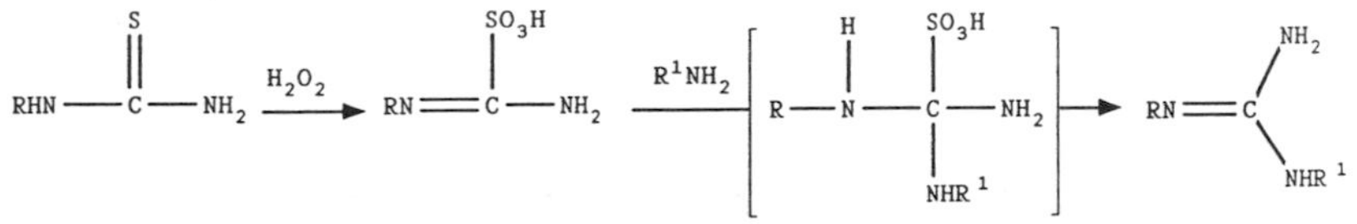

Cyclic and bicyclic guanidines. (1) A convenient route to bicyclic guanidines involves intramolecular Hg^{II}-induced amination of alkenylated monocyclic guanidines (Esser, F., Synthesis, 460–466, (1987)).
(2) Methods for the efficient synthesis of chiral C_2-symmetric bicyclic guanidines and unsymmetrically substituted guanidines from chiral α-amino acids have been demonstrated (Corey, E. J. and Ohtani, M., Tetrahedron Lett., **30**(39), 5227–5230, (1989); (Kurzmeier, H. and Schmidtchen, F. P., J. Org. Chem., **55**(12), 3749–3755, (1990); (Schmidtchen, F. P., Tetrahedron Lett., **31**(16), 2269–2272, (1990)).

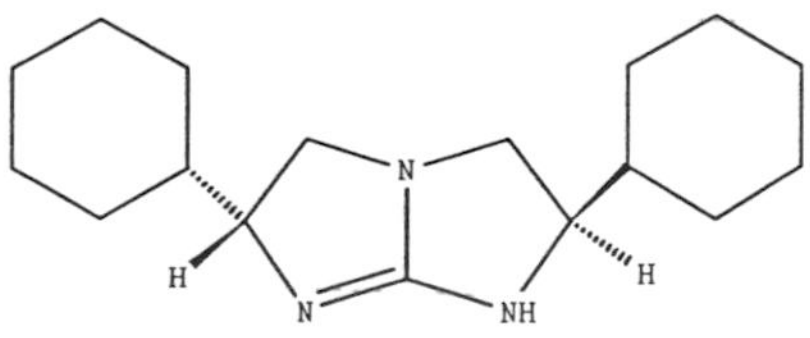

Di- and trisubstituted guanidines. (1) Di- and tri-substituted guanidines may be obtained from thioureas *via* S-oxidation followed by displacement with an amine nucleophile (Maryanoff, C. A., Stanzione, R. C. and Plampin, J. N., Phosphor. Sulfur. Relat. Elem., **27**(1/2), 221–232, (1986)).
(2) N,N′,N′′-trisubstituted guanidines have been prepared by

the action of iminophosphoranes on thioureas (Molina, P., Alajarin, M. and Saez, J., Synth. Commun., **13**(1), 67–72, (1983)) or S–methylation of monosubstituted thioureas with MeI and reaction of the resultant methyl carbamimidothioate hydroiodides with secondary amines in boiling *tert*–BuOH or CH_3CN (Rasmussen, C. R. et al., Synthesis, 460–466, (1988)).

$$Ar^3{-}N{=}PPh_3 + (Ar^1{-}NH)(Ar^2{-}NH)C{=}S \longrightarrow (Ar^1{-}NH)(Ar^2{-}NH)C{=}N{-}Ar^3 + S{=}PPh_3$$

$$H_2N{-}C({=}S){-}NHR^3 \xrightarrow{MeI/MeOH} H_2N{-}C(SMe){=}NR^3 \cdot HI \xrightarrow{HNR^1R^2} H_2N{-}C(NR^1R^2){=}NR^3 \cdot HI$$

(3) N,N′–Dialkyl carbodiimides react with H_2 in the presence of ruthenium clusters to give N,N′,N′′–trialkyl guanidines (Schmidt, G. F. and Sussfink, G., J. Organometal. Chem., **356**(2), 207–211, (1988)).

Tetrasubstituted guanidines. (1) Symmetrically 2,2,8,8–tetraalkyl–substituted bicyclic guanidines have been prepared stereoselectively *via* construction of open–chain triamines and subsequent cyclisation (Schmidtchen, F. P., Chem. Ber., **113**(6), 2175–2182, (1980)).
(2) A number of phenyl–substituted derivatives of 2–phenyl–1,1,3,3–tetramethylguanidine have been synthesised and the pKa of each measured in CH_3CN using a conventional, general purpose glass electrode (Leffek, K. T., Pruszynski, P. and Thanapaalasingham, K., Can. J. Chem., **67**(4), 590–595, (1989)). The influence of electronic and steric effects on the basicity of the compounds is discussed and the pKa values compared with those obtained in water.

Penta- and hexasubstituted guanidines. (1) Sterically hindered pentaalkylguanidines are prepared by the reaction of Vilsmeier salts, derived from tetraalkyl–ureas or –thioureas, with primary aliphatic amines (Barton, D. H. R., Elliott, J. D. and Gero, S. D., J. Chem. Soc., Perkin Trans. I, 2085–2090,

(1982)).
(2) The properties of these strong but highly hindered organic bases have been investigated (Barton, D. H. R., Elliott, J. D. and Gero, S. D., J. Chem. Soc., Chem. Commun., 1136–1137, (1981)) and several are superior catalysts for the preparation of aryl and aralkyl ethers, the methylation of phenols and the esterification of acids (Sennyey, G. et al., Tetrahedron, **46**(6), 1839–1848, (1990)).
(3) The preparation of penta- and hexasubstituted guanidinium salts and pentaalkylguanidines has been described (Kantlehner, W., Haug, E., Mergen, W. W., Spehe, P. and Maier, T., Liebigs Ann. Chem., 108–126, (1984)).

Spectroscopy. (1) A ^{1}H NMR study of hexasubstituted guanidine salts shows that restricted rotation exists for the three major C–N bonds of the guanidine nucleus (Santoro, A. V. and Mickevicius, G., J. Org. Chem., **44**(1), 117–120, (1979)) and provides evidence for the existence of a non-planar structure.
(2) ^{15}N NMR spectroscopy has been used to investigate the barriers to isomerisation about the C–N bond in guanidinium and guanidino groups of arginine (Kanamori, K. and Roberts, J. D., J. Amer. Chem. Soc., **105**(14), 4698–4701, (1983)) and the results compared with those previously reported for tetramethylguanidine derivatives.
(3) SCF calculations have been reported for the guanidinium ion ($C(NH_2)_3^+$) (Sapse, A. M. and Massa, L. J., J. Org. Chem., **45**, 719–721, (1980)) which is an important fragment in a variety of compounds (Subramanian, S., Sarma, T. S., Balasubramanian, D. and Ahluwalia, J. C., J. Chem. Phys., 75, 815, (1971); Bally, T., Diehl, P., Haselback, E. and Tracey, A., Helv. Chim. Acta., **58**, 2398, (1975); Capitani, J. F. and Pedersen, L., Chem. Phys. Lett., **54**, 547, (1978)).

Reactions. (1) Pentafluoroguanidine reacts with isocyanic acid in the presence of a catalyst to form bis(difluoroamino)-fluoraminomethyl isocyanate (Firth, W. C., Frank, S. and Schriffert, E. J., J. Org. Chem., **38**(6), 1080–1083, (1973)).
(2) A method for the introduction of guanidinium groups into macrocycles has been developed (Lehn, J. M. et al., J. Chem. Soc., Chem. Commun., 934–936, (1978)). The macrocycles display enhanced binding of phosphate anion.
(3) The conversion of guanidines into diaziridinimines on treatment with *t*-BuOCl and *t*-BuOK has been applied to the synthesis of sulphonyl-, phosphoryl-, and cyano-substituted derivatives bearing two additional *tert*-butyl groups (Labbe,

G. et al., J. Org. Chem., **46**(22), 4478–4481, (1981)).

Bu^t—NH HN—Bu^t C=N—X $\xrightarrow[\text{t-BuOK}]{\text{t-BuOCl}}$ Bu^t—N—N—Bu^t C=N—X

(4) 1,3-Diaryl-2-propen-1-ones react with guanidine hydrochloride in the presence of 3 moles of NaOH to give 2-amino-4,6-diarylpyrimidines (Alhajjar, F. H. and Sabri, S. S., J. Heterocycl. Chem., **19**(5), 1087–1092, (1982)).

(5) Acrylonitrile and guanidine react in DMF to yield 3,4,6,7-tetrahydro-2H-pyrimido[1,2-a]pyrimidine-2,8(1H)-diimine and 2-amino-4-imino-1,4,5,6-tetrahydropyrimidine-1-propionitrile as the main products (Wendelin, W. and Riedl, R., Monatsh Chem., **115**(4), 445–453, (1984)).

(6) The reactions of guanidine and thiourea with α, β, γ and δ-unsaturated ketones have been reported (Wendelin, W., Schramm, H. W. and Blasirabassa, A., Monatsh Chem., **116**(3), 385–400, (1985)).

(7) Monosubstituted guanidines react with methyl acrylate in DMF or EtOH preferentially to afford 1-substituted 2-amino-5,6-dihydro-4(1H)-pyrimidinones. N,N-disubstituted guanidines give N^2,N^2-disubstituted 2-amino-5,6-dihydro-4-(1H)-pyrimidinones (Wendelin, W. and Riedl, R., Monatsh Chem., **116**(2), 237–251, (1985)).

(8) N-Aryl- and N-aralkyl-N′,N′′-dicarbomethoxyguanidines on treatment with thiophosgene undergo fragmentation to yield N,N′-dicarbomethoxyurea and isocyanates (Viswanathan, N. and Sidhaye, A. R., Indian J. Chem., Sect. B, **25**(6), 659–660, (1986)).

(9) Base-catalysed 1,3-aryl migrations of N-substituted 2-arylamino-4,5,6,7-tetrahydro-1H-1,3-diazepines proceed smoothly at room temperature (Esser, F. et al., J. Chem. Soc. Perkin Trans. I, 3311–3316, (1988)). The reaction is related to the Chapman-type isourea rearrangement (Suttle, N. A. and Williams, A., J. Chem. Soc. Perkin Trans. II, 1369–1372, (1983)).

(10) A synthesis of melamine from guanidines and hexamethyldisilazane has been reported (Kim, Y. H., Lim, B. U. and Kim, K., Chem. Ind., London, 622–633, (1990)).

(11) The condensation products of aldoses with aminoguanidine exist in aqueous solution at pH 6 as cyclic pyranosyl-aminoguanidines, and at pH 12 and in DMSO as acyclic carboximidamidehydrazones (Szilagyi, L., Gyorgydeak, Z. and Duddeck, H., Carbohyd. Res., 158, 67–79, (1986)). The sites of protonation and tautomerism at the aminoguanidine moiety have been studied by ^{1}H–, ^{13}C–, and ^{15}N– NMR methods and the mechanism of the pH-dependent cyclic–acyclic interconversion studied.

Naturally-occurring guanidines. Compounds containing the guanidine unit such as the cyclic marine-derived guanidines saxitoxin, ptilocaulin and tetrodotoxin are of considerable biological interest (Cotton, F. A., Day, V. W., Hazen, E. E. and Larsen, S., J. Amer. Chem. Soc., **95**, 4834, (1973); Corey, E. J. and Cheng, X.-M., in "The Logic of Chemical Synthesis", p366, Wiley, New York, 1989).

Tetrodotoxin. (1) Long known as a food toxin caused by the puffer fish, *Spheroides rubripes*, tetrodotoxin blocks the sodium ion current in the nervous system of mammals (Catterall, W. A., Ann. Rev. Pharmacol. Toxicol., **20**, 15, (1980)). The origin of the toxin has since been attributed to gram negative bacteria (Yasumoto, T., Yasumura, D., Yotsu, M., Michishita, T., Endo, A. and Kotaki, Y., Agric. Biol. Chem., **50**, 793, (1986); Shimizu, Y., Ann. New York Acad. Sci., **479**, 24, (1986)).

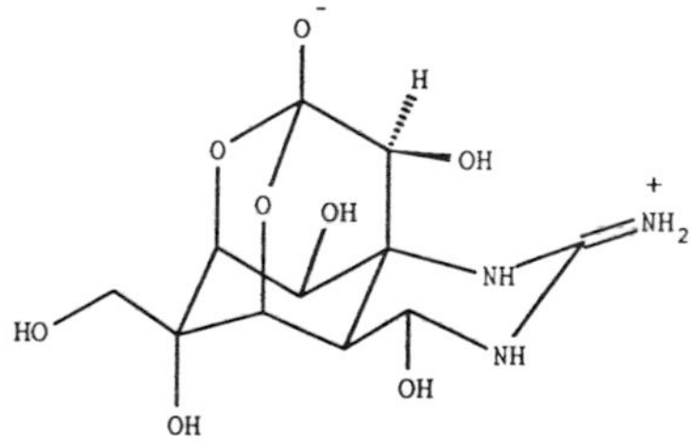

O^- H OH O O OH NH_2^+ NH NH HO OH OH

(2) Total synthesis and numerous synthetic studies have been reported (Kishi, Y. et al., Tetrahedron Lett., 5127–5128, 5129–5132, (1970); Kishi, Y. et al., J. Amer. Chem. Soc., **94**, 9217–9219, 9219–9221, (1972); Kishi, Y. et al., Tetrahedron Lett., 335, (1974); Keana, J. F. W. et al., J. Org. Chem., **48**(21), 3621–3626, (1983); Keana, J. F. W. et al., J. Org. Chem., **48**(21), 3627–3631, (1983); Isobe, M., et al., Tetrahedron

Lett., **28**(51), 6485–6488, (1987)).

Saxitoxin. (1) Saxitoxin is the paralytic agent of the Alaska butter clam, *Saxidomas giganteus*, but in fact originates in dinoflagellates such as *Gonyaulax catenella* and *Gonyaulax tamarensis*. It is also produced by a freshwater cyanobacterium *Aphanizomenon flos-aquae*. Like tetrodotoxin, saxitoxin binds to the sodium channel in nerve and muscle cells with high affinity and specificity. Initially purified in 1957, it was not until 1975 that X-ray analysis (Clardy, J. et al., J. Amer. Chem. Soc., **97**(5), 1238–1239, (1975); Rapoport, H. et al., J. Amer. Chem. Soc., **97**, 6008–6012, (1975)) confirmed the structural assignment (Wong, J. L., Oesterlin, R. and Rapoport, H., J. Amer. Chem. Soc., **93**(26), 7344–7345, (1971)).

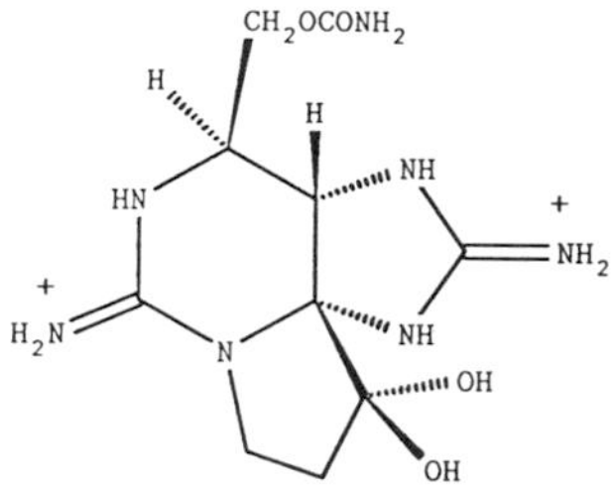

CH_2OCONH_2 H H HN NH $^+NH_2$ H_2N^+ N NH OH OH

(2) Total synthesis and other synthetic studies have been reported (Kishi, Y. et al., J. Amer. Chem. Soc., **99**(8), 2818–2819, (1977); Kishi, Y., Heterocycles, **14**, 1477, (1980); Schantz, E. J. et al., Bioorganic Chem., **10**, 412–428, (1981); Hannick, S. M. and Kishi, Y., J. Org. Chem., **48**(21), 3833–3835, (1983); Jacobi, P. A., Martinelli, M. J. and Polanc, S., J. Amer. Chem. Soc., **106**(19), 5594–5598, (1984)).
(3) Numerous analogues (gonyautoxins) have been isolated (Shimizu, Y. et al., J. Amer. Chem. Soc., **100**(21), 6791–6793, (1978); Shimizu, Y. and Hsu, C. P., J. Chem. Soc., Chem. Commun., 314–315, (1981); Shimizu, Y., et al., Tetrahedron, **40**(3), 539–544, (1984); Hall, S., et al., Tetrahedron Lett., **25**(33), 3537–3538, (1984)).

Ptilocaulin. (1) This antimicrobial and cytotoxic compound is isolated from a Caribbean sponge, *Ptilocaulis aff. P. spiculifer*, and has been identified as a cyclic guanidine

(Rinehart, K. L., Jr. et al., J. Amer. Chem. Soc., **103**(18), 5604–5606, (1981)).

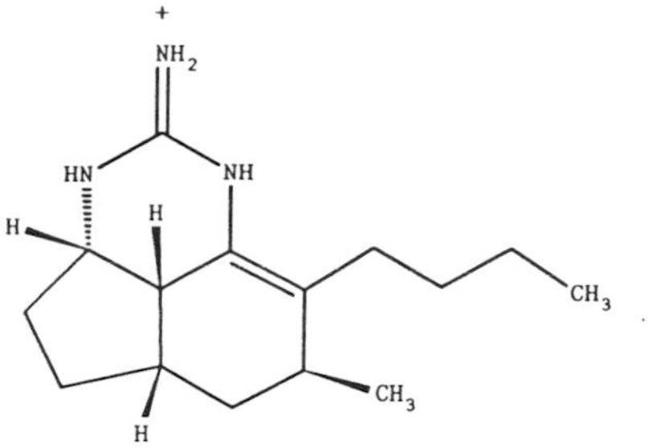

(2) Several total syntheses have been reported (Snider, B. B. and Faith, W. C., J. Amer. Chem. Soc., **106**(5), 1443–1445, (1984); Roush, W. R. and Walts, A. E., J. Amer. Chem. Soc., **106**(3), 721–723, (1984); Walts, A. E. and Roush, W. R., Tetrahedron, **41**(17), 3463–3478, (1985); Uyehara, T. et al., J. Chem. Soc., Chem. Commun., 539–540, (1986); Asaoka, M., Sakurai, M. and Takei, H., Tetrahedron Lett., **31**(33), 4759–4760, (1990)).

Anatoxin-a(s). (1) A potent anticholinesterase (Mahmood, N. A. and Carmichael, W. W., Toxicon, **25**, 1221, (1987); Cook, W. O., Beasley, V. R., Dahlem, A. M., Dellinger, J. A., Harlin, K. S. and Carmichael, W. W., Toxicon, **26**, 750, (1988)), anatoxin-a(s) has been isolated from the blue-green alga *Anabaena flos-aquae* and identified as the phosphate ester of a cyclic N-hydroxyguanidine (Moore, R. E., et al., J. Amer. Chem. Soc., **111**(20), 8021–8023, (1989)).

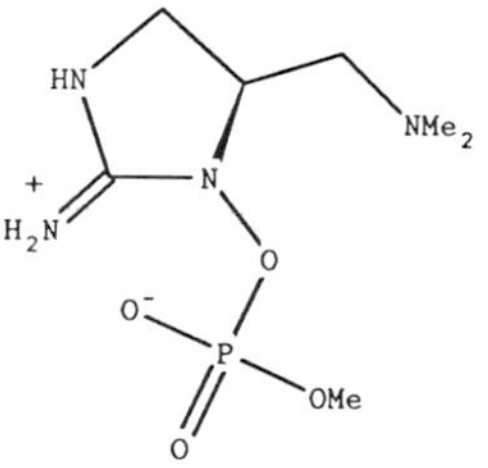

Argiotoxin. One of a number of toxins (Shiba, T. et al., Tetrahedron Lett., **28**(30), 3509–3510, (1987); Hashimoto, Y. et al., Tetrahedron Lett., **28**(30), 3511–3514, (1987)) isolated from spiders of the genera *Argiope*, *Araneus* and *Nephila*, argiotoxin contains an unusual arrangement of two amino acids (asparagine and arginine) linked *via* a polyamine moiety. It is reported to have glutamate receptor channel activity and has been synthesised (Shih, T. L., Ruiz–Sanchez, J. and Mrozik, H., Tetrahedron Lett., **28**(48), 6015–6018, (1987)).

Guanidino-acids, L-arginine. (1) The guanidino side–chain of arginine is an extremely strong base which remains protonated under the normal conditions of peptide synthesis. However, for a number of reasons, it is generally preferable to use some sort of protecting group (Bodanszky, M. and Martinez, J., Synthesis, 333–356, (1981); Bodanszky, M., in "Principles of Peptide Synthesis" Springer–Verlag, p137–141, (1984); Jones, J., in "The Chemical Synthesis of Peptides" Clarendon Press: Oxford, p77–80, (1991)).
(2) The 2,4,6–triisopropylbenzenesulphonyl group has been used in the protection of the guanidino function of arginine (Echner, H. and Voelter, N., Z. Naturforsch., B, **42**(12), 1591–1594, (1987)). The group is introduced by reaction with the chloride and is cleaved by treatment with methanesulphonic or trifluoromethanesulphonic acid.
(3) N^G–Methoxytrimethylbenzenesulphonyl (Mtr) (Atherton, E., Sheppard, R. C. and Wade, J. D., J. Chem. Soc., Chem. Commun., 1060–1062, (1983) and pentamethylchromansulphonyl (Pmc) (Ramage, R. and Green, J., Tetrahedron Lett., **28**(20), 2287–2290, (1987) derivatives have been used to protect the guanidino group of arginine during peptide synthesis. Bis(O–*t*–butoxycarbonyltetrachlorobenzoyl) (Btb) also protects the guanidino group and has the advantage of being cleaved in a mild two–stage process by treatment with F_3CCO_2H and then very dilute acid (Johnson, T. and Sheppard, R. C., J. Chem. Soc., Chem. Commun., 1605–1607, (1990)).

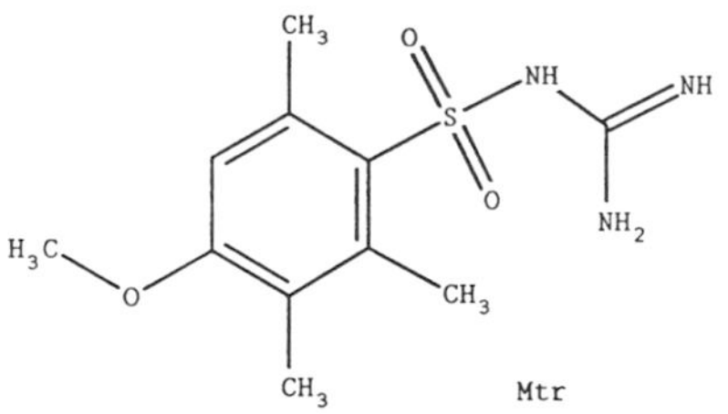

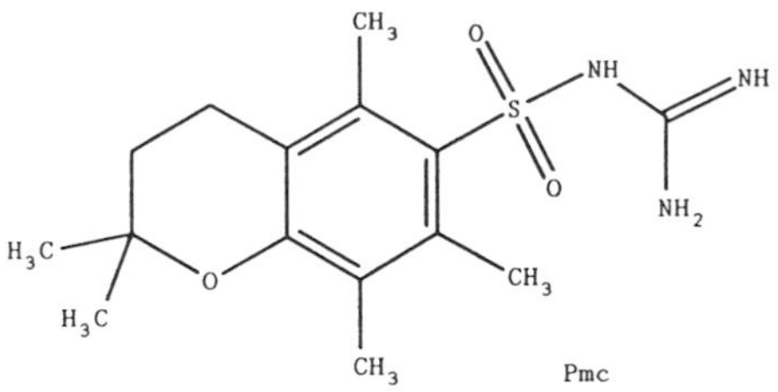

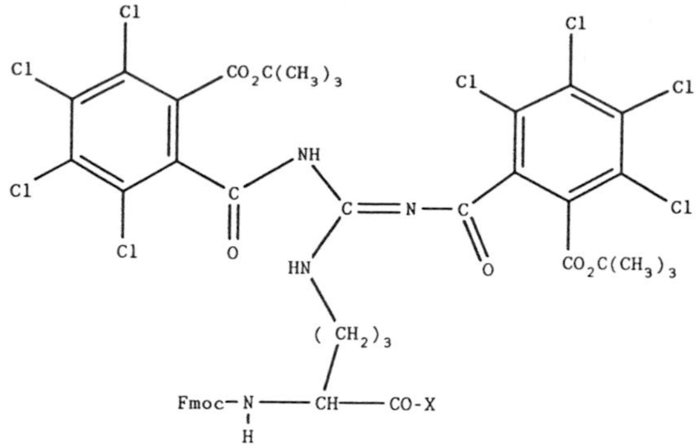

(4) The 9-Anthracenesulphonamide (Ans) group has been used in the protection of the guanidino function of arginine residues (Arzeno, H. B. and Kemp, D. S., Synthesis, 32-36, (1988)). The Ans group is removed either by hydrolysis with trifluoroacetic acid or by a variety of novel, mild reducing conditions such as reaction with aluminium amalgam or photoinduced ruthenium-catalysed reduction with 1-benzyl-1,4-dihydronicotinamide.
(5) The synthesis and bioactivity of N^{ω}-hydroxyarginine has been reported (Wallace, G. C. and Fukuto, J. M., J. Med. Chem., **34**(5), 1746-1748, (1991)).
(6) L-Canavanine [2-amino-4-(guanidinooxy)butyric acid], a potent arginine antimetabolite, has been incorporated into proteins and the expression of its effects analysed (Rosenthal, G. A. and Dahlman, D. L., J. Agr. Food Chem., **39**(5), 987-990, (1991)).
(7) The synthesis of N^6-[bis(ethylamino)methylene)-N^2-(1,1-dimethylethoxycarbonyl)lysine (Boc-diethylhomoarginine) by regiospecific amidination of lysine with ethylamino-ethylimino-methanesulphonic acid has been described (Arzeno, H. B., Bingenheimer, W. and Morgans, D. J., Syn. Commun, **20**(22), 3433-3437, (1990)).

Cyanoguanidines.

Preparations. (1) The synthesis of cyclic N-cyanoguanidines has been reported (Baltzer, C. M. and McCarty, C. G., J. Org. Chem., **38**(1), 155-156, (1973)).
(2) Cyanoguanidines may be derived from the corresponding thiourea/urea by a variety of methods, most of which involve isolation of the intermediate carbodiimide or S-alkylisothio-uronium salt (Petersen, H. J., Neilson, C. K. and Arrigoni-Martelli, E., J. Med. Chem., **21**(8), 773-781, (1978)).

Arylcyanoguanidines cannot be prepared from amines and thioureas *via* the S-methylisothiouronium salt.
(3) A synthesis of cyanoguanidines from carbodiimides has been reported (Mestres, R. and Palomo, C., Synthesis, 755–757, (1980)).

$$R^1{-}N{=}C{=}N{-}R^2 + H_2N{-}C{\equiv}N \xrightarrow{Et_3N} N{\equiv}C{-}N{=}C(HN{-}R^1)(HN{-}R^2)$$

(4) Cyanoguanidines are synthesised from the corresponding thioureas using 1-(3-dimethylaminopropyl)-3-ethylcarbodiimide hydrochloride (water-soluble carbodiimide) (Atwal, K. S., Ahmed, S. Z. and O'Reilly, B. C., Tetrahedron Lett., **30**(52), 7313–7316, (1989)). The reaction is cleaner and faster than that with DCCI (Hansen, E. T. and Petersen, H. J., Synth. Commun., **14**(13), 1275–1283, (1984)).

$$R^1NH{-}C({=}S){-}NH{-}CN + R^2{-}NH{-}R^3 \xrightarrow[RT]{WSC,\ DMF} R^1NH{-}C({=}NCN){-}NR^2R^3$$

(5) A polymer-supported synthesis of disubstituted N-cyanoguanidines has been reported (Zupan, M., Sket, B., Vodopivec, J., Zupet, P., Molan, S. and Japel, J. M., Synth. Commun., **11**(2), 147–156, (1981)).
(6) N-[(Methylthio)carbonyl]-N-trimethylsilylcyanamide (TMS-MCC) dimerises immediately in contact with water to form N-cyano-N,N'-bis[(methylthio)carbonyl]guanidine (D-MCC). When TMS-MCC is allowed to react with an amine, addition to the cyano groups occurs and the corresponding guanidine derivative is obtained (Suyama, T., Okuno, S. and Ichikawa, E., J. Chem Soc. Jap., Chem. Ind. Chem., 1202–1205, (1986)).

Reactions. (1) The preparation of phosphinylguanidine derivatives from cyanoguanidine or guanylurea has been described (Morris, C. E. and Chance, L. H., J. Chem. Eng. Data, **20**(4), 452–453, (1975)).
(2) 2,4,6-Triguanidino-1,3,5-triazine and 2,4,6-tri(3-

methylguanidino)-1,3,5-triazine are obtained by the cyclotrimerisation of cyanoguanidine and 1-methyl-3-cyanoguanidine respectively, using DMF in the presence of HCl. 1-Phenyl-3-cyanoguanidine, 1,2-dimethyl-3-cyanoguanidine, and 2-imidazolidinylidenecyanamide, are cyclodimerised to the corresponding guanidino-1,3,5-triazines in a similar manner (Iio, K. and Ichikawa, E., Bull. Chem. Soc. Jap., **58**(9), 2735-2736, (1985)).

3 RHN—C(=NH)—NH—C≡N → 2,4,6-tris(RHN—C(=NH)—NH)-1,3,5-triazine

(3) The condensation product in the reaction of N-cyanoguanidine with paraformaldehyde under strongly acidic conditions is N,N'-methylenebis[N'-(diaminomethylene)urea] (Shiba, R., Takahashi, M., Ebisuno, T. and Takimoto, M., Bull. Chem. Soc. Jpn., **61**(6), 2197-2198, (1988)).

Dicyanaoguanidine. Methods have been described for the prepartion of 1,3-dicyanoguanidine (reaction of dicyandiamide with cyanogen bromide) and 1,2,3-tricyanoguanidine (reaction of 1,3-dicyano-2-methylisothiourea or dimethyl N-cyanodithioimidocarbonate with cyanamide) (Suyama, T. and Odo, K., J. Syn. Org. Chem. Jap., **34**(6), 427-430, (1976)).

Silylguanidines. (1) N-Silylated guanidines are prepared by treatment of the corresponding imine with Me_3SiCl in the presence of Et_3N (Dorokhova, E. M., Levchenko, E. S. and Pel'kis, N. P., Zh. Org. Khim., **11**, 762, (1975)).
(2) Attempts to obtain oxo-substituted 2-azoniaallene hexachloroantimonate by treating silylated guanidine with phosgene and then $SbCl_5$ leads instead to the dimer (Hamed, A., Jochims, J. C. and Przybylski, M., Synthesis, 400-402, (1989)).

Hydroxyguanidines. (1) The synthesis and reactions of tri-substituted hydroxyguanidines has been investigated (Voss, G., Fischer, E. and Werchan, H., Z. Chem., **13**(2), 58, (1973)).

(2) Acyclic tri- and tetrasubstituted hydroxyguanidines have been prepared from C-chloroformamidinium chlorides (available from ureas or thioureas) *via* reaction with O-(tetrahydro-2-pyranyl)-hydroxylamine, followed by removal of the protecting group. Cyclic tri- and tetrasubstituted hydroxyguanidines have been prepared by the reaction of phosgene-O-(tetrahydro-2-pyranyl)oxime or phosgene-O-(N-methylcarbamoyl)oxime with a diamine, followed by removal of the protecting group (Ziman, S. D., J. Org. Chem., **41**(20), 3253-3255, (1976)).

Spectroscopy. Chemical shifts and coupling constants have been determined for several 2-hydroxyguanidines; the synthesis of 1-monosubstituted 2-hydroxyguanidines has been improved (Clement, B. and Kampchen, T., Chem. Ber., 3481-3491, (1985)).

$$R'-NH-C\equiv N \longrightarrow (R'HN)(H_2N)C=N\sim OH$$

Biguanides. (1) A synthesis of biguanide has been reported (Joshua, C. P. and Rajan, V. P., Chem. Ind.,- London, 497-498, (1974)).
(2) N'-Cyano-S-methylisothioureas react with amines in the presence of CuCl. Under an atmosphere of nitrogen, the methylthio group is substituted by amines to give N''-cyanoguanidines. Upon exposure to air, addition of amines to the cyano group occurs to form the Cu^{II} complex of N'-amidino-S-methylisothioureas. In the presence of CuCl and air, the N''-cyanoguanidines react with $BuNH_2$ to give the corresponding biguanides (Suyama, T., Iwaoka, T., Yatsurugi, T. and Ichikawa, E., J. Chem. Soc. Jap., Chem. Ind. Chem., 1192-1195, (1986)).
(3) Various rhenium complexes with biguanide and methyl biguanide have been synthesised and characterised (Roychowdhury, S., Sipani, M. K. and Sur, B., J. Indian Chem. Soc., **65**(10), 718-720, (1988)).

$ReO(OH)(Big)_2$ $ReO(OH)(MeBig)_2$ Big - -HN–C(=NH)–HN–C(=NH)–NH_2

$ReO(Cl)(Big)_2$ $ReO(Cl)(MeBig)_2$ MeBig - -HN–C(=NH)–NH–C(=NH)–NHMe

4. Isocyanates and derivatives

Reviews. (1) The chemistry (Richter, R. and Ulrich, H., Chem. Cyanates Their Thio Deriv., **2**, 619–818, (1977), Patai, S. (ed.) Wiley: Chichester, U. K.) and biochemistry (Brown, W. F., Curr. Top. Pulm. Pharmacol. Toxicol., **1**, 200–225, (1986)) of isocyanates, the chemistry of six-membered heterocyclic isocyanates and isothiocyanates (L'Abbe, G., Synthesis, 6, 525–531, (1987)) and of heteroallenes in general (carbodiimides, isocyanates and isothiocyanates) has been reviewed (Reichen, W., Chem. Rev., **78**(5), 569–588, (1978)).
(2) Aromatic di- and polyisocyanates (Twitchett, H. J., Chem. Soc. Rev., **3**(2), 209–230, (1974)), use of isocyanates in organometallic chemistry (Cenini, S., Inorg. Chim. Acta., **18**(3), 279–293, (1976)), organosilicon synthesis of isocyanates (Mironov, V. F., J. Organomet. Chem., **271**(1–3), 207–224, (1984)) and transition-metal-mediated reactions of organic isocyanates (Braunstein, P. and Nobel, D., Chem. Rev., **89**(8), 1927–1945, (1989)) have been described.
(3) The chemistry of α-haloalkyl isocyanates (Gorbatenko, V. I. and Samarai, L. I., Synthesis, 85–110, (1980)), 1,1-dihaloalkyl heterocumulenes (isocyanates, isothiocyanates and carbodiimides) (Matveyev, Y. I., Gorbatenko, V. I. and Samarai, L. I., Tetrahedron, 47(9), 1563–1601, (1991)) and the use of chlorocarbonyl isocyanate (ClCONCO) for the synthesis of heterocyclic compounds (Kamal, A., Heterocycles, **31**(7), 1377–1391, (1990)) has been reported.

Toxicity. The toxicology of isocyanates has been reviewed (Carney, I. F., Eur. J. Cell. Plast., **3**(3), 78–81, (1980); Carney, I. F., Cell. Non Cell. Polyurethanes, Int. Conf., 669–680, (1980); Zwi, A. B., SAMJ, **67**(6), 209–211, (1985)) including an account of respiratory effects of inhaled isocyanates (Karol, M. H., CRC Crit. Rev. Toxicol., **16**(4), 349–379, (1986)).

Methyl isocyanate. The hazardous properties of methyl isocyanate have been documented (Dangerous Prop. Ind. Mater. Rep., **5**(2), 68–70, (1985); Dodd, D. F., Rev. Environ. Contam. Toxicol., **105**, 71–98, (1988); Dangerous Prop. Ind. Mater. Rep., **9**(3), 68–74, (1989)) including the mechanism of toxicity of inhaled methyl isocyanate both during and after exposure (Nemery, B., Dinsdale, D. and Sparrow, S., Clin. Respir. Physiol., **23**(4), 315–322, (1987)).

Preparations. (1) Isocyanates may be prepared using Me_3SiN_3 (Kricheldorf, H. R., Synthesis, 551, (1972); Washburne, S. S. and Peterson, W. R., Synth. Commun., 227, (1972); MacMillan, J. H. and Washburne, S. S., J. Org. Chem., **38**, 2982, (1973)) or Bu_3SnN_3 (Kricheldorf, H. R. and Leppert, E., Synthesis, 329–330, (1976)). The preparation of tertiary alkyl isocyanates from alkanoyl chlorides and Me_3SiN_3 is successful only with Lewis acid catalysis (Olah, G. A. et al., J. Org. Chem., **55**(14), 4282–4283, (1990))
(2) Phosgene–free methods have been used for the preparation of isocyanates (Mironov, V. F., Kozyukov, V. P., Kirilin, A. D., Sheludyakov, V. D., Dergunov, Y. I. and Vostokov, I. A., Zh. Obshch. Khim., **45**(9), 2007–2010, (1975)) and diisocyanates (Lesiak, T. and Seyda, K., J. Prakt. Chem., **321**(1), 161–163, (1979)).

$$HCONH(CH_2)_nHNCHO \xrightarrow{Cl_2\ /\ DABCO} O{=}C{=}N{-}(CH_2)_n{-}N{=}C{=}O$$

(3) Reaction of carbodiimides with carboxylic acids (Guldener, D. B. V. and Sikkema, D. J., Chem. Ind.,–London, 628, (1980)), substituted ureas with chlorosilanes (Kozyukov, V. P., Orlov, G. I. and Mironov, V. F., Zh. Obshch. Khim., **50**(4), 960, (1980)), or alkyl halides with silver nitrocyanamide (Boyer, J. H., Manimaran, T. and Wolford, L. T., J. Chem. Soc., Perkin Trans. I, 2137–2140, (1988)) affords isocyanates.
(4) Oxidation of isothiocyanates with $PdCl_2$ and atmospheric oxygen (Paraskewas, S. M. and Danopoulos, A. A., Synthesis, 638–640, (1983)) or vinylogous aminoisocyanides with $Hg(OAc)_2$ (Herdeis, C. and Dimmerling, A., Arch. Pharm., **317**(1), 86–89, (1984)) yields isocyanates.

$$\text{N(Me), COOEt, N}{\equiv}\text{C} \xrightarrow[H_2O]{Hg(OAc)_2} \text{N(Me), COOEt, N}{=}\text{C}{=}\text{O}$$

(5) Methods for the synthesis of α–chloroalkyl– (Fetyukhin, V. N., Gorbatenko, V. I. and Samarai, L. I., Zh. Org. Khim., **11**(11), 2440, (1975)), α–azidoalkyl– (Fetyukhin, V. N.,

Gorbatenko, V. I., Koretskii, A. S. and Samarai, L. I., Zh. Org. Khim., **12**(2), 464–465, (1976)), α-acyloxyalkyl-(Fetyukhin, V. N., Vovk, M. V. and Samarai, L. I., Zh. Org. Khim., **18**(12), 2614–2615, (1982)), α-arylazoalkyl-(Schantl, J. G., Hebeisen, P. and Minach, L., Synthesis, 315–317, (1984)) and α-bromo-isocyanates (Reck, R. and Jochims, J. C., Chem. Ber., **115**(3), 860–870, (1982)) have been reported.

(6) Carbamoylisocyanates (Kozyukov, V. P. and Mironov, V. F., Zh. Obshch. Khim., **53**(6), 1434–1435, (1983)) and N-alkyl-N-chloromethylcarbamoylisocyanates have been synthesised (Gorbatenko, V. I. and Lure, L. F., Zh. Org. Khim., **22**(3), 677, (1986)).

(7) The use of pyridines as leaving groups has been investigated for the preparation of isocyanates (Katritzky, A. R. et al., Tetrahedron Lett., 2691–2694, (1976)).

(8) Aliphatic isocyanates are prepared *via* a two-phase Hofmann reaction using phase-transfer catalysis (Sy, A. O. and Raksis, J. W., Tetrahedron Lett., **21**(23), 2223–2226, (1980)).

(9) Isocyanates may be obtained from amines (Molina, P., Alajarin, M. and Arques, A., Synthesis, 596–597, (1982); Danopoulos, A., Avouri, M. and Paraskewas, S., Synthesis, 682–684, (1985)).

$$R{-}NH_2 \xrightarrow[(C_2H_5)_3N]{(C_6H_5)_3PBr_2} R{-}N{=}P(C_6H_5)_3$$

$$R{-}N{=}P(C_6H_5)_3 \xrightarrow{CO_2} R{-}N{=}C{=}O + (C_6H_5)_3P{=}O$$

$$R{-}N{=}P(C_6H_5)_3 \xrightarrow{CS_2} R{-}N{=}C{=}S + (C_6H_5)_3P{=}S$$

(10) *Vic*-iodoisocyanates are conveniently prepared by treatment of alkenes with I_2/TlOCN (Cambie, R. C., Hume, B. A., Rutledge, P. S. and Woodgate, P. D., Aust. J. Chem., **36**(12), 2569–2574, (1983)).

(11) A route from thiocarbamates to isocyanates has been utilised for the preparation of 2,2,2-trinitroethylisocyanate (Sitzmann, M. E. and Gilligan, W. H., J. Org. Chem., **50**(26), 5879–5881, (1985)).

(12) Di(1-isocyanatoalkyl) esters of dicarboxylic acids have been synthesised (Vovk, M. V., Momot, V. V., Dorokhov, V. I. and Samarai, L. I., Zh. Org. Khim., **22**(3), 673–674, (1986)).

(13) A key intermediate in the palladium-catalysed carbonylation of nitrobenzene into phenylisocyanate has been

identified (Metz, F. et al., J. Chem. Soc., Chem. Commun., 1616-1617, (1990)).

Thermolysis. (1) Thermolysis of oxazolidin-2-ones (Saito, N., Hatakeda, K., Ito, S., Asano, T. and Toda, T., Heterocycles, **15**(2), 905-910, (1981); Saito, N., Hatakeda, K., Ito, S., Asano, T. and Toda, T., Nippon Kagaku Kaishi, 11, 1881-1885, (1989)) or N-silylated oxazolidin-2-ones and oxazolidine-2-thiones (Kricheldorf, H. R., Ann. Chem., 772-792, (1973)) affords the corresponding isocyanates or isothiocyanates. N-silylated thiazolidin-2-ones and thiazolidine-2-thiones do not undergo an analogous reaction.
(2) Thermolysis of disilylated hydroxamic acids (Rigaudy, J. et al., Tetrahedron Lett., **21**(35), 3367-3370, (1980)), 2,4,6-trimethylbenzonitrile N-oxide (Taylor, G. A., J. Chem. Soc., Perkin Trans. I, 1181-1184, (1985)) or cyclic carbalkoxy-aminimides (Senet, J. P., Vergne, G. and Wooden, G. P., Tetrahedron Lett., **27**(52), 6319-6322, (1986)) leads to corresponding isocyanates.

$$R{-}C(=O)NHOH + H{-}N(SiMe_3)_2 \longrightarrow R{-}C(O{-}SiMe_3){=}N{-}OSiMe_3 \longrightarrow R{-}N{=}C{=}O$$

Spectroscopy. Microwave, infrared and Raman spectra, conformational stability, structure, dipole moment and vibrational assignments have been reported for cyclopropyl isocyanate (Durig, J. R., Berry, R. J. and Wurrey, C. J., J. Amer. Chem. Soc., **110**(3), 718-726, (1988)).

Vinylisocyanates. Vinylisocyanates are prepared *via* thermal dehydrohalogenation of 1-halogenoethylcarbamic acid halogenides with α-pinene (Konig, K. H., Feuerherd, K. H., Schwendemann, V. M. and Oeser, H. G., Angew. Chem., **93**(10), 915, (1981)) or bis(dibenzylideneacetone)palladium^0-catalysed conversion of azirines (Alper, H. and Mahatantila, C. P., Organometallics, **1**(1), 70-74, (1982)).

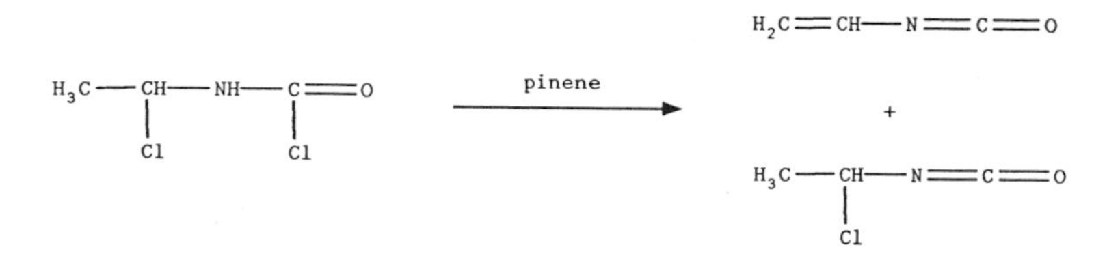

Silyl, Tin and Antimony isocyanates. Trimethylsiloxyisocyanate ($Me_3SiON{=}C{=}O$) (Sheludyakov, V. D., Dmitrieva, A. B., Kirilin, A. D. and Chernyshev, E. A., Zh. Obshch. Khim., **53**(3), 706–707, (1983); Mormann, W. and Leukel, G., Synthesis, 12, 990–992, (1988)), tributyltinisocyanate ($Bu_3SnN{=}C{=}O$) (Ratier, M., Khatmi, D., Duboudin, J. G. and Minh, D. T., Synth. Commun., **19**(11–12), 1929–1937, (1989)) and antimony triisocyanate $Sb(N{=}C{=}O)_3$ (Kijima, I., Wakeshima, I. and Sasaki, T., J. Chem. Soc. Jap., Chem. Ind. Chem., 1754–1757, (1986)) have been prepared.

$$((CH_3)_3SiO)((CH_3)_3Si)N{-}C(=O){-}OSi(CH_3)_3 \longrightarrow (CH_3)_3SiO{-}N{=}C{=}O + (CH_3)_3Si{-}O{-}Si(CH_3)_3$$

Reactions. (1) Aliphatic isocyanates react rapidly with aqueous bisulphite solutions to form stable water-soluble adducts which are salts of N-alkylcarbamyl sulphonic acids (Guise, G. B., Jackson, M. B. and Maclaren, J. A., Aust. J. Chem., **25**(12), 2583–2595, (1972)).
(2) Alkyl isothiocyanates are metallated by means of $KOBu^t$ or NaH to give anions which add to aldehydes or ketones. After neutralisation, di-, tri- or tetra-substituted 2-thioxo-oxazolidines are obtained (Hoppe, D. and Follmann, R., Chem. Ber., **109**(9), 3047–3061, (1976)).
(3) The hydrosilylation of isocyanates is promoted by palladium catalysts and affords N-silylformamides or C-silylamides (Ojima, I. and Inaba, S., J. Organometal. Chem., **140**(1), 97–111, (1977)).

(4) Diphenyl-*tert*-butylsilyllithium and Cl_3SiH-Et_3N reduce isocyanates to isocyanides (Baldwin, J. E. et al., J. Chem. Soc., Chem. Commun., 942–943, (1982)).
(5) Various reactions of trifluoroacetylisocyanate ($CF_3C(=O)N=C=O$) have been investigated (Kiemstedt, W. and Sundermeyer, W., Chem. Ber., **115**(3), 919–925, (1982)).
(6) Isocyanates have been used as universal reagents for derivative formation in gas chromatographic enantiomer separation (Benecke, I. and Konig, W. A., Angew. Chem., **94**(9), 709–710, (1982)).
(7) The reaction of ethoxy(thiocarbonyl)isocyanate $C_2H_5OC(=S)NCO$ with various nucleophiles has been studied (Goerdeler, J. and Schulze, A., Chem. Ber., **115**(3), 1252–1255, (1982)).
(8) The hydrogenation of isocyanates has been discussed (Howell, H. G., Synth. Commun., **13**(8), 635–637, (1983)).
(9) The reaction of methylisocyanate with mono- or dihydrogenphosphates yields solutions of mono- and bis-N-methylcarbamoyl phosphates (Seel, F. and Klein, N., Z. Naturforsch, Sect. B, **38**(7), 797–803, (1983)).
(10) N-Methyl- or N-phenylcyanothioformamide reacts with methyl or phenylisocyanate to provide 5-imino-1,3-disubstituted-4-thioxo-2-imidazolidinones. Methyl or phenyl isothiocyanates similarly give 5-imino-1,3-disubstituted-2,4-imidazolidinedithiones or 5-(phenylimino)-4-imino-2-thiazolidinethiones, depending upon reaction conditions and the reactants (Ketcham, R. et al., J. Org. Chem., **50**(19), 3431–3434, (1985)).
(11) Substituted aliphatic and aromatic dithiols add to isocyanates under catalysis with Triton B to yield biscarbamothioates (Hanefeld, W. and Schulzeweisschu, P., Arch. Pharm., **319**(4), 310–317, (1986)).
(12) Cyclisation of thiosemicarbazones and isocyanates gives 1,3,4-thiadiazolines (Graubaum, H., Nadolski, K. and Andreae, S., Z. Chem., **26**(3), 99–100, (1986)).
(13) Sonication of aryl halides, *t*-butylisocyanate and sodium in THF leads to an organosodium intermediate which is easily and regioselectively *ortho*-lithiated and reacted with electrophiles (Einhorn, J. and Luche, J. L., Tetrahedron Lett., **27**(4), 501–504, (1986)).
(14) Metal-catalysed synthesis of unsaturated carboxylic acid anilides from ethene and phenylisocyanates has been described (Hoberg, H. and Hernandez, E., J. Chem. Soc., Chem. Commun., 544–555, (1986)).

Ph—N=C=O + H_2C=CH_2 ⟶ (Ph_3P)—Ni (ring) N–Ph, =O

⟶ (Ph_3P)—Ni(H)—N(Ph)—CO—CH=CH_2 ⟶ H_2C=CH—CO—N(H)Ph

(15) The reaction of 4-chlorobutylisocyanate with anthranilonitrile gives a urea, which undergoes a double cyclisation to form 2,3,4,5-tetrahydro[1,3]diazepino[1,2-c]quinazolin-7(8H)-one (Kamal, A., Reddy, N. A. V. and Sattur, P. B., Heterocycles, **26**(3), 751-754, (1987)). Similarly, reaction with 2-hydroxybenzonitrile affords 2,3,4,5-tetrahydro[1,3]diazepino[1,2-c][1,3]benzoxazin-7-one.

(16) Isocyanates are converted to nitro-compounds with dimethyldioxirane in wet acetone (Eaton, P. E. and Wicks, G. E., J. Org. Chem., **53**, 5353-5355, (1988)).

O=C=N, N=C=O ⟶ (Me_2CO / H_2O) NO_2, O_2N

(17) 4,4-Dialkoxy-2-azetidinones are prepared in high yields by reaction between ketene acetals and isocyanates (Graziano, M. L. and Cimminiello, G., Synthesis, 54-56, (1989)).

R^1, R^2, OR^3, OR^3 + R^4—N=C=O ⟶ (4 - 85°C, 3 - 72 hrs) R^1, R^2, OR^3, OR^3, O, N, R^4

(18) Treatment of isocyanates with sulphonamides in DMF in the presence of CuCl affords sulphonylureas (Cervello, J. and Sastre, T., Synthesis, 221–222, (1990)).
(19) Mixed boron halide-pseudohalides $BX_n(NCS)_{3-n}$ (n = 1 or 2) have been prepared except for X = F (Dazord, J., Mongeot, H., Atchekzai, H. and Tuchagues, J. P., Can. J. Chem., **54**(13), 2135–2140, (1976)). The boron atom is bonded to the nitrogen of the NCS group and not to the sulphur.
(20) Mixed alkyl benzyl imidodicarbonates are prepared by reaction of benzyloxycarbonyl isocyanate with appropriate alcohols (Grehn, L., Almeida, M. L. S. and Ragnarsson, U., Synthesis, 992–994, (1988)).
(21) The formation of N-tributyltin heterocycles from bis(tributyltin) oxide and ω-haloalkylisocyanates allows one-pot syntheses of 2-oxazolidinones and tetrahydro-2H-1,3-oxazin-2-ones (Shibata, I., Nakamura, K., Baba, A. and Matsuda, H., Bull. Chem. Soc. Jpn., **62**(3), 853–859, (1989)).
(22) Epoxides may be stereoselectively converted to the thermodynamically less stable (Z)-2-oxazolidinones, regardless of the stereochemistry of the vinyl epoxides, by reaction with isocyanates (Trost, B. M. and Sudhakar, A. R., J. Amer. Chem. Soc., **110**(23), 7933–7935, (1988)).

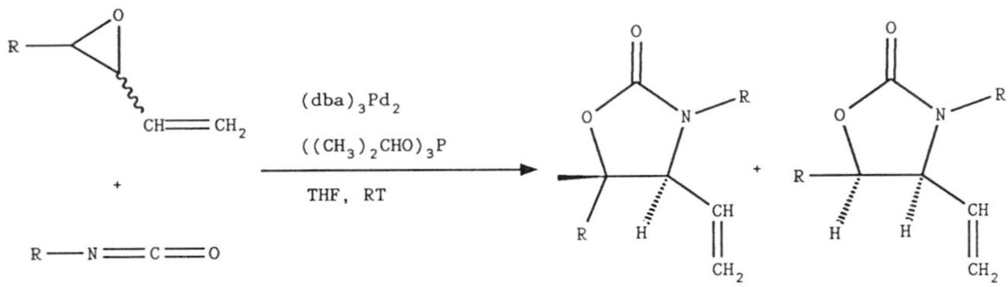

(23) The reactions of azirines with alkylisocyanates (Schaumann, E., Grabley, S. and Adiwidjaja, G., Justus Liebigs Ann. Chem., 264, (1981)), phenylisocyanate (Mukherjee-Muller, G., Heimgartner, H. and Schmid, H., Helv. Chim. Acta, **62**, 1429, (1979)), tosylisocyanate (Schaumann, E. and Grabley, S., Chem. Ber., **113**, 934, (1980)) or isothiocyanates (Schmid, U., Heimgartner, H., Schmid, H. and Oberhansli, W. E., Helv. Chim. Acta, **59**, 2768, (1976); Schaumann, E., Kausch, E. and Walter, W., Chem. Ber., **110**, 820, (1977)) have produced many heterocyclic systems, resulting from opening of either the 1,2- or 1,3-bond in the azirine. Reaction with trimethylsilylisocyanate or isothiocyanate gives imidazolin-2-ones and -thiones respectively (Handke, I., Schaumann, E. and Ketcham, R., J.

Org. Chem., **53**(22), 5298–5300, (1988)).

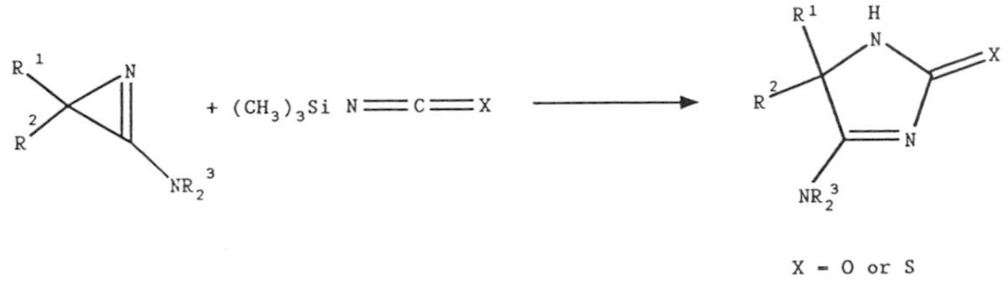

X = O or S

Trimerisation. (1) Trimerisation of isocyanates gives isocyanurates (N,N,N–trisubstituted 1,3,5–triazinetriones) by phase–transfer catalysed (Broda, W., Dehmlow, E. V. and Schulz, H. J., Isr. J. Chem., **26**(3), 222–224, (1985)) or electro–catalysed processes (Carelli, V., Liberatore, F., Moracci, F. M., Tortorella, S., Carelli, I. and Inesi, A., Synth. Commun., **15**(3), 249–258, (1985)).

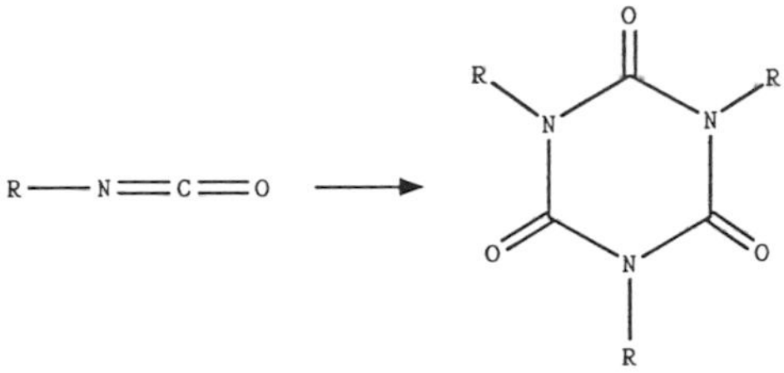

(2) Phenylisocyanate may be trimerised using an imidazole–epoxide–water catalytic system (Koyama, T. and Narahara, T., J. Chem. Soc. Jap., Chem. Ind. Chem., 1758–1764, (1986)) or in the presence of Et_3N under high pressure (Taguchi, Y., Shibuya, I., Yasumoto, M., Tsuchiya, T. and Yonemoto, K., Bull. Chem. Soc. Jpn., **63**(12), 3486–3489, (1990)).

Cycloaddition. (1) The cycloaddition reactions of hetero–cumulenes have been extensively investigated including the reaction of persubstituted guanidines with isocyanates, isothiocyanates and CS_2 (Schaumann, E., Kausch, E. and Rossmanith, E., Justus Liebigs Ann. Chem., 1543–1559, (1978)), the reaction of persubstituted isothioureas with isothio–cyanates and CS_2 (Schaumann, E. and Kausch, E., Justus Liebigs Ann. Chem., 1560–1567, (1978)), and the reaction of certain azirines with isocyanates and isothiocyanates (Schaumann, E. and Grabley, S., Justus Liebigs Ann. Chem., 1568–1585, (1978)).

(2) Cycloaddition reactions between acyl (or thioacyl) isocyanates (or isothiocyanates) and isocyanates (or isothiocyanates) have been investigated (Ratton, S., Moyne, J. and Longeray, R., Bull. Soc. Chim. Fr., Part 2 (1/2), 28–32, (1981)).
(3) The cycloaddition of alkynes with isocyanates (Hoberg, H. and Oster, B. W., Synthesis, 324–325, (1982); Bernat, J., Kniezo, L. and Kristian, P., Z. Chem., **25**(9), 324–325, (1985)) and vinyl isocyanates (Rigby, J. H., Holsworth, D. D. and James, K., J. Org. Chem., **54**(17), 4019–4020, (1989)) has been reported.
(4) 1,2-Cycloadditions of 3-methyl-1-(methylthio)-1-butenyl isocyanate with N-*t*-butyl-α-toluenimine N-oxide and with *in situ* generated N-*t*-butyl-α-methyl-α-lactam affords an oxadiazolidinone derivative and a hydantoin derivative respectively (Oshiro, Y., Komatsu, M., Okamura, A. and Agawa, T., Bull. Chem. Soc. Jap., **67**(3), 901–902, (1984)).
(5) Various types of five-membered heterocyclic compounds, such as 2-iminodioxolanes, 2-oxazolidinones and 2-iminoxazolidines, can be obtained from the cycloaddition of oxiranes with isocyanates or carbodiimides catalysed by organotin iodide-Lewis base complexes under very mild conditions (40°C, 2 hr) (Shibata, I. et al., Synthesis, 1144–1146, (1985)).

Transition Metals. Transition metal clusters such as $[HRu_3(CO)_{11}]^-$, $[H_3Ru_4(CO)_{12}]^-$, $[Ru_3(CO)_{11}]^{2-}$ and $[N(PPh_3)_2][HRu_3(CO)_{10}(SiEt_3)_2]$ catalyse coupling reactions of alkylisocyanates (Sussfink, G., Herrmann, G. and Thewalt, U., Angew. Chem., **95**(11), 899–900, (1983); Herrmann, G. and Sussfink, G., Chem. Ber., **118**(10), 3959–3965, (1985); Sussfink, G. and Herrmann, G., Angew. Chem., **98**(6), 568–569, (1986); Sussfink, G., Herrmann, G. and Schmidt, G. F., Polyhedron, **7**(22/23), 2341–2344, (1988)). With silanes, spiropentazadecanetetrons are obtained; with hydrogen, carbamylformamides are obtained; and with phenyl acetylene, benzylidene hydantoins.

$$5\ RN{=}C{=}O \xrightarrow{(HRu_3(CO)_{10}SiEt)_2^-}$$

Arylisocyanates. (1) O-Arylcarbamoyl-hydroxylamines react with arylisocyanates to give N,O-bis(arylcarbamoyl)-hydroxylamines and N,N,O-tris(arylcarbamoyl)hydroxylamines. Rearrangement of the former or splitting of isocyanate from the latter yielded the N,N-bis(arylcarbamoyl)hydroxylamines(1,5-diaryl-3-hydroxybiurets) (Ketz, E. U. and Zinner, G., Arch. Pharm., **309**(9), 741-747, (1976)).
(2) Ynediamines react with arylisocyanates to furnish diaminoquinolones which react with phosgene to give oxazolones (Bouvy, A., Janousek, Z. and Viehe, H. G., Bull. Soc. Chim. Belg., **94**(11/12), 869-871, (1985)).
(3) The iminophosphorane derived from 4-amino-6-methyl-3-(methylthio)-5-oxo-4,5-dihydro-1,2,4-triazine reacts with aromatic isocyanates to give 1,3-diazetidines (Molina, P. et al., J. Chem. Soc., Perkin Trans. I, 2037-2049, (1986)). Reaction of the same iminophosphorane with aliphatic isocyanates yields betaines (Molina, P. et al., J. Amer. Chem. Soc., **111**(1), 355-363 (1989)).

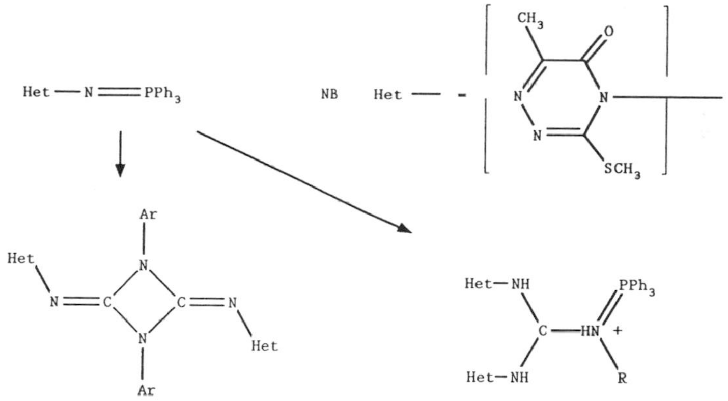

Het—N=PPh₃   NB Het — [triazinone ring: CH_3, O, N, N, N, SCH_3]

(Ar, Het, N=C, C=N, N, Ar, Het — 1,3-diazetidine) (Het—NH, Het—NH, C—HN⁺, PPh_3, R — betaine)

(4) Imidazolidine-2,4-diones are obtained from the reactions of CH_3NO_2 with arylisocyanates in the presence of Et_3N (Shimizu, T., Hayashi, Y. and Teramura, K., Bull. Chem. Soc. Jap., **59**(6), 2038-2040, (1986)).
(5) Reaction of tetraethylammonium nitrite with excess aryl isocyanate at 0°C rapidly produces 1,3-diaryltriazenes (Botting, N. P. and Challis, B. C., J. Chem. Soc., Chem. Commun., 1585-1586, (1989)).

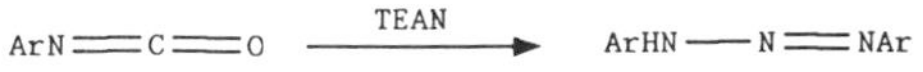

$$ArN{=}C{=}O \xrightarrow{TEAN} ArHN{-}N{=}NAr$$

Sulphonylisocyanates and isothiocyanates.

Reviews. Preparations and reactions of sulphonylisocyanates and sulphonyl isothiocyanates have been reviewed (McFarland, J. W., Sulfur Reports, **1**(4), 215–269, (1981)) as has the chemistry of chlorosulphonylisocyanate ($ClSO_2NCO$) (Rasmussen, J. K. and Hassner, A., Chem. Rev., **76**(3), 389–408, (1976); Szabo, W. A., Aldrichimica Acta, **10**(2), 23–29, (1977); Dhar, D. N. and Murthy, K. S. K., Synthesis, 437–449, (1986); Kamal, A. and Sattur, P. B., Heterocycles, **26**(4), 1051–1076, (1987)).

Preparations. Fluorosulphinylisocyanate has been prepared (Krannich, H. J. and Sundermeyer, W., Angew. Chem. **88**(3), 88, (1976)).

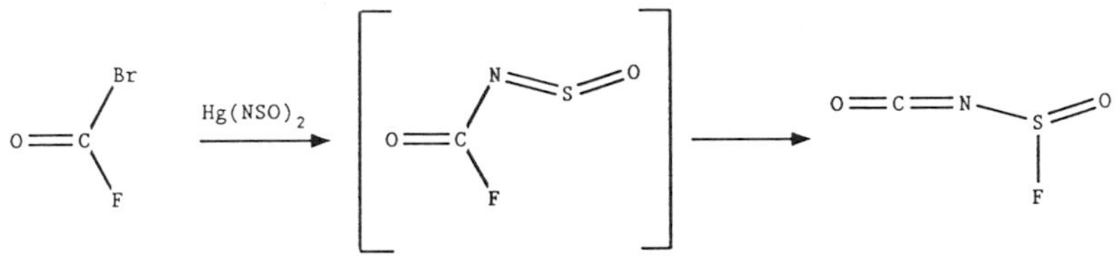

Reactions. (1) Fluorosulphurylisocyanate reacts with HgF_2 and CdF_2 in CH_3CN solution to form fluoroformylfluorosulphurylimide salts, $Hg[N(SO_2F)C(O)F]_2$ and $Cd[N(SO_2F)C(O)F]_2$ (Noftle, R. E., Green, J. W. and Yarbro, S. K., J. Fluorine Chem., **7**(1–3), 221–227, (1976)).
(2) Free-radical reactions of chlorosulphonylisocyanate (CSI) with alkanes can be initiated with either light or thermal initiators. The major products are chlorides, with low yields of isocyanates and sulphonyl chlorides (Mosher, M. W., J. Org. Chem., **47**, 1875–1879, (1982)).
(3) Reaction of CSI with 6-aryl-3(2H)-pyridazinones yields 3-chloro-substituted pyridazines (Srinivasan, T. N., Rao, K. R. and Sattur, P. B., Synth. Commun., **16**(5), 543–546, (1986)).
(4) Reaction of vinylepoxides with *p*-toluenesulphonylisocyanate in the presence of a Pd^0 catalyst is an effective approach to vinyloxazolidin-2-ones (Trost, B. M. and Sudhakar, A. R., J. Amer. Chem. Soc., **109**(12), 3792–3794, (1987)).
(5) N-Adamantyl sulphamates are obtained by the reaction of 1-adamantanol with alkoxy and aroxysulphonylisocyanates (Hedayatullah, M. and Beji, M., Bull. Soc. Chim. Belg., **97**(3), 219–225, (1988)).

(6) Cyclisation of sulphonyl isocyanates and isothiocyanates with epoxides yields oxazolidinones and oxazolidinethiones (McFarland, J. W., Beaulieu, J. J., Arrey, L. N. and Frey, L. M., J. Heterocycl. Chem., **25**(5), 1431–1434, 1435–1436, (1988).
(7) The reaction of 4-arylthiosemicarbazones with CSI yields Δ^1–[1,2,4]triazoline–5–thiones (Tripathi, M. and Dhar, C. N., Synthesis, 1015, (1986)).

Thiocyanates and isothiocyanates.

Reviews. (1) The chemistry of thiocyanates (Guy, R. G., Chem. Cyanates, their Thio Deriv., **2**, 819–886, (1977), Patai, S. (ed.) Wiley: Chichester, U. K.), isothiocyanates (Drobnica, L., Kristian, P. and Augustin, J., Chem. Cyanates, their Thio Deriv., **2**, 1003–1221, (1977), Patai, S. (ed.) Wiley: Chichester, U. K.); Hartmann, A., Houben-Weyl, 4th ed., Vol. E.4, Georg Thieme Verlag, Stuttgart, 834–883, (1983)), isotopically–labeled cyanates, isocyanates, thiocyanates and isothocyanates (Ceccon, A., Chem. Cyanates, their Thio Deriv., **1**, 507–548, (1977), Patai, S. (ed.) Wiley: Chichester, U. K.) and acyl and thioacyl derivatives of isocyanates, thiocyanates and isothiocyanates (Tsuge, O., Chem. Cyanates, their Thio Deriv., **1**, 445–506, (1977), Patai, S. (ed.) Wiley: Chichester, U. K.) has been reviewed.
(2) The synthesis and preparative applications of monosaccharide thiocyanates and isothiocyanates has been described (Witczak, Z. J., Heterocycles, **20**(7), 1435–1448, (1983); Witczak, Z. J., Adv. Carbohydr. Chem. Biochem., **44**, 91–145, (1986)).
(3) The application of thiocyanates and isothiocyanates in the synthesis of nitrogen– and sulphur–containing organophosphorus compounds has been reported (Kamalov, R. M., Stepanov, G. S., Khailova, N. I. and Pudovik, M. A., Phosphorus, Sufur Silicon Relat. Elem., **49-50**(1–4), 93–96, (1990)).

Thiocyanates.

Preparations. (1) Dithiocarbonic ester chlorides react with NaSCN to give thiocyanates which rearrange thermally to the isomeric isothiocyanates (Goerdeler, J. and Hohage, H., Chem. Ber., **106**(5), 1487–1495, (1973)).
(2) Two methods have been reported for the synthesis of aroyl thiocyanates (Christophersen, C. and Carlsen, P., Tetrahedron, **32**(6), 745–747, (1976)). One method is based on the thermal breakdown of 5–aroylthio–1,2,3,4–thiatriazoles, the other is

based on the reaction between thioacid salts and cyanogen halides.

(3) Thiocyanates have been prepared using phase-transfer catalysis (Reeves, W. P., Simmons, A. and Keller, K., Synth. Commun., **10**(8), 633–636, (1980)), polymer-supported thiocyanates (Harrison, C. R. and Hodge, P., Synthesis, 299–301, (1980)) and silica gel-supported thiocyanates (Kodomari, M., Kuzuoka, T. and Yoshitomi, S., Synthesis, 141–142, (1983)).

$$P^+SCN^- + R\text{-}X \longrightarrow R\text{-}SCN + P^+X^-$$

(4) (E)-Alkenylthiocyanates are formed stereoselectively from the oxidative cleavage of (E)-alkenylpentafluorosilicates with $Cu(SCN)_2$ in DMF (Tamao, K., Kakui, T. and Kumada, M., Tetrahedron Lett., **21**(1), 111–114, (1980)).

(5) Diethyl phosphorocyanidate has been used for the simple one-step transformation of sulphinic acids to thiocyanates (Harusawa, S. and Shioiri, T., Tetrahedron Lett., **23**(4), 447–448, (1982)).

$$RSO_2Na \xrightarrow[THF]{(C_2H_5O)_2P(O)CN} RSCN$$

(6) Primary alcohols are converted to thiocyanates by electrolysis of thiocyanate ion in DCM containing an alcohol, $(PhO)_3P$ and 2,6-lutidinium perchlorate or tetrafluoroborate (Maeda, H., Kawaguchi, T., Masui, M. and Ohmori, H., Chem. Pharm. Bull.-Tokyo, **38**(5), 1389–1391, (1990)).

(7) Tertiary, allylic and benzylic halides react with zinc thiocyanate prepared *in situ* in DCM to yield predominantly the corresponding thiocyanates (Gurudutt, K. N., Rao, S. and Srinivas, P., Indian J. Chem., Sect. B, 30(3), 343–344, (1991)).

Spectroscopy. The preparation and HeI photoelectron spectra of halogenothiocyanates, XSCN (X = Cl, Br) have been reported (Westwood, N. P. C. et al., J. Amer. Chem. Soc., **103**(15), 4423–4427, (1981)). Selenium analogues, XSeCN (X = CN, Cl, Br) also take this form (Jonkers, G., Mooyman, R. and de Lange, C. A., Mol. Phys., **43**, 655, (1981)) though the oxygen analogues, XNCO (X = CN, Cl, Br, I) bond only through nitrogen (Westwood, N. P. C. et al., Chem. Phys., **47**, 111–124, (1980)).

Isothiocyanates.

Preparations. (1) A novel method for breaking amide bonds involves the conversion of amides to thiol acids and isothiocyanates (Shahak, I. and Sasson, Y., J. Amer. Chem. Soc., **95**(10), 3440–3441, (1973)).
(2) Isothiocyanates are prepared by pyrolysis of alkyl dithiourethanes (Moharir, Y. E., J. Indian Chem. Soc., **52**(2), 148–149, (1975)).

$$RNH\overset{S}{\overset{\|}{C}}SEt \xrightarrow[160\text{-}175°C]{pyrolysis} RN{=}C{=}S + EtSH$$

(3) Alkylisothiocyanates are obtained in good yield by the reaction of As_2O_3 with monoalkylamine and CS_2 in a molar ratio of 1:3:3. When PbO is used in this reaction, lead^{II} dithiocarbonimidates are formed in high yield, which on standing or on heating give pure isothiocyanates. Alkylisothiocyanates are formed in good yield by the reaction of triethylammonium dithiocarbonimidate with $AsCl_3$ (Okuda, I. and Sugiyama, H., J. Syn. Org. Chem. Jap., **36**(12), 1090–1094, (1978)).
(4) Thermal isomerisation of sulphur dicyanide yields cyanogen isothiocyanate (King, M. A. and Kroto, H. W., J. Chem. Soc., Chem. Commun., 606, (1980)).
(5) 2–Oxoisothiocyanates are prepared from hydrochlorides of amino ketones by reaction with thiophosgene in the presence of $CaCO_3$ (Bobosik, V., Piklerova, A. and Martvon, A., Collect. Czech. Chem. Commun., **48**(12), 3421–3425, (1983)). At room temperature the enol form of these compounds undergoes slow cyclisation to the isomeric 4–oxazoline–2–thiones.
(6) Addition of I_2/thiocyanogen to alkenes in the dark proceeds by a regioselective ionic reaction to give mainly *vic*–iodoisothiocyanates, and under irradiation with u.v. light by a radical reaction (except in the case of 1–methylene–4–*t*–butylcyclohexane) to give mainly *vic*–iodothiocyanates (Woodgate, P. D. et al., J. Chem. Soc., Perkin Trans. I, 553–565, (1983)).
(7) A new procedure for the synthesis of glycosyl isothiocyanates has been reported (Delasheras, F. G., Mendezcastrillon, P. P. and Sanfelix, A., Synthesis, 509–510, (1984)).
(8) Reaction of amines with di–2–pyridylthionocarbonate affords isothiocyanates, while reaction of N,N′–disubstituted thioureas with the same reagent in the presence of DMAP as a catalyst

gives the corresponding carbodiimides in high yield (Kim, S. and Yi, K. Y., Tetrahedron Lett., **26**(13), 1661–1664, (1985)).
(9) Isothiocyanates have been prepared from dithiocarbamates and diphenylthiophosphinic acid chloride (Bodeker, J., Cammerer, B. and Kockritz, P., Z. Chem., **25**(5), 173, (1985)).
(10) A fluorescent isothiocyanate has been prepared for use as a reagent for the detection of NH functional groups (Hallenbach, W. and Horner, L., Synthesis, 791, (1985)).
(11) A synthesis of vinyloxyethylisothiocyanate has been reported (Nedolya, N. A., Gerasimova, V. V. and Trofimov, B. A., Zh. Org. Khim., **21**(9), 2019–1020, (1985)).
(12) Diethyl alkylphosphoramidates react with CS_2 in boiling benzene after deprotonation with NaH to yield isothiocyanates. The presence of 5 mol% of Bu_4NBr greatly improves the reaction (Olejniczak, B. and Zwierzak, A., Synthesis, 300–301, (1989)).

$$(EtO)_2P(=O)\text{—NH—R} \xrightarrow[\text{2. } CS_2]{\text{1. NaH / n-}Bu_4N^+Br^- \text{ / } C_6H_6} \text{R—N=C=S}$$

Spectroscopy. (1) The microwave spectrum of cyclopropyl isothiocyanate has been investigated (Durig, J. R., Nease, A. B., Berry, R. J., Sullivan, J. F., Li, Y. S. and Wurrey, C. J., J. Chem. Phys., **84**, 3663, (1986)) and shows the molecule exists as an equilibrium mixture of trans and cis conformers at ambient temperature.
(2) An HeI photoelectron study of cyanogen isothiocyanate (NCNCS), produced by thermal isomerisation of sulphur dicyanide ($S(CN)_2$), has been reported (King, M. A. and Kroto, H. W., J. Amer. Chem. Soc., **106**(24), 7347–7351, (1984)).

Ethoxycarbonylisothiocyanate. This material has been useful for the synthesis of a variety of compounds (Esmail, R. and Kurzer, F., Synthesis, 301, (1975); Babu, G. and Popadopoulos, E. P., J. Heterocycl. Chem., **20**, 1127, (1983)). A process for the preparation of ethoxycarbonylisothiocyanate using pyridine or quinoline as a catalyst in an aqueous medium has been presented (Lewellyn, M. E., Wang, S. S. and Strydom, P. J., J. Org. Chem., **55**(18), 5230–5231, (1990)). Reaction *in situ* with a suitable nucleophile such as an alcohol or amine gives excellent yields of thionocarbamate or thiourea.

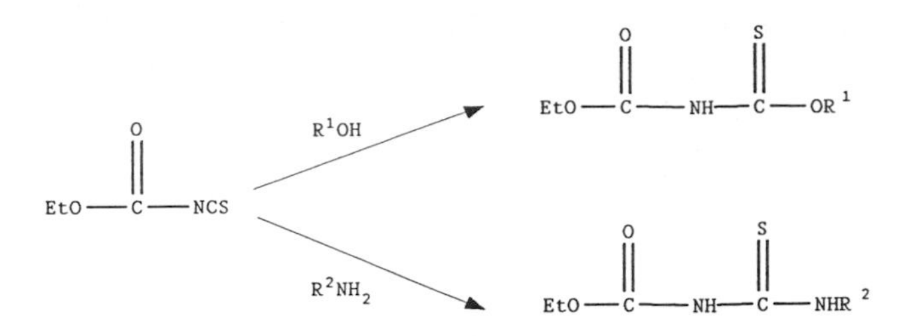

Ethoxy(thiocarbonyl)isothiocyanate has been prepared from ethyl chlorothioformate and NaSCN in CCl_4 using pyridine as a catalyst (Goerdeler, J. and Kunnes, W., Chem. Ber., **116**(5), 2044–2047, (1983)).

Arylisothiocyanates. (1) Arylisothiocyanates have been prepared using amines, CS_2 and BuLi (Sakai, S., Aizawa, T. and Fujinami, T., J. Org. Chem., **39**(13), 1970–1971, (1974)) or a Grignard reagent (Sakai, S., Fujinami, T. and Aizawa, T., Bull. Chem. Soc. Jap., **48**(10), 2981–2982, (1975)).
(2) Arylisothiocyanates (Pak, C. S., Youn, I. K. and Lee, Y. S., Synthesis, 969–970, (1982)) and diisothiocyanates (Garin, J., Melencez, E., Merchan, F., Merino, P. and Tejero, T., Bull. Soc. Chim. Belg., **98**(4), 289–290, (1989)) may be synthesised from dithiocarbamates. A method for the synthesis of sterically hindered arylisothiocyanates has been reported (Habib, N. S. and Rieker, A., Synthesis, 825–827, (1984)).

Vinylisothiocyanates. (1) A convenient method for the conversion of olefins to vinylicisothiocyanates involves the selective synthesis of the precursors, β-(phenylseleno)alkyl isothiocyanates, *via* the reaction of olefins with a mixture of PhSeCl and $Hg(SCN)_2$ (Toshimitsu, A. et al., J. Org. Chem., **48**(26), 5246–5251, (1983)).
(2) Thiocyanic acid (HSCN) adds to some acetylenic compounds through a two-step one-pot procedure which involves the generation of β-thiocyanato and/or β-isothiocyanato alkenyl mercuric compounds (Giffard, M. et al., Tetrahedron, 42(8), 2243–2252, (1986)). The reaction is thermodynamically controlled and allows vinylisothiocyanates to be obtained even when the isomeric vinylthiocyanates are kinetically favoured.

Acylisothiocyanates. Acylisothiocyanates are obtained by reaction of acylthiochlorosulphites with KSCN (Motoki, S. and Kagami, H., J. Org. Chem., **41**(16), 2759–2762, (1976)), phase-transfer catalysis (Reeves, W. P., Simmons, A., Rudis, J. A. and Bothwell, T. C., Synth. Commun., **11**(10), 781–785, (1981)) or reaction of phosphoryl isothiocyanate ($P(=O)(NCS)_3$) with carboxylic acids (Kniezo, L. and Bernat, J., Synth. Commun., **20**(4), 509–513, (1990)).

R CSSCl
KSCN
N≡C
R
S
O
S—S
O
R
N
C
S
O

Reactions. (1) Polycyclic aromatic isothiocyanates have been utilised as fluorescent agents for protein labelling and for use in microanalytical techniques (Sinsheimer, J. E., Jagodic, V., Polak, L. J., Hong, D. D. and Burckhalter, J. H., J. Pharm. Sci., **64**(6), 925–930, (1975)).
(2) 5–(Substituted)amino–1,2,3–thiadiazoles have been prepared from isothiocyanates and diazomethane (Uher, M., Rybar, A., Martvon, A. and Lesko, J., Collect. Czech. Chem. Commun., **41**(4), 1182–1187, (1976)).
(3) On heating in ethanolic KOH, 1–isothiocyanato–2–thiocyanates cyclise to 4,5–thiazolidine–2–thiones (Maxwell, R. J., Moore, G. G. and Silbert, L. S., J. Org. Chem., **42**(9), 1517–1520, (1977)).
(4) Trimethylsilylmethylisothiocyanate has been used as an isothiocyanatomethanide equivalent (Agawa, T. et al., Angew. Chem. Int. Ed. Engl., **20**(1), 126–127, (1981)). Reaction with carbonyl compounds in the presence of a catalytic amount of Bu_4NF gives oxazolidine–2–thiones.

$MeSiCH_2NCS \xrightarrow{n\text{-}Bu_4NF} [^-CH_2NCS] \xrightarrow[2.\,H_2O]{1.\,RCOR^1}$ (oxazolidine-2-thione bearing R and R^1 at C-5; O, NH, S labels)

Bis(trimethylsilyl)methylisothiocyanate under similar conditions with benzaldehyde gives β-styrylisothiocyanate as the major product together with 5-phenyl-4(trimethylsilyl)-oxazolidine-2-thione.

(5) 2-Benzylamino-2-thiazolines are synthesised from *vic*-iodoisothiocyanates (Cambie, R. C., Rutledge, P. S., Strange, G. A. and Woodgate, P. D., Heterocycles, **19**(10), 1903-1908, (1982)).

(6) N-Acetyl or benzoyl-amino acids, when heated with isothiocyanates in the presence of suitable aromatic aldehydes with pyridine as a catalyst, afford 1,2-disubstituted 4-arylmethylene-2-imidazolin-5-ones (Ashare, R. and Mukerjee, A. K., Indian J. Chem., Sect. B **25**(7), 762-764, (1986)). N-benzyloxycarbonylaminoacetic acid gives the corresponding anilide.

(7) Cinnamoylisothiocyanate reacts with amines to give the corresponding cinnamoylthiourea derivatives. These compounds undergo cyclisation when refluxed with NaOEt solution to give the corresponding perhydropyrimidine derivatives. (Gohar, A. M. N., Abdellatif, F. F. and Regaila, H. A. A., Indian J. Chem., Sect. B, **25**(7), 767-768, (1986)).

(8) Tri-*n*-butyltin ω-haloalkoxides are useful reagents for the preparation of five- and six-membered cyclic compounds from isothiocyanates (Baba, A., Shibata, I., Kashiwagi, H. and Matsuda, H., Bull. Chem. Soc. Jap., **59**(1), 341-343, (1986)).

(9) Nucleophilic addition reactions of α,β-unsaturated acyl isothiocyanates with aromatic and heteroaromatic amines have been studied (Abed, N. M., Elagamey, A. G. A., Sowellim, S. Z. and Harb, A. F. A., Rev. Roum. Chim., **33**(4), 393-398, (1988)).

(10) Substituted benzenes and phenols react with aliphatic and aromatic isothiocyanates in a CH_3NO_2 solution of $AlCl_3$ to yield secondary thiobenzamides (Jagodzinski, T., Synthesis, 717-720, (1988)).

(11) ω-halogenoalkylisothiocyanates react with tetraaza-pentalenes to give novel tetraazapentalene derivatives with fused cyclic systems (Matsumura, N., Tomura, M., Chikusa, H., Mori, O. and Inoue, H., Chem. Lett., 965-968, (1989)).

(12) Per-O-acylglycosylisothiocyanates react with ω-halo-alkylamines to yield glycosylamino- and N,N′-bis-glycosyl-heterocycles (Gonzalez, M. A., Caballero, R. B., Moreno, P. C., Mota, J. F., Requejo, J. L. J. and Albarran, J. C. P., Heterocycles, **29**(1), 1-4, (1989)).

(13) Thioformamides are obtained by reduction of isothio-cyanates with Bu_3SnH (Babiano, R. et al., Tetrahedron Lett., **31**(17), 2467-2470, (1990)).

(14) Treatment of 1,4-disubstituted 1,4-diaza-1,3-dienes with isothiocyanatotrimethylsilane yields crisscross addition products, 1,4-disubstituted tetrahydroimidazo[4,5-d]imidazole-2,5-dithiones (Takahashi, M. and Miyadai, S., Heterocycles, **31**(5), 883-888, (1990)).

R—N=CH—CH=N—R —(TMS—N=C=S)→ [TMS—N(R)—CH(N=C=S)—CH(N=C=S)—N(R)—TMS] → [tetrahydroimidazo[4,5-d]imidazole-2,5-dithione, 1,4-di-R]

Cycloaddition. (1) Isothiocyanates undergo [2+2] cycloaddition with DCCI to form 1,3-thiazetidine derivatives (Hritzova, P. J. and Kniezo, L., Collect. Czech. Chem. Commun., **48**(6), 1745-1748, (1983)) and with N-substituted keteniminylidene-triphenylphosphoranes to form four-membered ring phosphonium ylides (Bestmann, H. J., Siegel, B. and Schmid, G., Chem. Lett., 1529-1530, (1986)).
(2) The [4+2] cycloaddition of isothiocyanato-alkenes with tetracyanoethylene has been studied (Giffard, M. and Cousseau, J., J. Chem. Res., 5300-5301, (1985)).

Anodic oxidation. The anodic oxidation of isocyanates and isothiocyanates has been studied in CH_3CN (Becker, J. Y., Zinger, B. and Yatziv, S., J. Org. Chem., **52**(13), 2783-2789, (1987)) and DCM (Becker, J. Y. and Yatziv, S., J. Org. Chem., **53**(8), 1744-1748, (1988)). Isocyanates give low yield mixtures of products arising mainly from C-NCO bond cleavage. Isothiocyanates undergo three types of processes dependent on the nature of the alkyl group: (a) formation of five-membered heterocyclic products from primary alkyl isothiocyanates (b) formation of α-cleavage products ($RNHCOCH_3$) from secondary and tertiary RNCS (c) formation of isocyanates (RNCO) from secondary and tertiary RNCS.

Selenocyanates and isoselenocyanates.

Reviews. Selenenylselenocyanates, sulphenylselenocyanates and isoselenocyanates (seleno mustard oils) (Bulka, E., Chem. Cyanates, their Thio Deriv., **2**, 887–922, (1977), Patai, S. (ed.) Wiley: Chichester, U. K.; Kneen, G., Gen. Synth. Methods, **4**, 172–195, (1981)) have been reviewed.

Preparations. (1) 1:1 complexes of primary amines with $HgCl_2$ react with CSe_2 and Et_3N in CH_3CN to give alkyl and aryl isoselenocyanates (Henriksen, L. and Ehrbar, U., Synthesis, 519–521, (1976)).

$$RNH_2 \xrightarrow[Et_3N]{HgCl_2,\ CSe_2} R-N{=}C{=}Se$$

(2) The reaction of trimethylsilylisoselenocyanate ($Me_3SiCNSe$) with oxiranes and oxetane gives 2-trimethylsiloxyalkyl selenocyanates and 3-trimethylsiloxypropylselenocyanate respectively. The effects of solvents and catalysts on regioselectivity have been examined (Sukata, K., Bull. Chem. Soc. Jap., **63**(3), 825–828, (1990)).

5. Carbodiimide and its derivatives

Reviews. (1) Carbodiimide chemistry (Muraca, R. F., Treatise Anal. Chem., **2**(16), 403–525, (1980); Mikolajczyk, M. and Kielbasinski, P., Tetrahedron, **37**(2), 233–284, (1981); Williams, A. and Ibrahim, I. T., Chem. Rev., **81**, 589–636, (1981)) together with the wider chemistry of heteroallenes (carbodiimides, isocyanates and isothiocyanates) (Reichen, W., Chem. Rev., **78**(5), 569–588, (1978)) and 1,1-dihaloalkyl heteroallenes (Matveyev, Y. I., Gorbatenko, V. I. and Samarai, L. I., Tetrahedron, **47**(9), 1563–1601, (1991)) has been reviewed.

(2) Modification of proteins (Carraway, K. L. and Koshland, D. E., Jr., Methods Enzymol., **25**(B), 616–623, (1972)) and peptide synthesis (Rich, D. H. and Singh, J., in "The Peptides" Gross, E. and Meienhofer, J., (Eds.) Academic Press: New York, **1**, 241–261, (1979)) using carbodiimides has been described.

(3) α,ω-Diisocyanatocarbodiimides, polycarbodiimides and their derivatives (Dietrich, W. et al., Angew. Chem. Int. Ed. Engl., **20**(10), 819–830, (1981)) and phosphorus-containing N-(trifluoromethyl)carbodiimides (Martynov, I. V. et al., Phosphorus, Sulfur Silicon Relat. Elem., **49-50**(1–4), 223–226, (1990)) have been reviewed.

(4) Sulphoxide-carbodiimide oxidations (Moffatt, J. G., Oxidation, **2**, 1–64, (1971), Augustine, R. L. (ed.) Dekker: New York)) have been reviewed.

Preparations. (1) Carbodiimides are formed in the cycloaddition of azirines with nitrile oxides (Nair, V., Tetrahedron Lett., 50, 4831–4833, (1971)).

(2) A versatile method of carbodiimide synthesis involves oxidation of carbene–Pd^{II} complexes with Ag_2O (Ito, Y., Hirao, T. and Saegusa, T., J. Org. Chem., **40**(20), 2981–2982, (1975)).

(3) Perfluoroazapropene reacts with primary aromatic amines to give intermediate carbodiimides which dimerise or trimerise (Flowers, W. T., Franklin, R., Haszeldine, R. N. and Perry, R. J., J. Chem. Soc., Chem. Commun., 567–568, (1976)).

(4) Carbodiimides may be prepared by the reaction of dimetallothioureas with SO_2 (Fujinami, T., Otani, N. and Sakai, S., Synthesis, 889–890, (1977)) or oxidation of thioureas with hypochlorite (Broda, W. and Dehmlow, E. V., Isr. J. Chem., **26**(3), 219–221, (1985)). The latter reaction is improved by the use of phase-transfer catalysis.

(5) A convenient synthesis of unstable carbodiimides has been reported (Palomo, C. and Mestres, R., Synthesis, 373–374,

(1981)).
(6) Carbodiimides are obtained in high yields under essentially neutral conditions using di-2-pyridyl sulphite (Kim, S. and Yi, K. Y., Tetrahedron Lett., **27**(17), 1925–1928, (1986)).
(7) Cycloalkylene carbodiimides have been synthesised *via* a modified Tiemann rearrangement of cyclic amidoxime O-methanesulphonates (Richter, R., Tucker, B. and Ulrich, H., J. Org. Chem., **48**(10), 1694–1700, (1983)).
(8) Trifluoromethylaryl(alkyl)carbodiimides have been prepared (Knunyants, I. L. et al., J. Fluorine Chem., **15**(2), 169–172, (1980)).

$$F_3C{-}N{=}CF_2 + Ar{-}NH_2 \xrightarrow[Et_2O]{2\ KF} F_3C{-}N{=}C{=}N{-}Ar$$

(9) The preparation and reactions of carbamoyl-(Goerdeler, J. and Raddatz, S., Chem. Ber., **113**(3), 1095–1105, (1980)) and thiocarbamoylcarbodiimides (Goerdeler, J. and Losch, R., Chem. Ber., **113**(1), 79–89, (1980)) have been described.

$$R_2N{-}C({=}S){-}NH{-}C({=}S){-}NHR^1 \longrightarrow R_2N{-}C({=}S){-}N{=}C{=}NR^1$$

(10) N-α-Isocyanatoalkyl- (Gorbatenko, V. I., Melnichenko, N. V., Gertsyuk, M. N. and Samarai, L. I., Zh. Org. Khim., **12**(1), 231–232, (1976)) and N-α-[(N-sulphinylamino)alkyl]carbodiimides (Shermolovich, Y. G. and Gorbatenko, V. I., Zh. Org. Khim., **12**(5), 1129–1130, (1976)) have been synthesised.
(11) Chloral ureas have been used in the synthesis of N-substituted-N'-1,2,2,2-tetrachloroethylcarbodiimides ($Cl_3CCHClN{=}C{=}NCHClCCl_3$) (Sinitsa, A. D. and Parkhomenko, N. A., Zh. Org. Khim., **18**(3), 668–669, (1982)).

Reactions. (1) Fragmentation of [2+2] cycloadducts derived from methyl-*tert*-butylcarbodiimide and other heterocumulenes allows assignment of structure for the heterocyclic four-membered ring adducts (Ulrich, H., Tucker, B. and Sayigh, A. A. R., J. Amer. Chem. Soc., **94**(10), 3484–3487, (1972)). While arylisocyanates and arenesulphonylisocyanates add across the C=N double bond

of heterocumulenes, benzoylisocyanate adds across the cumulative C=O double bond. Reactive isothiocyanates, such as 4-nitrophenyl, tosyl, and ethyl isothiocyanatoformate, add across their C=S double bond, and N-sulphinylsulphonamides add across their N=S double bond.
(2) Carbodiimides react with Me_3SiCN in the presence of $AlCl_3$ to give N-trimethylsilyl-1-cyanoformamidines. These react with isocyanates and carbodiimides to give 4,5-diiminodiazoxolidine-2-ones and 2,4,5-triiminoimidazolidines respectively (Ojima, I., Inaba, S. I. and Nagai, Y., J. Organomet. Chem., **99**(1), C5-C7, (1975)).
(3) Reaction of carbodiimides with *p*-toluenesulphonyl azide (Yamamoto, I. et al., J. Chem. Soc., Perkin Trans. I, 1241-1243, (1977)) or bis-silylated phosphines (Issleib, K., Schmidt, H. and Meyer, H., J. Organometal. Chem., **192**(1), 33-39, (1980)) gives guanidines and phosphaguanidines respectively.
(4) Carbodiimides and N,N'-disubstituted ureas form adducts with (dichloromethylene)dimethylammonium salts ("phosgeniminium salts") (Elgavi, A. and Viehe, H. G., Angew. Chem. Int. Ed. Engl., **16**(3), 181-182, (1977)). These adducts are bis-electrophiles which yield biurets upon hydrolysis and biguanides upon aminolysis or ammonolysis.

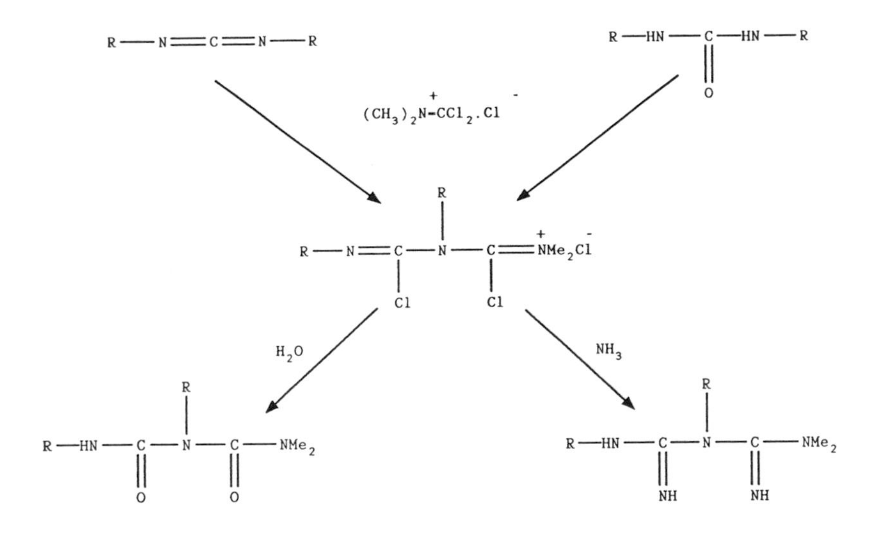

(5) The reaction of carbodiimides with a Reformatsky reagent gives 4-iminoazetidin-2-ones in 51-88% yield (Piotrowska, K.

and Mostowicz, D., J. Chem. Soc., Chem. Commun., 41, (1981)).
(6) Depending on the substituents, carbodiimides effect 1,2 or 1,3 ring opening of aminoazirines (Schaumann, E. and Grabley, S., Liebigs Ann. Chem., **198**(2), 290–305, (1981)).

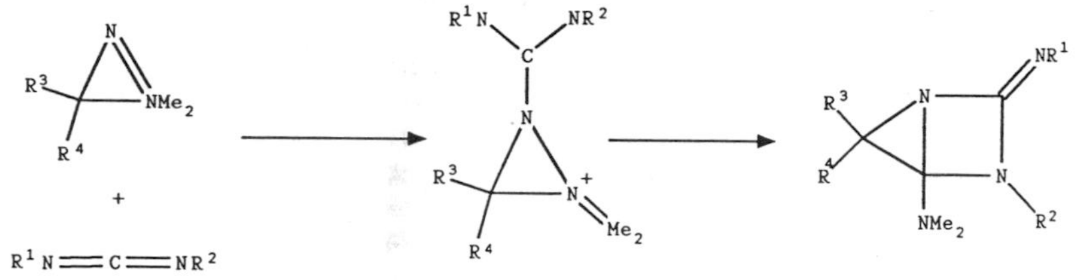

(7) Carbodiimides react with N–methyl–N–sulphinylmethanaminium tetrafluoroborate to give 1,2,4–thiadiazetium salts (Schwobel, A. and Kresze, G., Liebigs Ann. Chem., 900–903, (1984); Kresze, G., Schwobel, A., Hatjiissaak, A., Ackermann, K. and Minami, T., Liebigs Ann. Chem., 904–910, (1984)).
(8) The reaction of carbodiimides with phosphorothioic, –dithioic and –selenoic acids has been described (Mikolajczyk, M. Kielbasinski, P. and Basinski, W., J. Org. Chem., **49**(5), 899–908, (1984)).
(9) The cycloaddition of ketenes with cyclic carbodiimides yields β–lactams in good to excellent yields (Brady, W. T. and Shieh, C. H., J. Heterocycl. Chem., **22**(2), 357–360, (1985)).

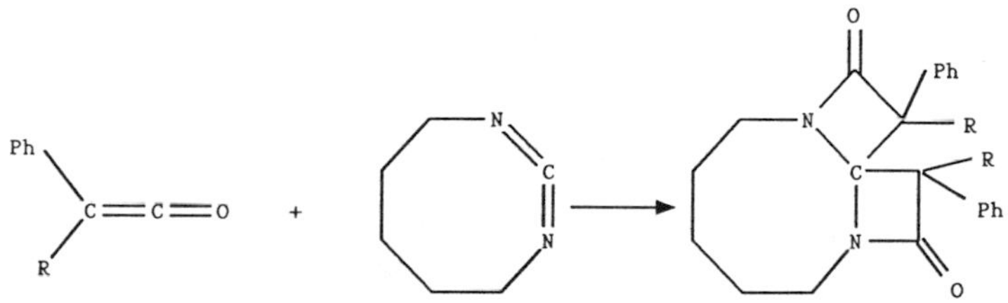

(10) The synthesis of some N,N–dialkylcarbamoyl azides in the presence of carbodiimides gives two types of products: cyclic ammonioamates (from N,N–dialkylaminoisocyanates, formed by a photo–Curtius rearrangement of the azides) and five–membered mesoionic 5–(dialkylamino)–1,2,4–triazoles (Yoshida, K. and Subbbaraj, A., J. Org. Chem., **51**(10), 1719–1723, (1986)).

Conjugated carbodiimides. The preparation and synthetic applications of C=C conjugated carbodiimides have been discussed (Molina, P., Fresneda, P. M. and Alarcon, P., Tetrahedron Lett., **29**(3), 379–380, (1988); Molina, P. et al., Chem. Ber., **122**, 307, (1989); Molina, P., Alajarin, M. and Vidal, A., J. Chem. Soc., Chem. Commun., 7–8, (1990)). A one-pot preparation of α-carbolines and quinindolines from conjugated carbodiimides has been described (Molina, P., Alajarin, M. and Vidal, A., J. Chem. Soc., Chem. Commun., 1277–1279, (1990)).

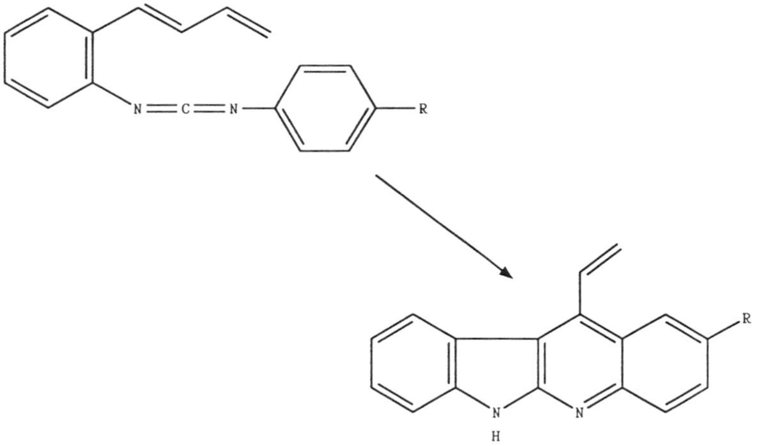

Tosylmethylcarbodiimides. The successful use of tosylmethyl isocyanide (TosMIC, $RSO_2CH_2N{=}C$) in organic synthesis (van Leusen, A. M., Lect. Heterocycl. Chem., **5**, S111, (1980) (a supplementary issue of volume 17 of J. Heterocycl. Chem.) has prompted the investigation of other potential synthons. A number of N-(tosylmethyl)carbodiimides ($R^1SO_2CH_2N{=}C{=}NR^2$) have been synthesised and used in the preparation of 2-amino-1,3-oxazoles (van Leusen, A. M., et al., J. Org. Chem., **46**(10), 2069–2072, (1981)).

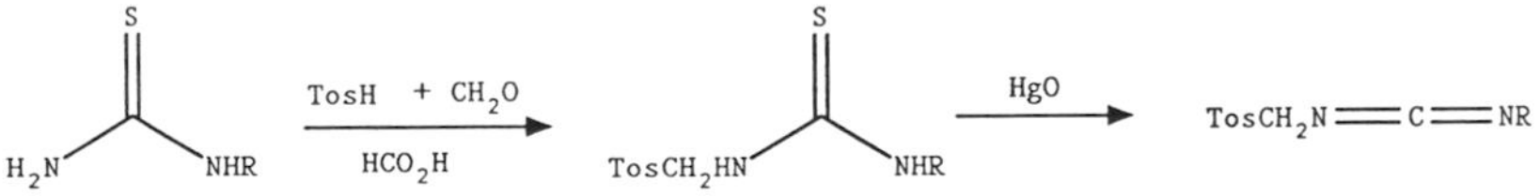

The trityl-substituted carbodiimide is a perfectly stable solid, which can be stored indefinitely, the *tert*-butyl substituted one may be subjected to bulb-to-bulb distillation but others decompose on attempted purification.

N,N'-Disubstituted carbodiimides.

Preparations. (1) N,N'-Disubstituted thioureas, on reaction with cyanuric chloride and then NaOH (Furumoto, S., J. Chem. Soc. Jap. Pure, **92**(11), 1005–1007, (1971)) or 2,4-dichloropyrimidine in the presence of HCl followed by NaOH (Furumoto, S., J. Syn. Org. Chem. Jap., **33**(10), 748–752, (1975)) give N,N'-disubstituted carbodiimides.
(2) Carbodiimides may be prepared by dehydration of ureas with arenesulphonyl chlorides under solid–liquid phase-transfer catalysis using K_2CO_3 as base and a lipophilic quaternary ammonium salt as catalyst (Toke, L. et al., Synthesis, 520–523, (1987)). The method is generally applicable for the synthesis of disubstituted carbodiimides, but is especially useful for unsymmetrically substituted carbodiimides.

$$R^1NH-CO-NHR^2 \xrightarrow[\text{TEBAC}]{R^3SO_2Cl\ /\ K_2CO_3} R^1-N=C=N-R^2$$

Dimethylcarbodiimide. Dimethylcarbodiimide has been synthesised and its chemical and physical properties studied (Rapi, G., Sbrana, G. and Gelsomini, N., J. Chem. Soc. C, Org., 3827–3829, (1971)). The compound spontaneously trimerises to give hexamethylisomelamine and reacts with esters of α-hydroxy acids to give 2-methylimino-3-methyloxazolidin-4-ones.

Diarylcarbodiimides. (1) Symmetrical diarylcarbodiimides may be prepared *via* iminophosphoranes (Molina, P. et al., Synth. Commun., **12**(7), 573–577, (1982)).

$$2\ Ph\text{-}N\text{=}PPh_3 + CS_2 \longrightarrow Ph\text{-}N\text{=}C\text{=}N\text{-}Ph + SPPh_3$$

(2) Polystyrene-anchored triphenylarsine oxide (prepared from brominated polystyrene beads) is a highly effective insoluble catalyst for the conversion of arylisocyanates to diarylcarbodiimides (Smith, C. P. and Temme, G. H., J. Org. Chem., **48**(24), 4681–4685, (1983)).

Reactions. (1) The ozonolysis of a series of N,N'-disubstituted carbodiimides has been studied (Kolsaker, P. and Joraandstad, O., Acta Chem. Scand., Ser. B, **29**(1), 7–12, (1975)). Gradually, two molar equivalents of O_3 are absorbed. The main products are the correspondingly substituted ketones, isocyanates, cyanamides and O_2. Minor amounts of nitrocyclohexane, N,N'-dicyclohexylurea and CO_2 are obtained from N,N'-dicyclohexylcarbodiimide, and unidentified peroxy functions observed to an extent of 10–15% from all carbodiimides.
(2) Alkyl diazoacetates react with N,N'-diisopropylcarbodiimide in the presence of transition metal salts to give aziridines (Hubert, A. J. et al., Tetrahedron Lett., 1317–1318, (1976)).

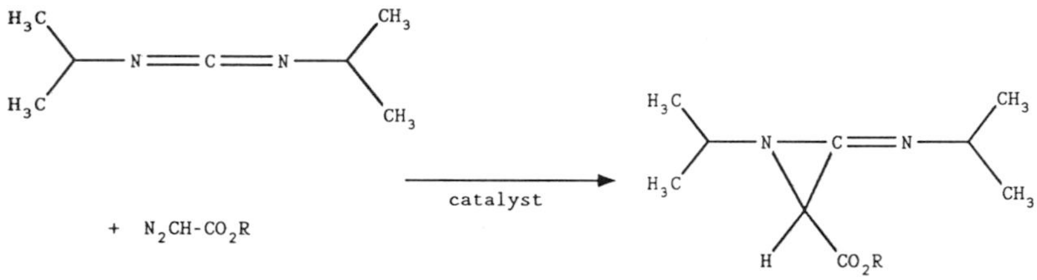

(3) N-Chlorosulphinyldiarylketimine reacts with Lewis acids to give an orange precipitate of 1-oxa-2-thia-3-azabutatrienium salts (Tashtoush, H. and Altalib, M., Rec. Trav. Chim.-J. Roy Neth. Chem., **108**(3), 117–119, (1989)). These readily react with two equivalents of dialkylcarbodiimides to afford 1-oxo-1,2,4,6-thiatriazinium salts.
(4) The role of the nitrogen lone pair in additions to carbodiimides has been investigated using di-(*p*-nitrophenyl)-carbodiimide (Amils, R. and Morenomanas, M., An. Quim., **67**(9/10), 901–902, (1971)).

N,N'-Dicyclohexylcarbodiimide (DCCI).

Reviews. The use of DCCI for activating carboxy groups in peptide synthesis has been reviewed (Jones, J., in "The Chemical Synthesis of Peptides" Clarendon Press: Oxford, p51–54, (1991)).

Preparations. (1) DCCI is obtained from the reaction of N,N'-dicyclohexylformamidine with N-bromosuccinimide in the presence of pyridine (Furumoto, S., J. Syn. Org. Chem. Jap., **34**(7), 499–502, (1976)).
(2) ^{14}C-DCCI has been prepared using ^{14}C-CS_2 (Perry, C. W. and

Burger, W., J. Label. Compound Radiopharm., **13**(1), 113–117, (1977)).

Reactions. (1) The reaction of 1,2- and 1,3-diols with DCCI has been studied (Vowinkel, E. and Gleichenhagen, P., Tetrahedron Lett., 2, 143–146, (1974)).
(2) DCCI is alkylated by prolonged heating in pure alkyl bromides or iodides (Scheffold, R. and Saladin, E., Angew. Chem. Int. Ed. Engl., **11**(3), 229–231, (1972)). N,N′-dicyclohexylcarbodiimidium iodide crystallises as colourless needles, m.p. 111–113°C and can convert alcohols into iodides. N-methyl- and N,N′-dicyclohexylcarbodiimidium tetrafluoroborate and fluorosulphonate are also formed by alkylation of DCCI. Both compounds crystallise as their DCCI-adducts and both dehydrate alcohols (Scheffold, R. and Mareis, U., Chimia, **29**(12), 520–522, (1975)).

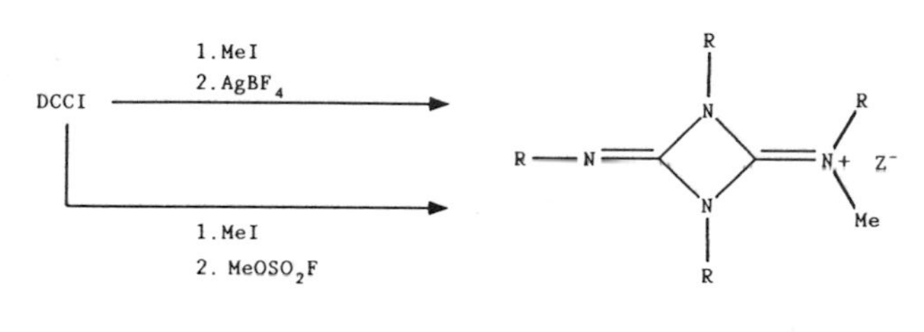

(3) 5-Hydroxythiazolidine-2-thiones react with DCCI to give 2-keto-isothiocyanates which cyclise to 4-oxazoline-2-thiones if an enolisable H atom is attached at the 1-position (Jochims, J. C. and Abutaha, A., Chem. Ber., **109**(1), 154–167, (1976)).

Silyl, Tin and Germanium carbodiimides.

Preparations. (1) Silylcarbodiimides have been prepared from ureas and guanidines (Kozyukov, V. P., Orlov, G. I. and Mironov, V. F., Zh. Obshch. Khim., **51**(1), 245–246, (1981)).
(2) Bis(trimethylsilyl)carbodiimide $Me_3Si{-}N{=}C{=}N{-}SiMe_3$ (Mai, K. and Patil, G., J. Org. Chem., **52**(2), 275–276, (1987)) and bis(triphenylsilyl)carbodiimide $Ph_3Si{-}N{=}C{=}N{-}SiPh_3$ have been synthesised. The structure of the latter has been determined by single crystal x-ray diffraction (Sheldrick, G. M. and Taylor, R., J. Organometal. Chem., **101**(1), 19–25, (1975)).

Reactions. (1) The reactions of organosilicon, organogermanium and organotin carbodiimides with selected compounds have been reported (Dergunov, Y. I., Gordetsov, A. S., Vostokov, I. A. and Galperin, V. A., Zh. Obshch. Khim., **45**(10), 2234–2237, (1975); Vostokov, I. A., Gordetsov, A. S. and Dergunov, Y. I., Zh. Obshch. Khim., **45**(10), 2237–2240, (1975)).
(2) The reaction of N,N′-bis(tributyltin)carbodiimide ($Bu_3SnN{=}C{=}NSnBu_3$) with alcohols and monocarboxylic acids has been investigated (Dergunov, Y. I., Gerega, V. F. and Mushkin, Y. I., Zh. Obshch. Khim., **42**(2), 378–380, (1972)).
(3) The reaction of organosilicon and organotin carbodiimides with hexafluoroacetone has been studied (Fetyukhin, V. N., Vovk, M. V., Dergunov, Y. I. and Samarai, L. I., Zh. Obshch. Khim., **51**(7), 1678–1679, (1981)).
(4) Bis(triphenyltin)carbodiimide reacts with 1,1,3-trisubstituted thioureas to give N,N,N′-trisubstituted N′-cyanoguanidines (Kupchik, E. J., Hanke, H. E., Dimarco, J. P. and Chessari, R. J., J. Chem. Eng. Data, **26**(1), 105–106, (1981)). Bis(triphenyltin)carbodiimide and bis(trimethyltin)-carbodiimide react with ethyl N-arylthiocarbamates to give the corresponding N-aryl-N′-(triorganotin)-N′,N′′-dicyano-guanidines.

Monophosphorus analogues of carbodiimides. (1) A general method for the preparation of monophosphorus analogues of carbodiimides [(iminomethylidene)phosphines, RP=C=NR] together with an initial study of their chemical reactions has been reported (Wentrup, C. et al., J. Amer. Chem. Soc., **105**(24), 7194–7195, (1983)). Prior to this, only the stable and sterically protected $(CH_3)_3CP{=}C{=}NC(CH_3)_3$ had been prepared (Kolodiazhnyi, O. I., Tetrahedron Lett., **23**(47), 4933–4936, (1982)).
(2) The transient formation of hitherto unknown cationic phosphacumulenes $R_2P^+{=}C{=}NR'$, in resonance with the nitrilium salts $R_2PC{\equiv}N^+R'$, has been postulated to rationalise the synthesis of the first 1-aza-3-phosphetine cations (Sanchez, M. et al., J. Org. Chem., **54**(23), 5535–5539, (1989)).

Guide to the Index

This index is constructed in a similar manner to the volume indexes of the first edition of the Chemistry of Carbon Compounds. However, to make the index easier to use, more descriptive entries have been made for the commonly occurring individual, and groups of chemicals.

The indexes cover primarily the chemical compounds mentioned in the text, and also include reactions and techniques, where named, and some sources of chemical compounds such as plant and animal species, oils, etc.

Chemical compounds have been indexed alphabetically under the names used by authors, editing being restricted to ensuring uniformity of entries under the same heading. In view of the alternative nomenclature that can often be used, a limited amount of cross-referencing has been done where it is considered to be helpful, but attention is particularly drawn to Convention 2 below.

For this and the succeeding volumes, the indexing conventions listed below have been adopted.

1. Alphabetisation

(a) A letter by letter alphabetical sequence is followed for entries, firstly for the main entry, followed by the descriptive entry.

(b) The following prefixes have not been counted for alphabetising:

n-	*o-*	*as-*	*meso-*	*C-*	*E-*
	m-	*sym-*	*cis-*	*O-*	*Z-*
	p-	*gem-*	*trans-*	*N-*	
	vic-			*S-*	
		lin-		*Bz-*	
				Py-	

Some prefixes and numbering have been omitted in the index, where they do not usefully contribute to the reference.

(c) The following prefixes have been alphabetised:

Allo	Epi	Neo
Anti	Hetero	Nor
Bis	Homo	Pseudo
Cyclo	Iso	

2. Cross references

In view of the many alternative trivial and systematic names for chemical compounds, the indexes should be searched under any alternative names which may be indicated in the main body of the text. Only a limited amount of cross-referencing has been carried out, where it is considered that it would be helpful to the user.

3. Derivatives

Simple derivatives are not normally indexed if they follow in the same short section of the text.

4. Collective and plural entries

In place of "– derivatives" the plural entry has normally been used. Plural entries have occasionally been used where compounds of the same name but differing numbering appear in the same section of the text.

5. Main entries

The main entry of the more common individual compounds is indicated by heavy type. Multiple entries, such as headings and sub-headings over several pages are shown by "–", e.g., 67–74, 137–139, etc.

Index